S0-AQI-815

BRAIN & BEHAVIOUR

Psychology:
Revisiting the Classic Studies

Series Editors:
S. Alexander Haslam, Alan M. Slater and Joanne R. Smith
School of Psychology, University of Exeter, Exeter, EX4 4QG

Psychology: Revisiting the Classic Studies is a new series of texts aimed at students and general readers who are interested in understanding issues raised by key studies in psychology. Volumes centre on 12–15 studies, with each chapter providing a detailed account of a particular classic study and its empirical and theoretical impact. Chapters also discuss the important ways in which thinking and research has advanced in the years since the study was conducted. Chapters are written by researchers at the cutting edge of these developments and, as a result, these texts serve as an excellent resource for instructors and students looking to explore different perspectives on core material that defines the field of psychology as we know it today.

Also available:
Cognitive Psychology: Revisiting the Classic Studies
Michael W. Eysenck and David Groome

Social Psychology: Revisiting the Classic Studies
Joanne R. Smith and S. Alexander Haslam

Developmental Psychology: Revisiting the Classic Studies
Alan M. Slater and Paul C. Quinn

BRAIN & BEHAVIOUR

REVISITING THE CLASSIC STUDIES

Los Angeles | London | New Delhi
Singapore | Washington DC | Melbourne

Los Angeles | London | New Delhi
Singapore | Washington DC | Melbourne

SAGE Publications Ltd
1 Oliver's Yard
55 City Road
London EC1Y 1SP

SAGE Publications Inc.
2455 Teller Road
Thousand Oaks, California 91320

SAGE Publications India Pvt Ltd
B 1/I 1 Mohan Cooperative Industrial Area
Mathura Road
New Delhi 110 044

SAGE Publications Asia-Pacific Pte Ltd
3 Church Street
#10-04 Samsung Hub
Singapore 049483

Editor: Luke Block
Editorial assistant: Lucy Dang
Production editor: Imogen Roome
Proofreader: Leigh C. Timmins
Marketing manager: Alison Borg
Cover design: Wendy Scott
Typeset by: C&M Digitals (P) Ltd, Chennai, India
Printed and bound by CPI Group (UK) Ltd, Croydon, CR0 4YY

First published 2017

Library of Congress Control Number: 2016935221

British Library Cataloguing in Publication data

A catalogue record for this book is available from the British Library

ISBN 978-1-4462-9651-6
ISBN 978-1-4462-9652-3 (pbk)

Contents

About the editors

Bryan Kolb is a native of Calgary, Canada, and is currently a professor in the Department of Neuroscience at the University of Lethbridge, Canada, where he has been since 1976. He received his PhD from Pennsylvania State University in 1973 and did postdoctoral work at the University of Western Ontario and the Montreal Neurological Institute. His recent work has focused on the development of the prefrontal cortex and how neurons of the cerebral cortex change in response to various developmental factors including hormones, experience, stress, drugs, neurotrophins, and injury, and how these changes are related to behavior. He has also published five books, including two textbooks with Ian Whishaw (*Fundamentals of Human Neuropsychology*, 7th edition; *Introduction to Brain and Behavior*, 5th edition), and more than 390 articles and chapters. Kolb is a Fellow of the Royal Society of Canada, the Canadian Psychological Association, the American Psychological Association, the Association for Psychological Science, and is currently a member of the Canadian Institute for the Advanced Research Program in Child Brain Development. He and his wife train and show horses in Western riding performance events.

Ian Whishaw received his PhD from Western University and is a Professor of Neuroscience at the University of Lethbridge. He has had visiting appointments at the University of Texas, the University of Michigan, Cambridge University, and the University of Strasbourg. He is a fellow of Clare Hall Cambridge, the Canadian Psychological Association, the American Psychological Association, and the Royal Society of Canada. He is a recipient of the Canadian Humane Society Medal for Bravery, the Speaker Medal for Research, the Alberta Science and Technology Leadership Award, the Donald O. Hebb Prize from the Canadian Society for Brain Behavior and Cognitive Science, and the distinguished teaching medal from the University of Lethbridge. He has received the keys to the City of Lethbridge and honorary degrees from Thompson Rivers University and the University of Lethbridge. His research addresses the neural basis of skilled movement and brain disease. The Institute of Scientific Information includes him in its list of most cited neuroscientists. He trains and shows horses in Western riding performance events.

About the contributors

Mary Baldwin is currently a postdoctoral fellow within the Center for Neuroscience at the University of California, Davis. She received both her BSc in Neuroscience and her PhD in Psychology from Vanderbilt University in Nashville, Tennessee. She completed her graduate work under the supervision of Dr. Jon Kaas with a research focus on the evolution of extrageniculate visual pathways, as well as the development and evolution of cortical visual areas in rodents, tree shrews, and primates. Currently, her postdoctoral research, with Dr. Leah Krubitzer, is focused on the evolution of the organization and function of frontoparietal networks involved in sensorimotor processing across a wide range of mammals including marsupials, rodents, tree shrews, and primates.

Antoine Bechara is a Professor of Psychology and Neuroscience at the University of Southern California and the Brain and Creativity Institute, and an adjunct Professor of Neurology at the University of Iowa Hospitals and Clinics. His research focuses on understanding the neural processes underlying how we make decisions and choices. Among the most influential work he did using the brain lesion method is the work he conducted with Antonio R. Damasio at the University of Iowa in patients who have suffered injury to the ventromedial sector of their prefrontal cortex. His development of what became known as the Iowa Gambling Task (IGT) enabled researchers, for the first time, to detect these patients' elusive impairment in the laboratory, measure it, and investigate its possible causes. Another influential work is his finding that lesions to the insula disrupt smoking addiction suddenly and effortlessly. Both lines of work have certainly drawn attention to the potential value in studying the neural basis of addiction to drugs and related neuropsychiatric disorders, including pathological gambling.

Kent C. Berridge, PhD, is the James Olds Collegiate Professor of Psychology and Neuroscience at the University of Michigan. His research focuses on answering questions such as: How is pleasure generated in the brain? How does wanting a reward differ from liking the same reward? What causes addiction? Does fear share anything with desire?

Richard E. Brown, PhD, is a Professor in the Department of Psychology and Neuroscience, Dalhousie University, Halifax, Nova Scotia, Canada. He received his BSc in Psychology from the University of Victoria, his MA and PhD in Psychology and Physiology from Dalhousie University and completed a 2-year PDF in Zoology at the University of Oxford, England. At Dalhousie he teaches courses in Hormones and Behaviour, Measuring Behaviour, the Neurobiology of Learning and Memory, and the History of Neuroscience. He was the Chairman of the Psychology Department at Dalhousie University from 1989–1996 and from 2002–2008, and Director of the Neuroscience Institute from 1996–1999. He was President of the International Behavioural and Neurogenetics Society (2009–2012) and President of the Canadian Society for Brain Behaviour and Cognitive Science (1998–2000). He was the Senior Visiting Research Fellow at St. John's College, Oxford, UK (2008–09); received the Faculty of Science Teaching Excellence Award (2006); and was a Faculty of Science Killam Research Professor (2002–2007). His research is on mouse models of Alzheimer Disease, Fragile X Syndrome, ADHD and other neurological disorders. He is currently examining the age-related hormonal changes in transgenic Alzheimer's mice. He has published over 130 journal articles and 42 book chapters; edited one book [R.E. Brown and D.W. Macdonald, *Social Odours in Mammals.* Oxford, 1985] and written one book [Brown, R.E. *An Introduction to Neuroendocrinology*, Cambridge University Press, 1994; second edition 2015]. Information on his laboratory, research projects and publications is on his lab website: http://myweb.dal.ca/rebrown/.

G. Campbell Teskey received his PhD from Western University in 1990 and then conducted postdoctoral work at McMaster University. He relocated to the University of Calgary in 1992, where he is a professor in the Department of Cell Biology and Anatomy and the Hotchkiss Brain Institute. His federally funded research program examines the development, organization and plasticity of the motor cortex as well as how seizures alter brain function. Teskey has won numerous teaching awards, developed new courses and co-created the Bachelor of Science in Neuroscience program at his home university. He currently serves as Education Director for the Hotchkiss Brain Institute and co-leads the Epilepsy Neuro Team. He recently published his first book, *An Introduction to Brain and Behavior* – 5th edition, with colleagues Bryan Kolb and Ian Whishaw. His hobbies include hiking, biking, kayaking, and skiing.

Michael Corballis was born and educated in New Zealand before completing his PhD in psychology at McGill University in Montreal, Canada. He taught at McGill from 1968 to 1977, before returning to the University of Auckland, where he is now Emeritus Professor. He works in cognitive neuroscience, including research on split-brain patients and, more recently, neuroimaging studies of brain asymmetry, language, and memory. His most recent books are *Pieces of Mind* (2011), *The Recursive Mind* (2011), and *The Wandering Mind* (2015).

Jody Culham is a Professor in the Department of Psychology and Brain and Mind Institute at the University of Western Ontario. Her research investigates how vision is used for perception and to guide actions in human adults. She uses a combination of cognitive neuroscience techniques – including functional magnetic resonance imaging (fMRI), behavioral psychophysics and kinematics, neuropsychology, and neurostimulation – to investigate hand actions such as grasping, reaching, and tool use. One theme of her work is bringing cognitive neuroscience closer to everyday life by investigating real actions upon real objects. This work has revealed interesting differences between more realistic approaches and commonly used proxies for actions (such as pantomimed or imagined movements) and stimuli (such as images). Dr. Culham received a Bachelor's degree from the University of Calgary, a PhD from Harvard University, and did a postdoctoral fellowship at the University of Western Ontario before becoming a faculty member there.

Jason Flindall is a senior PhD student in Neuroscience at the University of Lethbridge. Born and raised in Halifax, Nova Scotia, Jason received a BSc in Biology from Dalhousie University in 2007. He joined the University of Lethbridge in 2009, and received an MSc in Kinesiology in 2012. Jason will receive his doctorate in 2017. His research interests include task-dependent kinematics of reach-to-grasp actions, the functional lateralization of behaviour, and the dual influence of vision and hapsis on dextrous motor movements. His research has appeared in the *Journal of Neurophysiology*, *Laterality*, *PLoS One*, *Frontiers in Psychology*, and the *Journal of Experimental Child Psychology*, as well as at dozens of Canadian and international conferences. He is the recipient of over $100,000 in scholarships and awards, including an Alexander Graham Bell CGS prize for his proposal *Behavioural Correlates of Action Intent: from Development to Degeneration*. He lives in Lethbridge.

Robbin Gibb was born and raised in the Lethbridge area. She received her Bachelor's degree in Chemistry and her Master's and PhD in Neuroscience, all from the University of Lethbridge. She is currently an Associate Professor in the Department of Neuroscience at the University of Lethbridge. Her current research is focused on 1) how prenatal and preconception experience influence brain development and 2) how to improve outcomes for kindergarten children by enhancing early literacy, executive function and self-regulation, and motor skills in preschool children. In addition to the more than 75 articles and chapters she has published, Gibb is currently editing a book on the neurobiology of brain development. She is on the steering committee for the Canadian Centre for Behavioral Neuroscience and the Building Brains and Futures Initiative in Lethbridge and is an affiliate in the Institute of Child and Youth Studies at the University of Lethbridge.

Claudia Gonzalez is an Associate Professor and Canada Research Chair in the Department of Kinesiology at the University of Lethbridge. Gonzalez obtained a Bachelor's degree in Psychology from the National Autonomous University of Mexico (UNAM) and a MSc and a PhD in Neuroscience from the University of Lethbridge. Gonzalez investigates how the brain processes and integrates sensory and motor information, in particular how vision and hapsis (touch) guide arm and hand movements for reaching and grasping. She is interested in understanding the complex interactions of the motor system with cognitive processes such as language, memory and spatial abilities. She uses human psychophysics, behavioural measures including eye and hand kinematics, and cerebral blood flow to infer brain function. Research includes healthy and neurological populations. She has received funding from the Natural Sciences and Engineering Research Council (NSERC), the Canadian Foundation for Innovation (CFI), and Alberta Innovates: Health Solutions (AIHS). She has published over 50 papers and book chapters.

Theresa A. Jones is a Professor in the Psychology Department and Neuroscience Institute at the University of Texas in Austin (UT). She received her PhD from UT Austin in 1992, was a postdoctoral fellow at the University of Illinois' Beckman Institute and faculty of the University of Washington through 2001. Her research focus is on the neural, glial and vascular plasticity underlying learning and post-stroke recovery in adult animals.

Jenni M. Karl is an Assistant Professor in the Department of Psychology at Thompson Rivers University in Kamloops, British Columbia, Canada. She received the Governor General's Gold Medal for her PhD in Behavioural Neuroscience at the University of Lethbridge in Alberta, Canada and subsequently completed a Postdoctoral Fellowship on the neuroimaging of skilled hand movements at Western University in Ontario, Canada. Jenni uses high-speed frame-by-frame video analyses, motion capture technology, and linear kinematics to investigate how the human central nervous system generates skilled hand and mouth movements. She is especially interested in how these movements and their underlying neural substrates arose through evolution, are established during development, and break down in various neurological disorders. She is currently investigating the differential contribution of visual versus tactile sensory inputs to each of these processes.

Leah Krubitzer is currently a Professor in the Department of Psychology and Center for Neuroscience at the University of California, Davis. She received a BS at Penn State University in Communication Disorders and a PhD in Psychology at Vanderbilt University, Nashville Tennessee. Her graduate work, under the mentorship of Dr. Jon Kaas, focused on the evolution of visual cortex in primates. Her interest in the evolution of the neocortex was extended in her postdoctoral work at the University of Queensland, Australia to include a variety of mammals such as monotremes and marsupials. While in Australia she performed comparative analysis on the neocortex of a variety of different species and to date has worked on the brains of over 37 mammals. Her current research focuses on the impact of early experience on the cortical phenotype, and she specifically examines the effects of the sensory environment on the development of connections, functional organization and behavior in normal and visually impaired mammals. She also examines the evolution of sensory motor networks involved in manual dexterity, reaching and grasping in mammals. She received a MacArthur award for her work on evolution.

Marie Monfils is currently an Associate Professor in the Department of Psychology at the University of Texas at Austin. She received her PhD in Behavioral Neuroscience from the Canadian Centre for Behavioural Neuroscience, under the supervision of Bryan Kolb and Jeffrey Kleim, and her Master's degree from the University of Calgary with Cam Teskey. She conducted a postdoctoral fellowship at New York University in Joseph LeDoux's lab. Marie's lab is currently examining the mechanisms by which post-consolidation manipulations can persistently attenuate fear memories, and isolating the factors that underlie the social transmission of fear.

Morris Moscovitch, (BA, McGill, 1966; PhD, University of Pennsylvania, 1972; Postdoctoral, Montreal Neurological Institute, 1973–1974) holds the Glassman Chair in Neuropsychology and Aging at the University of Toronto. Born in Romania, he moved to Israel at four and to Canada at seven. In 1971, he joined the Mississauga (Erindale) Campus, and moved to the St. George Campus in 2000. He is a senior scientist at the Rotman Research Institute (1989–present), and was a visiting professor at the Hebrew University (1978–79; Institute for Advanced Studies 1985–1986), and at the University of Arizona (1996, 1999–2000). Best known for his work on the cognitive and brain basis of memory, he also has made important contributions to research on face-recognition, attention, and hemispheric specialization. His component process model of memory posits that the neural structures mediating memory encoding, retention and retrieval depend on interactions between the nature of memory representations and task demands. His Multiple Trace Theory and Trace Transformation Theory account for hippocampal-neocortical interactions in systems level consolidation (see Moscovitch et al., *Annual Review of Psychology*, 2016). He has published over 300 papers, edited five books, and served as Co-Editor-in Chief of *Neuropsychologia*. A Fellow of Divisions 3 and 6 of APA, of AAAS and of The Royal Society of Canada, Morris is the recipient of lifetime achievement/distinguished career awards for his research, including the Hebb Award (2007) and the William James Award (2008), and of teaching and mentorship awards from his department (2003), his University (2015) and Women in Cognitive Science (2005).

Sarah Raza is a PhD student in the Department of Pediatrics at the University of Alberta. Her research examines the developmental pathways and risk factors leading to the emergence of autism spectrum disorder (ASD) in a cohort of high-risk infants, with particular emphasis on the reciprocal relationships between attention control and emotional regulation. Sarah previously completed her MSc in behavioral neuroscience at the University of Lethbridge, where she investigated the role of early experiences on animal behavior, brain development, and plasticity. Her research pursuits have included examining the synergistic brain–behavior relationship and abnormalities underlying ASD utilizing a rodent model. As well, she has studied the effects of executive function training, antipsychotic drug exposure, and therapeutic interventions on the developing brain and subsequent behavioral outcomes.

Terry E. Robinson received his PhD in Psychology from the University of Western Ontario in 1978, and after postdoctoral training he joined the Department of Psychology at The University of Michigan in Ann Arbor. Dr. Robinson is known internationally for his research concerning the persistent behavioral and neurobiological consequences of repeated psychostimulant drug use, and the implications of these for addiction and relapse. His present research focuses on individual differences in the propensity to attribute incentive motivational properties to cues associated with rewards, and how this may predispose some individuals to develop impulse control disorders, such as addiction. He has published over 225 articles, and his papers have been cited over 36,000 times (h=85; he is listed on ISIHighlyCited.com as one of the highest cited [top 0.5%] scientists in Neuroscience). Awards include the D.O. Hebb Distinguished Scientific Contribution Award from the American Psychological Association (APA), the Distinguished Scientist Award from EBPS, the William James Fellow Award for Lifetime Achievement from APS, and he shared the Award for Distinguished Scientific Contributions from APA with his colleague, Kent Berridge. He is currently the Elliot S. Valenstein Distinguished University Professor of Psychology & Neuroscience at Michigan, and President of the European Behavioral Pharmacology Society (EBPS).

Melanie J. Sekeres obtained her PhD in Physiology from the University of Toronto, and is an Assistant Professor in the Department of Psychology & Neuroscience at Baylor University. She studies memory consolidation processes in humans and rodents using a combination of functional neuroimaging and molecular genetics.

Matthew Shapiro has been studying how the brain remembers for more than 20 years, starting with his graduate studies with David Olton at Johns Hopkins University. His work investigates the mechanisms that let neural circuitry encode momentary experiences into representations that can be retrieved selectively to guide behavior even after decades. Focusing on encoding mechanisms while in the Psychology department at McGill University he discovered that the same mechanisms required for rapid connection changes in the hippocampus, NMDA dependent long-term potentiation, are also needed to establish stable place field maps that support spatial working memory. His laboratory started investigating the brain mechanisms of memory retrieval mechanisms after moving to the Department of Neuroscience at the Icahn School of Medicine at Mount Sinai in 1999. His students found that hippocampal place fields represent more than current location, but encoded temporally extended spatial representations that include the recent past and future goals. They also demonstrated that internal context, defined by hunger or thirst, were stored along with perceptual information as part of memory representations that help distinguish past experiences in identical spatial contexts. His laboratory is now investigating how goals help form and retrieve memories, studying how distributed brain circuits communicate by simultaneously recording the activity of neuronal populations in the prefrontal cortex, circuits that compute abstract rules and outcome expectancies, together with the hippocampus, during decision-making.

Stephen J. Suomi, PhD is Chief of the Laboratory of Comparative Ethology at the Eunice Kennedy Shriver National Institute of Child Health & Human Development (NICHD), National Institutes of Health (NIH) in Bethesda, Maryland. He also holds research professorships at the University of Virginia, the University of Maryland, College Park, the Johns Hopkins University, Georgetown University, the Pennsylvania

State University, and the University of Maryland, Baltimore County. Dr. Suomi earned his BA in psychology at Stanford University in 1968, and his MA and PhD in psychology at the University of Wisconsin-Madison in 1969 and 1971, respectively. He then joined the Psychology faculty at the University of Wisconsin-Madison, where he eventually attained the rank of Professor before moving to the NICHD in 1983. Dr. Suomi's initial postdoctoral research successfully reversed the adverse effects of early social isolation, previously thought to be permanent, in rhesus monkeys. His subsequent research at Wisconsin led to his election as Fellow in the American Association for the Advancement of Science 'for major contributions to the understanding of social factors that influence the psychological development of nonhuman primates.' His present research at the NICHD focuses on three general issues: the interaction between genetic and environmental factors in shaping individual developmental trajectories, the issue of continuity versus change and the relative stability of individual differences throughout development, and the degree to which findings from monkeys studied in captivity generalize not only to monkeys living in the wild but also to humans living in different cultures. Throughout his professional career Dr. Suomi has been the recipient of numerous awards and honors, the most recent of which include the Donald O. Hebb Award from the American Psychological Association, the Distinguished Primatologist Award from the American Society of Primatologists, and the Arnold Pffeffer Prize from the International Society of Neuropsychoanalysis. To date, he has authored or co-authored over 400 articles published in scientific journals and chapters in edited volumes.

Robert J. Sutherland is a Board of Governors' Research Chair in Neuroscience, Director of the Canadian Centre for Behavioural Neuroscience, and a Professor of Neuroscience at The University of Lethbridge, Alberta, Canada. Sutherland obtained his BSc from the University of Toronto and PhD from Dalhousie University, as well as his postdoctoral training in Neuropsychology at The University of Lethbridge. In addition, he has served as a faculty member at the University of New Mexico, the University of Colorado, and Norwegian University of Science of Technology in Trondheim. Much of his work is focused on the neurobiology of cognition, especially on neural processes involved in normal and pathological memory. He has authored more than 300 publications on his research results. His work has recently been funded by the Alberta Heritage Foundation for Medical Research, Canadian Institutes of Health Research, Natural Science and Engineering Research Council of Canada, the Canada Foundation for Innovation, and the National Institutes of Health.

Gordon Winocur is Senior Scientist at the Rotman Research Institute, Professor of Psychiatry and Psychology at the University of Toronto, and Professor of Psychology at Trent University. His research is concerned with cognitive changes associated with selective brain damage, normal aging, neurodegenerative disease, and chemotherapy, in humans and animal models. His work is supported by the Canadian Institutes of Health Research, Natural Sciences and Engineering research Council, and the Breast Cancer Research Foundation.

Preface

When Michael Carmichael at SAGE Publishing first approached us about this book the idea was to provide background reading as a companion to existing materials on a brain and behavior course. This was quite a challenge given the breadth of studies in brain and behavior. Beginning with the field of physiological psychology in the 1930s and 1940s the emphasis was on the basic biological mechanisms of regulatory behaviors (e.g. feeding, drinking), and later expanded to nonregulatory behaviors (e.g. sex). However the field has expanded over the past 30 years as a general field of behavioral neuroscience has emerged and now includes cognitive neuroscience as well. Some current courses would focus on biological psychology, including some cognitive neuroscience, whereas others would be more molecular in nature. This book is in the first camp and we chose to identify classic studies that reflected our thinking about historical issues in brain and behavior. We believe that it could be used as companion information in a lower-level course in either behavioral neuroscience or cognitive neuroscience, but would also act as a stand-alone volume in more senior courses in which discussions would go well beyond the chapters here.

Bryan Kolb and Ian Q. Whishaw

1 An introduction to classic studies in behavioral neuroscience

Bryan Kolb and Ian Q. Whishaw

There have been literally tens of thousands of papers on brain organization and function published over the past 100 years, but the majority have focused on reducing complex phenomena of brain function to simpler (and often molecular) terms yet do not help our understanding of brain organization and function, and few stand out as "classics." The classics are distinguished by introducing new ideas that shape the direction of subsequent research.

Such was the book *The Treatise of Mind*, written by the French philosopher René Descartes in the 1600s, that proposed that the brain makes an important contribution to behavior. In contrast to earlier ideas about the control of behavior, Descartes placed the control of behavior in the brain, although he made a distinction between "mind" and "brain". For Descartes, the nonmaterial mind controlled the body through the brain and received information about the body from the brain, a position now referred to as dualism. Cartesian ideas were influential, and although there were difficulties with the details of his theory, it was dominant until the theory of materialism emerged in the mid-nineteenth century with the writings of Wallace and Darwin. Materialism held that rational behavior did not require a separate mind but instead could be explained simply by the activities of the brain and nervous system.

Yet a major problem remained – how was the brain organized and how did it work? Early ideas dating back to the phrenology of Gall and Spurtzheim in the 1800s suggested that functions were localized in the cerebral hemispheres. Although Gall placed no emphasis on the study of cases with brain damage, the neurologists of the latter part of the nineteenth century began to describe evidence from brain-damaged people purporting to show that at least language was localized, epitomized by Broca's claim that language was localized in the third frontal convolution, which came to be known as Broca's area. Although many neurologists were seduced by the idea of localization of functions in the late nineteenth and early twentieth centuries, there were other writers – most notably John Hughlings Jackson and Karl Lashley – who argued that many functions were distributed and the brain could not be understand by any strict theory of localization.

The emergence of noninvasive imaging techniques in the late twentieth century has reinforced this view, and the emphasis has shifted more to networks of cerebral functioning and away from earlier debates about localization of function. Nonetheless, big questions regarding how the brain is organized remained to be formalized.

This book identifies 17 papers and one book that we consider should be considered classics because they provide key insights into the big questions related to brain and behavior. We wanted to show how these classics have influenced current ideas about brain and behavior, so we selected active researchers or research groups whose research interests were related to specific classics and asked them to write a chapter on the relevant classic study, explaining its influence on subsequent research.

Selecting the classic studies by anyone is obviously biased, and in our case this is related to our experience both as researchers and textbook writers. Undoubtedly some people may wonder why some studies were not included and why others were. We did our best to include those writings that in our view had the most influence on subsequent research and on current thinking about the big questions in behavioral neuroscience. We divided our search into four general categories: cerebral organization, cortical functions, chemicals and behavior, and brain plasticity. Although it was tempting to include topics that occupied many researchers at specific times over the past 100 years – such as the fundamental regulatory mechanisms of feeding and drinking and the organization of the synapse – we decided to focus on those topics that we felt had had the largest impact on our understanding of the broad questions of psychology and behavioral neuroscience. What is striking about these chapters is the unique interpretations on the development of brain research that feature in each of the contributions.

CEREBRAL ORGANIZATION

As we set about writing the original edition of *Fundamentals of Human Neuropsychology* in the late 1970s, we were struck by the absence of a coherent theory of how the brain was organized as a whole. There were early attempts to do this but most had little lasting impact. Several more modern writers, however, did have an impact. In a series of books and papers Alexander Luria provided the first clear theory of the functional organization of the brain, and especially the cerebrum. Although his model may seem simple today, we found it empowering both because students could understand it easily and because it led others to consider this very big question. The idea that there are functional maps in the cerebrum is obvious today, but the original studies by Wilder Penfield and his colleagues were groundbreaking, and although they were wrong in details they revolutionized how we think about the organization of the sensory and motor systems.

Later work, especially by Jon Kaas and his many colleagues using more sophisticated electrophysiological techniques, showed that Penfield's early maps were far more complex than they first appeared, leading to a shift in thinking that remains today on how cerebral maps represented the external world. But how is the map information related to the rest of the brain? This is where the anatomical and behavioral studies of Leslie Ungerleider and Mort Mishkin led to an influential shift towards parallel streams of cortical processing in the visual system, a story that has had a major influence on the past 25 years of thinking on cerebral organization. Finally, although Broca suggested that there was lateralization of language in the brain, it was the split-brain studies of Sperry and his students that revolutionized our thinking about cerebral asymmetry and the lateralization of cognitive functions.

CORTICAL FUNCTIONS

Although behavioral neuroscience encompasses both cortical and subcortical regions, it is our view that it is the organization of the cerebral cortex, and its relationship to subcortical regions, that are central to our understanding of brain organization. We begin with Donald Hebb's book, the *Organization of Behavior*. We would be hard pressed to identify any other single paper or book that has so influenced our thinking about brain and behavior in the past 100 years. Owing to the importance of Hebb and his book, Richard Brown's chapter is somewhat broader than the others as it includes historical information related to Hebb, his book, and his legacy. Arguably Hebb's most influential student, Brenda Milner, combined with William Scoville to present the case of H.M. – a case that changed our understanding of memory. At about the same time Paul MacLean's writings on the limbic system and motivation had an equally influential impact on how we currently view the control of emotion. While many of the details of the Scoville and Milner and MacLean papers are now known to be wrong, these publications have had an amazing impact, with Scoville and Milner being cited over 5,400 times.

Perhaps the only case in behavioral neuroscience that is as well known as H.M. is that of Phineas Gage. It is descriptive rather than scientific but the impact on our understanding and study of emotional behavior and the frontal lobes is extraordinary, leading to a rich literature on neurology and psychology. Finally, the 2014 Nobel Prize in physiology or medicine went for research on the brain's Global Positioning System, which began with the studies of John O'Keefe and his colleagues.

CHEMICALS AND BEHAVIOR

This is a huge topic but there are two papers that stand out as classics. The first is by Charles Phoenix and colleagues who transformed how sex differences in sexual behavior were believed to develop. Looking at the field more than 50 years later, this paper clearly led to extensive work that has had an enormous impact on the fundamental topic of sexual differentiation, which continues to fascinate students and researchers. The second is Roy Wise and colleagues' paper on catecholamine theories of reward. Like the Phoenix et al. paper, the Wise et al. paper stimulated a lot of new research that has moved the field to a completely different place than it started from – that it was not just the structure of a specific brain region but also its chemical function that shapes behavior. The question of why things are rewarding remains central to the whole field of motivated behavior.

BRAIN PLASTICITY

One of the fundamental paradigm shifts in behavioral neuroscience was the development of the concept, and now the field, of brain plasticity. A PubMed search under brain plasticity has over 29,000 papers, which is truly remarkable. It was a challenge to choose classic papers, but we settled on five. Harry Harlow was ranked by a 2002 survey as being the 26th most cited psychologist of the twentieth century. While he also did influential work on learning, it was his work on the role of maternal deprivation in infant development that had a transformational role in our understanding the role of attachment in brain and behavioral development. Some of Harlow's early work may now seen to be dramatic, but developmental influences are often dramatic. His original findings clearly were serendipitous and classic.

About the same time that Harlow was studying development in infant monkeys, the Berkeley group including David Krech, Mark Rosenzweig, Edward Bennett, and Marion Diamond made the remarkable finding that experience produced neurochemical and neuroanatomical changes in the brain. Although initially greeted with skepticism, their work was seminal in showing that experience can chronically change the brain. This research did not include electrophysiological studies, but the paper by Timothy Bliss and Terje Lømo revealed a different way to study plasticity, a phenomenon now known as long-term potentiation (LTP). LTP has become the most widely-studied experimental model of how the brain stores memories.

We noted above that Penfield's studies of brain stimulation revolutionized the study of the organization of cerebral maps. The study by Timothy Pons and his colleagues took advantage of the brain stimulation technique to demonstrate unexpectedly large changes in the somatosensory maps of monkeys (known as the Silver Spring monkeys) with chronic deafferentation of an arm. Like Harlow's work, the Silver Spring monkey studies were controversial, but the final chapter in their legacy has had an important and lasting impact on ideas of brain plasticity.

Finally, Per Roland and colleagues were among the first to show plastic changes in the activity of the human brain during the performance of a simple motor task. While primitive by today's standards, this work set in motion a new way to examine plastic changes in brain functioning – the brain could be directly observed as it did its business. Neuroimaging has advanced rapidly since the Roland studies of the 1980s, and no doubt there will be more advances as current techniques are improved and new techniques are developed, but the key point remains that cerebral blood flow and metabolism change with experience.

CONCLUSION

Behavioral neuroscience is an exciting and rapidly changing field. However, one central point is that big questions in behavioral neuroscience are not new and not likely to change very much in the near future. The classic papers that we chose range from 1937 to 1988 but most are from the 1960s and 1970s. The key issues revolve around how the brain is organized, how the cerebral cortex works as a unit, and how the brain changes with experience. If we fast forward ahead 30 years, there will be thousands more papers but most will be focused on specific issues and not on the big questions of brain and behavior, and those that are on those questions will have originated from the classic studies highlighted here.

One of the fundamental pieces of advice for students of the brain is that understanding the big questions means reading old books and papers. Many if not all of these papers are not easily available as PDF files, but that is no reason not to revisit them. Their wisdom will continue to guide us for a long time. While most of us will do experiments looking at smaller questions, understanding what the big questions are is essential both to understanding the organization of the brain and behavior.

PART 1

Cerebral Organization

2 Revisiting Luria: The organization of higher cortical functions

Bryan Kolb

The fundamental question in neuroscience at the beginning of the twentieth century and now is how the brain, and especially the cerebral cortex, is organized. For students new to studying the brain this is especially daunting because there are so many parts to the brain, both at a neuroanatomical level as well as cellular and molecular levels. The challenge is to see the big picture without being buried in the details. The writings of the Russian neuropsychologist Alexander Luria provided a simple, yet elegant model to conceptualize cerebral organization. Before getting to Luria, however, we need to consider earlier conceptualizations.

THE DIAGRAM MAKERS

The two extreme historical views were that of localization versus nonlocalization of function. Early examples of localization were provided by Broca, Wernicke, and Fritsch and Hitzig. They identified specific regions (e.g. Broca's area) that were associated with discrete functions. In order to explain how these regions worked together, the investigators drew arrows connecting one region to the next, and so they came to be known as "diagram makers."

Early nonlocalizatists, who rejected diagrams, included Hughlings Jackson, Fredrick Goltz, and later Karl Lashley and Norman Geschwind. They saw functions as being more distributed and Lashley went so far as to say that the cortex was "equipotential," meaning that all areas could take over functions of other areas. By the twenty-first century the question changed from the extreme positions, especially with the emergence of noninvasive imaging and the new field of connectonomics. Nonetheless, the more general question of how the brain is organized remains central to understanding the basis of psychological phenomena.

HIERARCHY

Perhaps the most influential idea of how the nervous system is organized came from Herbert Spencer's mid-nineteenth-century speculations that the brain is organized in functional levels, likely resulting from evolutionary change in which new levels were added, reflecting new levels of behavioral complexity. This general idea formed the basis of John Hughling-Jackson's writings, and by the mid-twentieth century underlay Paul MacLean's triune brain hypothesis. In its simplest form, MacLean's idea was that the organization of the human brain began with a reptilian brain, which added another functional layer (the limbic system) in the primitive mammalian brain (MacLean's Paleomammalian level), and finally the neomammalian complex, which consisted of the much-expanded neocortex found in more recent larger-brained mammals – and especially humans. MacLean's hypothesis was not supported by comparative neuroanatomical work in the past 50 years (e.g. see Heimer et al., 2008), but more importantly for our current discussion, it did not address the question of how the cerebral cortex was organized.

WHY DO WE NEED A CORTEX?

Before considering how the cortex might be organized, it is worth considering the broader question – what does the cortex do? One way to address this is to consider what a mammal without a cerebral cortex *cannot* do. Although one might suspect that the rat with the cortex removed would be paralyzed and unable to perceive very much, this is incorrect. Rats that have the cerebral cortex removed shortly after birth are very difficult to differentiate from their unoperated littermates (Kolb and Whishaw, 1981). In fact, we have a laboratory exercise in which undergraduates are challenged to identify which rats are decorticates versus normal when a mixture are placed together in a box. It is not obvious, but they are different. Decorticates eat and drink, their general motor abilities develop seemingly normally, they startle from unexpected sensory events, they mate and females can raise litters of pups, they interact socially, they have the rudiments of play behavior, and they can learn simple problems. In fact, David Oakley (1979) has shown that decorticate rats are nearly as good at classical conditioning, operant conditioning, and simple cue learning as normal animals.

Upon careful study, however, it is apparent that decorticate rats are severely handicapped at a wide range of behaviors. For example, they do not build nests, although they can move nesting material around. They also do not hoard food, but can carry it around, only to drop it randomly. They have great difficulty making sequences of skilled movements of the tongue or forelimbs, such as would be seen in protruding the tongue to obtain small food items or by reaching for food objects with one forelimb. They are also severely impaired at all but very simple pattern discriminations, sequential motor learning, or spatial learning (e.g. see Kolb and Whishaw, 1983). Further, the deficits in decorticate rats become especially clear as their spatial world grows. For example, if they are living in a standard laboratory shoebox cage they appear remarkably normal, but if they are released to live in a 9×12 foot room, they will not survive (Whishaw et al., 1981). They cannot organize their behavior in a larger space and would likely starve to death. Living in a more spatially demanding environment clearly requires the cortex.

In sum, the cortex is not necessary for simple sensory or motor behaviors and is also not necessary for learning per se. But if either a sensory or motor problem is the least bit complex, or a spatial map is required, or if any sort of planning or organization of movement sequences is needed, the cortex must be present. Yet how is it organized to do these tasks? This is where Luria provided some insight.

LURIA'S THEORETICAL ORGANIZATION OF THE CORTEX

Although Donald Hebb (1949) wrote about the principles underlying cortical functioning (see also *Beyond Hebb*), it was Alexander Luria (1962, 1973) who provided the first comprehensive model of the cortex as a series of functional systems. Although there had been general agreement for decades that the posterior cortical regions were more involved in sensory functions than the anterior cortex, which in turn was more involved in motor functions, Luria conceived the anterior and posterior regions as two distinct functional systems (see Figure 2.1). The first was the *sensory unit*, which included the parietal, temporal, and occipital cortex. Its function was to receive sensory impressions, process these, and store them as information. The second was the *motor unit*, which was located in the frontal lobe. The motor unit formulated intentions, organized them into programs of action, and executed the programs. Both cortical units had a hierarchical structure with three cortical zones arranged in a linear organization. In each case, the functional zones were based upon the idea that cortical regions could be described (and identified by Brodmann's cytoarchitectonic numbers) as primary, secondary, and tertiary.

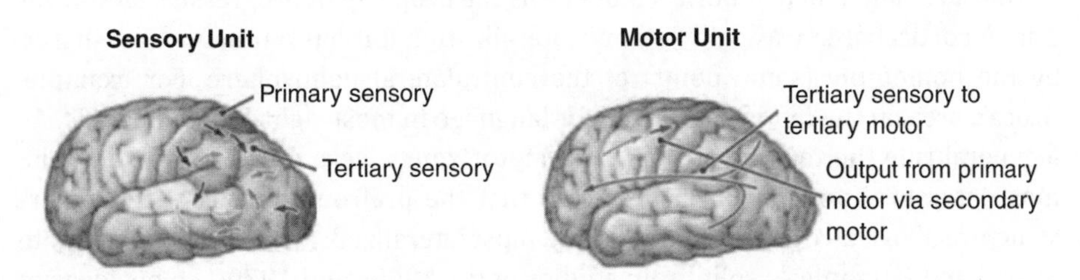

Figure 2.1 Luria's functional units of the cortex (after Kolb and Whishaw, 2015)

Primary sensory regions were those that received direct projections (via the thalamus) from the visual (Brodmann's area (BA 17), auditory (BA 41), and somatosensory (BA 1, 2, 3) receptors. In each of these zones the general features of sensory input were organized in an array representing the topography, intensity, and pattern of stimulation. The secondary zones were the adjacent cortical regions that received the outputs of the primary regions (e.g. BAs 18 and 19 in the visual cortex). The secondary zones retained the sensory modality (such as vision) but had less fixed topographic organization. The tertiary zones were the next level of sensory projections, largely receiving connections from the secondary zones, and represented about one-quarter of the posterior zone's total area. It is here that the sensory inputs were translated into symbolic processes and abstract thinking. Luria suggested that primary sensory zones showed specificity in that they analyzed only one sensory modality. The higher zones showed diminishing specificity as the different senses began to be blended to synthesize more abstract features of sensory experience. Luria's formulation suggested that our perceptions of the world are unified and coherent, which was in accord with the commonsense view of perceptions. The obvious place to form this perception was in the tertiary cortex.

The Polish neuroscientist Jerzy Konorski expanded on this idea in 1967 by proposing that the tertiary regions formed gnostic (knowing) areas that contained gnostic neurons that enabled stimuli to be categorized in spite of variations in the details of their presentation. He was not proposing that individual neurons represented specific stimuli, but rather that there is a redundant set of neurons that are most active when exposed to stimuli from the category they represent. Luria does not appear to have written about Konorski's ideas, although they were contemporaries in the Soviet Bloc, but the gnostic neurons provided a mechanism for Luria's tertiary sensory cortex to operate. Konorski's ideas have recently been revisited by Christopher Kanan (2013), who described a computational model of Konorski's theory.

Luria's primary motor zone was the motor strip (BA 4), which Luria saw as the final cortical command area that controlled the spinal cord. The secondary zone was the premotor cortex (BA 6, 8, 44), which prepared motor programs for execution. The tertiary zone comprised the prefrontal and anterior cingulate cortex, which includes at least 50% of the frontal lobe. Luria described this as the "superstructure above all other parts of the cerebral cortex."

One additional idea of Luria's model was the property of progressive lateralization. A cortical area was said to show lateralization if it has a function not shared by the homotopic (same point) of the contralateral hemisphere. For example, Broca's area on the left is lateralized for language in most right-handed people. As a general rule the tertiary sensory and motor zones were predicted to show the most lateralization, and Luria predicted that the prefrontal cortex – the superstructure above all others – would be the most lateralized. Luria did not anticipate Sperry and Gazzaniga's split brain studies of the 1960s and 1970s, so his ideas of the organization of laterality may now appear a bit too simple (see *Beyond Sperry*).

As the brain has evolved there is an obvious premium on efficiency because connections take up space. In the primate cortex these occupy about 40% of the entire volume. Klyachko and Stevens (2003) determined that when all possible connections between different brain regions are considered, adjacent areas have more connections than areas that are not adjacent, which is implicit in Luria's formulation.

Luria's model provided a theoretical basis to understand cortical organization and was a useful heuristic tool to move away from the strict localizational ideas of the brain diagram makers such as Broca, Wernicke, and the others that followed, in part owing to Brodmann's cytoarchitectonic map that begged for functions to be attached to different discrete anatomical areas. The beauty of Luria's formulation is that it used the known neuroanatomical organization of the cortex to provide a simple explanation for observations that Luria made daily in his neuropsychological clinic in Moscow.

THE BINDING PROBLEM

One serious difficulty with Luria's model is that newer data have shown that cortical processing is much less hierarchical than was believed in the 1960s. Only about 40% of the possible connections among regions within a sensory modality are actually found, meaning that tertiary areas, for example, do not receive input from all other areas. This makes it difficult to form a unified percept in a single location. This lack of connections to tertiary areas forms the basis of what has become known as the binding problem. That is, how can information from different parts of a sensory modality (e.g. color, form, movement) be bound together to form a unified sensory experience that we believe we have? A related problem is that even within the connections within a sensory modality, there is more than a simple feed forward arrangement that Luria assumed. Rather, cortical regions have reciprocal connections - a property known as re-entry - such that any cortical area can influence the area from which it receives input. Thus, if area A sends a connection to area B, area B reciprocates and returns a message to area A. This allows area B to influence the inputs it gets from area A.

A more philosophical issue with Luria's model is that although it is easy to think of functions as sensory or motor, it is very difficult to make a clear sensory/motor distinction. Movements produce changes in sensation, and sensation is partly a motor function because it provokes movement. Indeed, motor and sensory functions, at least at the primary levels, are intertwined and difficult to distinguish.

DISTRIBUTED HIERARCHICAL PROCESSING

In Luria's formulation the connections from primary to secondary to tertiary in the sensory unit were sequential. Luria's idea was corroborated by the evidence of Hubel and Wiesel (1962, 1965) that the there was a progressive increase in the complexity of the properties of neurons in the visual cortex of cats. They saw this as the result of a serial feed forward scheme going from primary to secondary regions. By the 1980s this idea was difficult to reconcile with emerging neuroanatomical evidence. Felleman and van Essen (1991) summarized research on the connectivity of somatosensory and visual regions in the monkey. First, there were more discrete sensory regions than was predicted from Brodmann's maps. In the visual system they identified 32 cortical regions with extensive visual inputs (compared to Brodmann's 3) and 13 regions with somatosensory inputs (compared to Brodmann's 3). They identified 306 connections among the visual areas and 62 connections among somatosensory regions. They estimated that these connections represented only about 40% of the connections that would be expected if all the regions were connected, and identified a significant hierarchical organization within these connections, but also noted that many connections were parallel, leading them to propose the *distributed hierarchical model* illustrated in Figure 2.2. In contrast to a serial processing model, this model features several levels of processing with considerable processing within levels, likely reflecting different elements of the sensory experience. Thus, their model features multiple intertwined processing streams. Note that some connections skip levels and the number of areas expands as the hierarchy unfolds, which is the opposite of what Luria would have predicted.

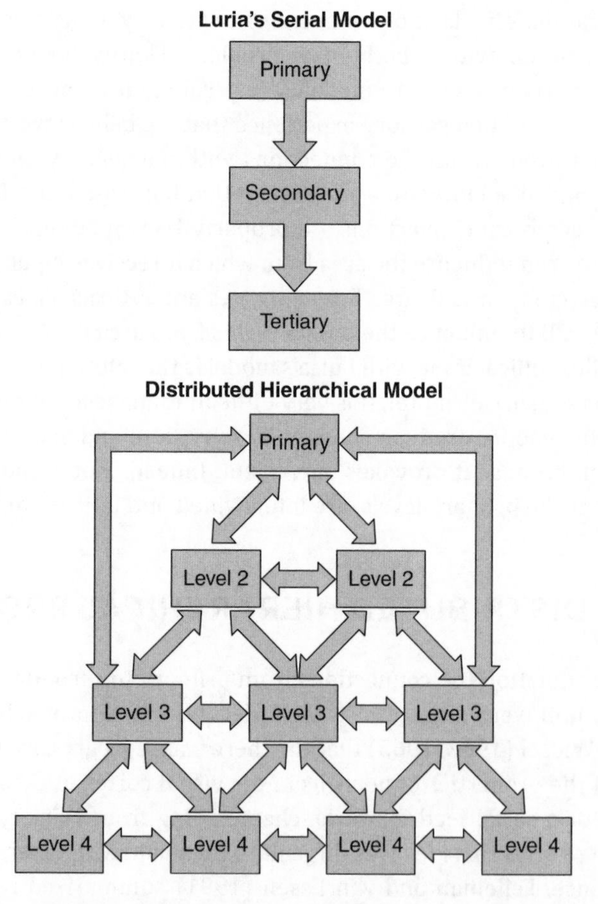

Figure 2.2 Two hierarchical models: (a) Luria's simple serial hierarchical model of cortical processing; (b) Felleman and van Essen's distributed hierarchical model featuring multiple levels of association areas interconnected with one another at each level (after Kolb and Whishaw, 2015)

STREAMS OF PROCESSING

Ungerleider and Mishkin (1982) originally conceived of two parallel visual pathways: a dorsal pathway projecting into the parietal cortex and a temporal pathway projecting into the temporal cortex. Their original conception has changed considerably (see *Beyond Ungerleider and Mishkin*), but the key point here is that there are two fundamentally different forms of visual processing. The dorsal pathway, which is unconscious and controls movements in space, contrasts with the ventral pathway, which is conscious and processes object quality. Like Luria's model, these pathways were originally conceived as serial pathways, but in keeping with the general idea of Felleman and van Essen's distributed hierarchical model, they are now believed to include several distinct cortical and subcortical systems (e.g. Kravitz et al., 2011, 2013). The dorsal stream is still reminiscent of Luria's general formulation, however, as there are subpathways to the premotor and prefrontal cortex that eventually influence motor output. Similarly, portions of the ventral pathway eventually terminate in the prefrontal cortex to be involved in object–reward associations and object working memory, two functions that Luria would have predicted to be in the "superstructure above all other parts of the cerebral cortex." Thus, in a general sense, Luria's model remains heuristically useful.

THE CONNECTOME

The Felleman and van Essen model and the evolution of the dual streams of visual processing model reflect a dynamic interplay between and among the operations of cortical regions. In this view, cortical regions should not be regarded as independent processors of specific information but rather as areas that act conjointly, forming large-scale neural networks (e.g. Meehan and Bressler, 2012). The challenge is to identify which connections such as those described by Felleman and van Essen form functional networks. Because the living brain is always active, it is possible to correlate activity in different cortical regions when the brain is engaged in specific tasks or even when the brain is not engaged in any task, the latter process known as resting-state functional magnetic resonance imaging (rsfMRI). Yeo et al. (2011) utilized such data from 1000 participants and were able to parcellate the cortex into 17 networks, including sensory and motor networks as well as networks largely comprised of the tertiary cortex. Although the primary sensory and motor networks were largely local, as adjacent areas tended to show strong functional coupling with one another, the higher order networks were widely distributed, including regions in prefrontal, posterior parietal, anterior temporal, and midline areas. This type of work has led to an ambitious project known as the Human Connectome Project (www.humanconnectome.org). When complete, the project will have generated functional connections between every 1-cubic-millimeter gray matter location in the cortex of 1200 humans (van Essen and Ugurbil,

2012). Given the human brain's size, the Connectome involves about 100 investigators who anticipate working on it until 2020.

Another way to analyze functional cortical organization is to cross correlate the activity across humans who are given specific information to process. Hasson et al. (2004) individually placed five participants in an MRI and had them view a 30-minute clip of a movie, *The Good, The Bad, and The Ugly*. Two features of the results are germane here. First, the ventral stream was coincidently active in all participants, as were the auditory processing regions. Second, although there was a general activation of the temporal lobe throughout the clip, there were selective activations related to the precise moment-to-moment film content. For example, the fusiform face area was especially active when participants viewed close-ups of faces. In contrast, the parahippocampal place area was especially active when viewing large scenes. There was no coherence in other brain regions, especially in the frontal lobe. We can conclude, therefore, that the ventral stream acts like an extensive visual network for analyzing the content of the movie, with subregions engaged in selective analyses of specific content. This activity was likely sent on to the frontal lobe, as Luria would predict, but because people differ in how they would interpret the information, there was no frontal coherence across people. Curiously, when the same participants viewed the same film clip on different occasions, there was also no coherence within the same people from time to time, likely because the frontal lobe interpretation of the content varied over time.

A parallel example of distributed processing is found in a recent study by Huth et al. (2016). These authors used fMRI to track cerebral activity when participants listened to natural narrative speech. They found that semantic concepts are represented in multiple cerebral regions and each region represented multiple semantic concepts. For example, one area responded to words related to people whereas another responded to numbers. These semantic maps were largely consistent across people (see an interactive version of the atlas at http://gallantlab.org/huth2016).

THE SOCIAL BRAIN

Luria's formulations were focused on cognitive and motor behaviors, but the same general contemporary principles of cortical organization should apply to socioaffective behaviors as well. Daniel Kennedy and Ralph Adolphs (2012) reviewed studies of brain-injured patients and fMRI activation in healthy participants to search for social networks. They identified four separate networks that were essentially of tertiary cortex in the parietal, temporal, and frontal lobes. In contrast to the resting state networks described by Yeo et al. (2011), the social networks did not include any primary sensory cortex, nor motor or premotor cortex. The absence of primary sensory regions makes sense in Luria's formulations because the cognitive processes would be fairly abstract. The absence of any motor or premotor activation, in the presence of a lot of prefrontal activation, makes sense too because the participants were not making movements but rather were responding to social and emotional materials. In Luria's formulation the sensory information went to the tertiary motor area (prefrontal cortex) but no further because no movements needed to be selected or made.

CONCLUSION

In sum, Luria proposed a universal theoretical model of cortical organization. It is difficult to determine how influential it was beyond behavioral neuroscience, but certainly within behavioral neuroscience it provided a simple heuristic model. When I began teaching neuropsychology in the mid-1970s I embraced his model because the students (and I) could understand it as a way of conceptualizing cortical organization. Ian Whishaw and I used it as a central concept in the first edition of our neuropsychology text, *Fundamentals of Human Neuropsychology* (in 1980), and over 35 years later it is still useful, although as we have seen, much more is known about the details of cortical connectivity than was available to Luria in 1962.

Of course, the final model of cerebral organization has yet to be written. New studies using Diffusion Tensor Imaging and rsfMRI are giving us a much more different view of the cerebral connections. An obvious problem is that as we describe more connections, it becomes more difficult to conceptualize the organization. A real challenge for the Connectome Project will be to identify simple pedagogical principles of cerebral organization that will make it possible for students and researchers to grasp the big picture of cortical organization.

REFERENCES

Felleman, D.J. and van Essen, D.C. (1991) Distributed hierarchical processing in the primate cerebral cortex. *Cerebral Cortex*, 1: 1–47.

Hasson, U., Nir, Y., Levy, I., Fuhrmann, G. and Malach, R. (2004) Intersubject synchronization of cortical activity during natural vision. *Science*, 303: 1634–40.

Hebb, D.O. (1949) *The Organization of Behavior*. New York: McGraw-Hill.

Heimer, L., Van Hoesen, G.W., Trimble, M. and Zahm, D.S. (2008) *Anatomy of Neuropsychiatry: New Anatomy of the Basal Forebrain and its Implications for Neuropsychiatric Illness*. Amsterdam/Boston, MA: Academic Press/Elsevier.

Hubel, D.H. and Wiesel, T.N. (1962) Receptive fields, binocular interaction and functional architecture in the cat's visual cortex. *Journal of Physiology*, 160: 106–54.

Hubel, D.H. and Wiesel, T.N. (1965) Receptive fields and functional architecture in two nonstriate visual areas (18 and 19) of the cat. *Journal of Neurophysiology*, 28: 229–89.

Huth, A.G., de Heer, W.A., Griffiths, T.L., Theunissen, F.E. and Gallant, J.L. (2016) Natural speech reveals the semantic maps that tile human cerebral cortex. *Nature*, 532: 453–8.

Kanan, C. (2013) Recognizing sights, smells, and sounds with Gnostic Fields. *PLOS ONE*, 8: e54088.

Kennedy, D.P. and Adolphs, R. (2012) The social brain in psychiatric and neurological disorders. *Trends in Cognitive Sciences*, 16: 559–72.

Klyachko, V.A. and Stevens, C.F. (2003) Connectivity optimization and positioning of cortical areas. *Proceedings of the National Academy of Sciences, USA*, 100: 7937–41.

Kolb, B. and Whishaw, I.Q. (1980) *Fundamentals of Human Neuropsychology*. San Francisco, CA: W.H. Freeman & Co.

Kolb, B. & Whishaw, I.Q. (1981) Decortication of rats in infancy or adulthood produced comparable functional losses on learned and species typical behaviors. *Journal of Comparative and Physiological Psychology*, 95: 468–83.

Kolb, B. & Whishaw, I.Q. (1983) Dissociation of the contributions of the prefrontal, motor and parietal cortex to the control of movement in the rat. *Canadian Journal of Psychology*, 37: 211–32.

Kolb, B. and Whishaw, I.Q. (2015) *Fundamentals of Human Neuropsychology*, 7th edition. New York: Worth.

Konorski, J. (1967) *Integrative Activity of the Brain*. Chicago, IL: University of Chicago Press.

Kravitz, D.J., Dadharbatcha, S.S., Baker, C.I. and Mishkin, M. (2011) A new neural framework for visuospatial processing. *Nature Reviews Neuroscience*, 12: 217–30.

Kravitz, D.J., Saleem, K.S., Baker, C.I. Ungerleider, L.G. and Mishkin, M. (2013) The ventral visual pathway: An expanded neural framework for the processing of object quality. *Trends in Cognitive Sciences*, 17: 26–49.

Luria, A.R. (1962) *Higher Cortical Functions in Man.* Moscow: Moscow University Press.

Luria, A.R. (1973) *The Working Brain*. New York: Basic Books.

MacLean, P. (1990) *The Triune Brain in Evolution: Role in Paleocerebral Functions*. New York: Plenum.

Meehan, T.P. and Bressler, S.L. (2012) Neurocognitive networks: Findings, models, and theory. *Neuroscience and Biobehavioral Reviews*, 36: 2232–47.

Oakley, D.A. (1979) Cerebral cortex and adaptive behavior. In D.A. Oakley and H.C. Plotkin (eds), *Brain, Evolution and Behavior.* London: Methuen.

Ungerleider, L.G. and Mishkin, M. (1982) Two cortical visual systems. In D.J. Ingle, M. Goodale and R.J.W. Mansfield (eds), *Analysis of Visual Behavior*. Cambridge, MA: MIT Press. pp. 549–86.

Van Essen, D.C. and Ugurbil, K. (2012) The future of the human connectome. *NeuroImage*, 62: 1299–1310.

Whishaw, I.Q., Nonneman, A.J. and Kolb, B. (1981) Environmental changes can improve grooming, swimming and eating in decorticate rats. *Journal of Comparative and Physiological Psychology*, 95: 792–804.

Yeo, B.T., Fienen, F.M., Sepulcre, J., Sabuncu, M.R., Lashkari, D. et al. (2011) The organization of the human cerebral cortex estimated by intrinsic functional connectivity. *Journal of Neurophysiology*, 106: 1125–65.

3 Revisiting Penfield and Boldrey: Somatic motor and sensory representation in the cerebral cortex of man as studied by electrical stimulation

Ian Q. Whishaw

THE MYSTERY OF THE MOTOR CORTEX

The search for the function of the neocortex, and the motor cortex in particular, resembles the search for the guilty party in an old fashioned whodunnit. No sooner is convincing evidence assembled against one suspect than evidence against another comes to light, and so forth, until suspicion once again falls on one or other of the previous suspects. Even then, some suspects seem to have been cleared while others remain under suspicion, and the case remains unsolved.

Wilder Penfield and Edwin Boldery's (1937) paper, *Somatic motor and sensory representation in the cerebral cortex of man as studied by electrical stimulation*, provided evidence for one theory of the organization of the motor cortex. As a testament to its influence over the 75 years since the paper was published, it has received about 2,500 citations and its central idea is highlighted in every textbook that discusses cortical function.

This chapter will first give a brief review of the history of motor cortex stimulation to provide a context for Penfield and Boldery's findings. It will then present a new interpretation of the effects of motor cortex stimulation that stresses natural actions, as well as some findings that show how the motor cortex interacts with the parietal cortex and how it participates in streams of action that begin in the visual cortex. It will then conclude with studies and curious findings that suggest that there are still many unknowns related to motor cortex function and that these will serve as the basis for future investigations.

A SHORT HISTORY OF MOTOR CORTEX STIMULATION

The first investigators to find that electrical stimulation of the motor cortex could produce movement were Gustav Fritsch and Eudard Hitzig (1960 [1870]). Prior to their study, the neocortex was thought to be but a covering, a bark, protecting underlying brain regions. For example, in René Descartes' (1664) theory of brain function, the cortex was a protective covering for the pineal body, which served as the interface between the mind and the body. Even much later, as the cortex was accepted to be an important part of the brain, it was thought to be unresponsive to electrical stimulation, whereas such stimulation produced movement when applied to other regions of the brain. The cortex was also thought to function as a whole, such that when a part was damaged the remaining cortex could take over the functions of the missing part. These features of cortical function were seen as being consistent with its role in representing the mind, which, as a nonmaterial and indivisible entity, would be immune to electrical stimulation or injury.

When Fritsch and Hitzig electrically stimulated the frontal lobes of a dog, they found that they could elicit movements on the opposite side of the animal's body. Furthermore, they found that stimulation in one location produced movement of the neck, in another location movement of the foreleg, in another location movement of the hind leg, and in still another location movement of the face. At a stroke, they demonstrated that the cortex was electrically excitable and that motor functions were localized.

These findings were replicated in numerous studies of other species of animals. In an illustrative study, using a monkey, David Ferrier (1874) described a map-like arrangement of movement locations of the cortex, with the feet located more medially and the face more laterally, with an upside-down image of a monkey when viewed from the side of the brain. Ferrier represented his findings with numbered zones. Still other researchers obtained similar results and represented their findings with labels of body parts pasted onto the respective parts of the cortex from which movement of those parts was produced. As study followed study, experimenters refined their electrical stimulation procedures, with evermore stimulation sites and with briefer and briefer electrical pulses. The result was an extremely detailed description of a part of the cortex that came to be recognized as the motor cortex.

Along with these experimental changes in delivering electrical stimulation came more refined descriptions of the changes in behavior. In the earlier studies of Fritsch and Hitzig and Ferrier, somewhat longer and less controlled pulses of electrical current produced rather gross movements of the body, for example the closing of a hand or the movement of an arm accompanied by a seemingly grasping behavior. The more refined experimental methods produced twitches of body parts, of a part of the tongue, the lips, or a finger. In order to represent the behavior elicited at different cortical locations, experiments typically indicated body parts (e.g. shoulder, elbow, wrist, thumb, and so forth) by attaching more labels for such body parts to a drawing of the cortex. Eventually, Grunbaum and Sherrington (1901) proposed that different cortical regions were related to the dorsal roots of the spinal cord, each of which projected to different groups of muscles in the body. In light of these results, the motor cortex came to be viewed as a repository for elementary actions. The speculation was that more complex behavior would require configurations of the elemental actions.

THE HOMUNCULUS

It was in this tradition that Penfield and Boldery conducted their experiment on human elective surgery patients. Their paper is a landmark study not only in its detail but also in its methodology. To avoid producing epileptic seizures in patients, they kept their current voltage low and the duration of the electrical stimulation short and they explored a large number of points in many patients. As soon as the stimulation evoked a response, they turned it off. They also checked the responses they obtained by sometimes administering stimulation to patients without warning.

Not only were they conducting a study on human patients, their patients were also conscious. Therefore, in addition to observing the movements that the stimulation produced, the experimenters had the patients report on the induced sensations: the tingling and numbness and so forth that the stimulation produced. They were able to document that regions of the brain that produced movement were largely different from regions of the brain that produced sensation. The movement-evoking regions mainly were located anterior to the central fissure, whereas those areas that evoked body sensations were mainly posterior to the central fissure.

The way that Penfield and Boldery summarized all of their data, as a *homunculus*, was a *tour de force*. This homunculus was illustrated in their Figure 28. A homunculus is a misshaped caricature of a person, and this one features very large hands, lips and feet, and actually has two parts (a motor part and a sensory part). The motor homunculus is largely located on the precentral gyrus of the frontal lobe, with the feet up and the head down, while the body sensory homunculus is largely located on the adjacent postcentral gyrus of the parietal lobe (and is also standing on its head, so to speak). They described a second, smaller motor homunculus that was located medially and in front of the main motor homunculus. To distinguish these two motor areas, the cortical region of the main motor homunculus was called the primary motor region whereas the frontal medial region was called the supplementary motor region.

When the homunculus was superimposed upon a drawing of the cortical surface, with feet up and head down, it not only replaced the traditional method of labelling body parts, it also illustrated at a glance the functional relationship between cortical regions and body parts. The lips, hands and feet occupy a disproportionate area of the cortex. Thus, the homunculus reveals that our sensitivity in these body parts and our dexterity in their use are enabled by the relatively large amount of cortex devoted to their control. In short, we use tools and talk because we have so much cortex devoted to our hands and mouth.

Investigators who followed Penfield and Boldery described homologues to the homunculus in other animals, e.g. a *simunculus* for the monkey, a *ratunculus* for the rat, and so forth. They also used ever more refined techniques, including microelectrodes (tiny wires with micron-sized tips) over which millisecond duration pulses were delivered, so that ever smaller regions of the cortex could be stimulated selectively. These electrodes also penetrated the cortex so that the large neurons in layers 5 and 6, which send projections to other brain regions and the spinal cord, could be directly stimulated. The effects of the stimulation were also more objectively monitored using electrodes that recorded the twitches of single muscles or single muscle fibers.

The results of these studies were debated within the framework of whether the stimulation was activating movements or muscles. What comprised a movement was usually not specifically defined, but it likely meant the movement of a defined body part around a joint, a movement of a hand or a finger. The results favored the idea that the cortex was organized in terms of muscles and its ultimate role was to configure the contraction and relaxation of many muscles to produce movement (Chang et al., 1947). In addition, many more homunculi than had been described by Penfield and Boldery were found. For example, Galea and Darian-Smith (1994) proposed that rather than two motor humunculi there were as many as nine.

MOTOR CORTEX STIMULATION AND POSTURE

Michael Graziano, in his book *The Intelligent Movement Machine* (2009), describes how a group in his laboratory at Princeton University rediscovered the findings of Fritsch and Hitzig and of Farrier and then elaborated on these to propose a new model of cortical motor function. One of this group, Tiren Moore, stimulated the frontal eye fields, which are located further forward in the frontal lobes than is the motor cortex, using rather long trains of electrical stimulation, as was required for eliciting eye movements. He placed one electrode further back in the cortex and from this location he elicited a remarkable movement. The monkey reached out as if attempting to take hold of something. The same movement could also be repeatedly obtained, to the surprise of both monkey and experimenters. The animal eventually sat on its hand in order to prevent the movement from reoccurring.

Graziano and Moore, along with a graduate student Charlotte Taylor, then went on to do a systematic study of the frontal lobes to see what other movements they could elicit (Graziano et al., 2002). They studied the movements that they could elicit from each cortical site and gauged how the movements changed in relation to the posture of the animals. They found, for example, that from a particular cortical site that produced a reaching movement, this always took the hand to the same position irrespective of its starting point, as if it were goal-directed. If the hand began in a more medial position, extensor muscles were activated to take it to a lateral destination. If the hand was more laterally placed, flexor muscles were activated to bring it more medially. Some stimulation sites produced movements that positioned the hand to the mouth, as if the monkey were feeding, although there was nothing in its hand. As long as the stimulation was on, the hand remained at the mouth. If the hand was blocked on its way, the hand remained at the block as long as the stimulation was on. The movement brought the limb to its final posture and it had nothing to do with the behavioral context, i.e. it was mechanical.

By exploring the lateral surface of the cortex in monkeys, Graziano and his co-workers were able to identify movements that took the body to eight general positions. Along the central fissure, medially located stimulation brought the hand into lower body space, central stimulation brought the hand into a central position relative to the body, and ventral stimulation produced chewing and licking movements. Further forward in the cortex, medially located stimulation elicited postures of climbing and leaping, medial stimulation elicited a reaching position of the arm, ventral stimulation elicited defensive movements of the hand and face, and stimulation in the most ventral region brought the hand to the mouth (see Figure 3.1).

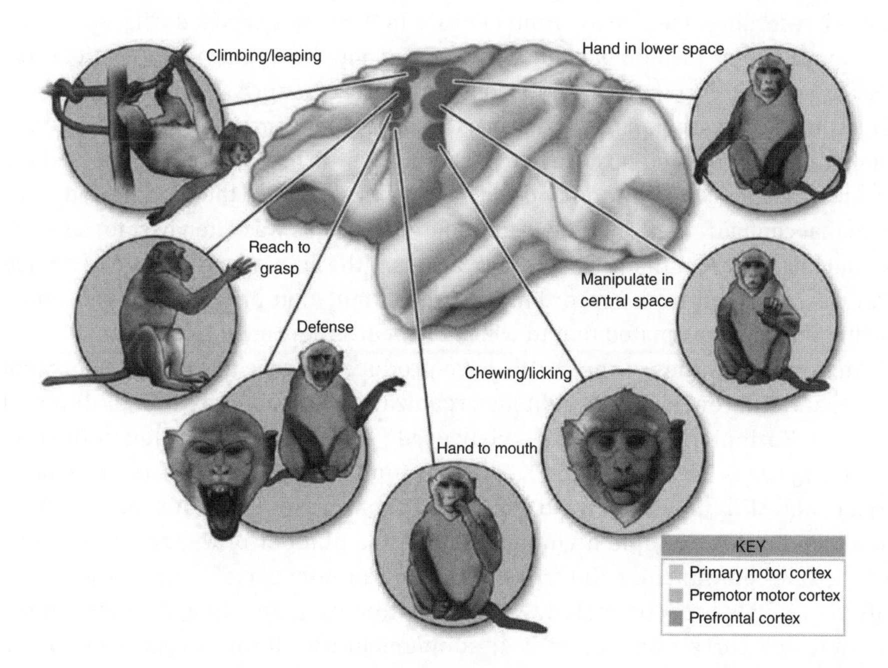

Figure 3.1 The brain's representation of biological action: topography of organized movements elicited from the frontal cortex of the rhesus monkey (adapted from Graziano and Afalo, 2007: 243, Figure 5.)

Although the Graziano rendition of motor cortex organization is different from all of the preceding maps of motor cortex, it does have a topographic organization. Sites that position the hand in lower body space are located more medially on the motor cortex and sites that position the hand in upper body space are located more laterally. Postures that involve more body joints (e.g. climbing) are more anterior, and those involving fewer joints (e.g. hand movement) are more posterior. Movements that might ordinarily go together, as represented by the idea "like attracts like", are closer to each other than are very different postures. For example, sites from which licking and chewing postures were obtained are close to sites that bring the hand to the mouth, and sites that evoke climbing are close to sites that evoke reaching movements. It is interesting that a number of previous studies, including those by Penfield and Boldrey, had remarked on defensive movements, but had interpreted these movements as the product of epileptic seizures elicited by the stimulation rather than as movements per se.

That the motor cortex represented a small set of organized movements was supported with two other lines of investigation. First, the behavior of monkeys in a more natural setting was examined, with the result that the postures elicited by brain stimulation were also those most frequently used by the monkeys. For example, reaching for objects, manipulating them in central body space, and placing them in the mouth were easily recognized and frequently occurring natural behaviors. Second, recordings were made from neurons in the motor cortex during spontaneous movements, including spontaneous arm and hand movements. Neuronal discharges were correlated with the joint angles that comprised different postures, for example eight joint angles of the arm during arm movements. The neurons were found to have preferred postural correlations. Their firing was high when the limb was in a certain configuration and their firing was lower as the limb moved away from that configuration. The tuning for different neurons was often best for a subset of joint angles, rather than all of the joint angles of the limb. When electrical stimulation was delivered to the recording site, the stimulation brought the limb into a position that approximated that to which the neuron was most closely tuned.

Anatomical studies of the cortex have produced a large number of maps that represent differences in the cellular organization and composition of different areas. In Korbinian Brodmann's widely used (1909) map, the region of the precentral gyrus is designated area 4, the region in front of area 4 is designated area 6, and still further forward is area 8. Other anatomical maps make further subdivisions, for example dividing area 6 into a number of regions. It was recognized in homuncular interpretations of motor cortex stimulation that different body parts correlated with the anatomical maps: finger regions of the homunculus correspond to area 4; supplementary motor cortex from which more gross body movements were obtained corresponds to the dorsal portion of area 6; and the frontal eye fields from which eye movements are obtained correspond to area 8.

Graziano and his co-workers suggest that what determines the cellular configuration of motor cortex regions is not body parts but body movements. The behavior of climbing requires a different composition of neurons and projections from that required for making hand manipulations in central body space or chewing. The cortical region adjacent to the central fissure features many large layer 5 neurons that make direct connections with the motor neurons of the brainstem and spinal cord. The layer 5 neurons in area 6 more densely innervate interneurons of the spinal cord. These anatomical features are presumably those required to produce the species typical movements of the hand displayed by an animal. Whereas in earlier thinking the primary motor cortex had the final say in action and the supplementary motor cortex worked its effects through the primary motor cortex, in the Graziano model each region directly activates a distinctive behavior. The behavioral studies now allow the various regions of the cortex to be named by the behaviors with which they are associated, in addition to having numerical designations.

THE PARIETAL CORTEX

Penfield and Boldrey, in addition to describing the motor homunculi, also described a sensory homunculus. The stimulation sites along the postcentral gyrus that evoked body sensation displayed a pattern that was similar to the pattern for evoked movement responses. However some locations also evoked movements. Omar Gharbawie and his co-workers at Vanderbilt University have followed up the evoked movement line of investigation by stimulating a wide region of parietal cortex and by anatomically tracing the relationship between different regions of the parietal cortex and movement-eliciting regions of the motor cortex. The regions of the parietal cortex that produced various movements, and the connections of those regions to motor cortex regions from which similar movements are produced, are illustrated in Figure 3.2.

They found that short duration trains of electrical stimulation did not evoke movement in the parietal cortex, but if they used long trains (500 msec) of electrical stimulation, similar to the stimulation duration used by Graziano, they were able to evoke movements. In addition, the movements that they evoked were very similar to the movements evoked from the motor cortex. Different movements were also elicited from different regions of the parietal cortex. A more medial zone elicited hindlimb and trunk movements, and lateral to this zone, stimulation elicited reach movements and arm defensive movements. Further lateral in parietal cortex stimulation elicited hand to body movements and grasping movements. Face defensive and aggressive movements as well as eye movements were elicited in the most lateral region of the parietal cortex.

This topography and the elicited movements resemble the topography and movements elicited in the motor cortex. In addition, when dyes were injected into the parietal cortex in order to trace the connections of parietal neurons, each parietal cortex region preferentially projected to the motor cortex region from which similar movements were obtained. Thus, the connections between the parietal cortex and motor cortex formed functional streams connecting a small number of zones representing canonical movements. The movement zones for the most part did not lie within the primary sensory cortex zone from which Penfield and Boldery reported most of their sensations, but rather were found in the region of the parietal association cortex, largely in Brodmann's area 5.

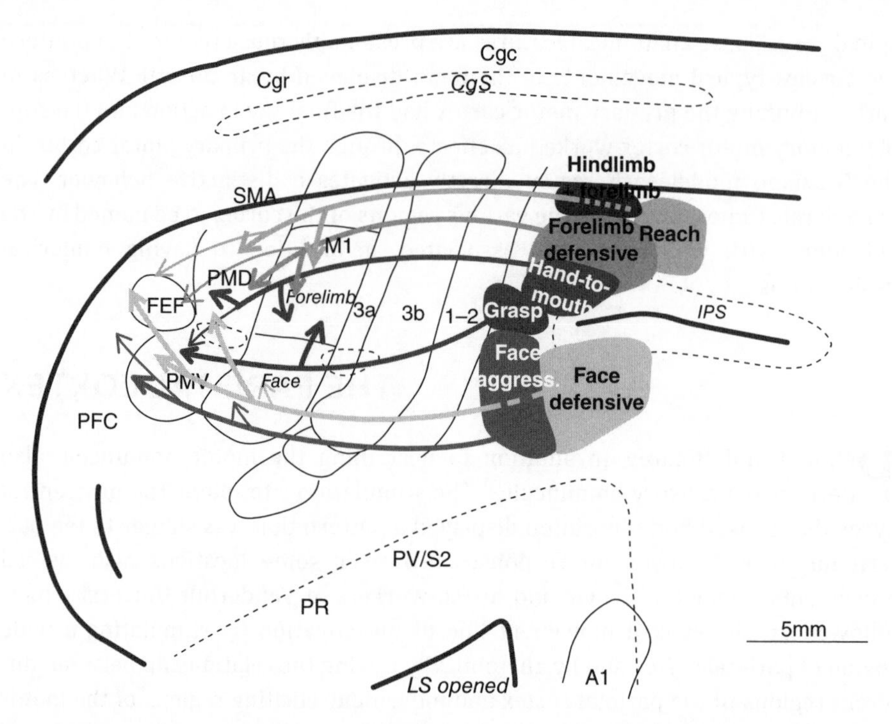

Figure 3.2 Schematic of parietal–frontal connections of the specific movement zones in PPC of galagos. Functionally distinct movement zones and their connections are marked on the flattened view of left cerebral hemisphere. Thick lines represent strong connections, and thin lines represent weak connections. Opened sulci are marked with dashed lines. (From Kaas et al., 2011.)

That both the motor cortex and parietal cortex have similar motor functions suggests not that these are equivalent but that they have a hierarchical arrangement, such that the parietal cortex contributes to adaptive movements by providing information about somatosensory events. Following this hierarchical model, other senses are envisioned as projecting into these circuits to produce relevant movements to odors, sounds, and visual stimulation. For purposes of illustration, a large body of research is directed toward how visual information interacts with the somatosensory-motor circuits for movement. The following section will provide a brief account of how the visual system is proposed to contribute to motor events.

VISUOMOTOR STREAMS AND MOTOR RESPONSES

In 1994 Leslie Ungerleider and James Haxby provided the first modern synthesis of visuomotor function (see the chapter by Gonzalez). They suggested that there were two main groups of pathways, which they called streams, leading from the primary visual cortex to the rest of the brain: the dorsal stream projects through the parietal cortex to the frontal lobes; the ventral stream projects through the temporal cortex and limbic cortex to the frontal lobes.

David Milner and Melvyn Goodale (2006) provided functions for the two streams in their action-perception theory. They suggested the dorsal stream's function was to produce action requiring online control. For example, reaching for a cup requires visual guidance of the hand to the cup and then finger shaping to grasp it. Even though we reach for objects thousands of times in our lifetime, our ability to describe that movement of reaching is not good because of that behavior's automatic nature. Milner and Goodale also suggest that the ventral stream uses sensory information for the perception of objects. We do not need to reach for a cup in order to know that it is a cup. After seeing a cup, we can recall having seen it, we can describe it to others and we can return to get it, if need be. The function of this perceptual process of the ventral stream is to produce a conscious representation of the things we see and hear.

In an early study on pointing at an object, Woodworth (1899) had suggested this action involved two movements, i.e. orienting the limb followed by directing the finger to the target. Marc Jeannerod and colleagues (1995) used this idea to formulate his Dual Visuomotor Channel theory of reaching. He proposed that reaching was composed of two movements, i.e. directing the hand to a target and shaping the fingers to grasp that target. Different sensory features of the target guided each movement. The extrinsic properties of the target (its location) guided the reach and its intrinsic properties (its size and shape) guided the finger shaping to grasp. Support for the idea that the two movements were enabled by different neural pathways from the visual cortex through the parietal cortex to the motor cortex comes from a wide range of studies using brain imaging with functional magnetic resonance imaging (fMRI), both with experimental participants reaching for objects and patients who have lesions in different portions of the dorsal stream. The theory is also supported by behavioral studies that dissociate the Reach and the Grasp by manipulating sensory information.

Jenni Karl and co-workers have behaviorally dissociated the Reach and the Grasp in a number of ways (Karl and Whishaw, 2013). For example, they have compared reaching for food in sighted experimental participants and participants who were blindfolded. As illustrated in Figure 3.3, blindfolded subjects reached with an open hand to locate the object, and then after touching it adjusted their hand and fingers to grasp. Thus, the Reach occurred first and was followed by the Grasp. This dissociation is strikingly similar to that obtained by electrical stimulation of different movement areas of the frontal cortex:

> In our stimulation studies, when we stimulated in the cortical zone that we termed the "reach-to-grasp" zone, we tended to evoke the first process of extension of the arm, pronation of the forearm, extension of the wrist, and opening of the grip. In this sense the term reach-to-grasp is misleading because there was no grasp at the end of the reach. Rather we seemed to evoke a reach in preparation for a grasp, with the

grip open. In contrast, when we stimulated in the cortical areas that we termed the "manipulation" zone and the "hand-to-mouth" zone, we tended to evoke the second process of retraction of the arm toward the body, supination of the forearm, flection of the wrist, and closure of the grip. (Grazanio, 2009: 145)

Taken together, the behavioral studies using human subjects and the stimulation results obtained from monkeys suggest that complex behavior such as reaching for food and placing it in the mouth is produced by a process of combination and addition. Reaching for a piece of food involves activation of the visuomotor channels that project from visual regions, including the visual cortex and superior colliculus, through the parietal cortex to the motor cortex. One channel includes the reach-to-grasp region of the motor cortex and directs the hand to the target. The almost simultaneous activation of a second channel to the hand manipulation region of the motor cortex shapes the hand for grasping. To then bring the food to the mouth involves a third pathway, one from the parietal cortex to the motor cortex that relies solely on somatosensory information and does not require visual control. Touch and proprioception guide the hand to the mouth to release the food to the lips. (These three pathways are illustrated in Figure 3.4.) Presumably the flow of complex movement is produced by the similar combining and addition of activity in many other channels from the visual, somatosensory, auditory, and olfactory cortex all projecting through the parietal cortex to the motor cortex.

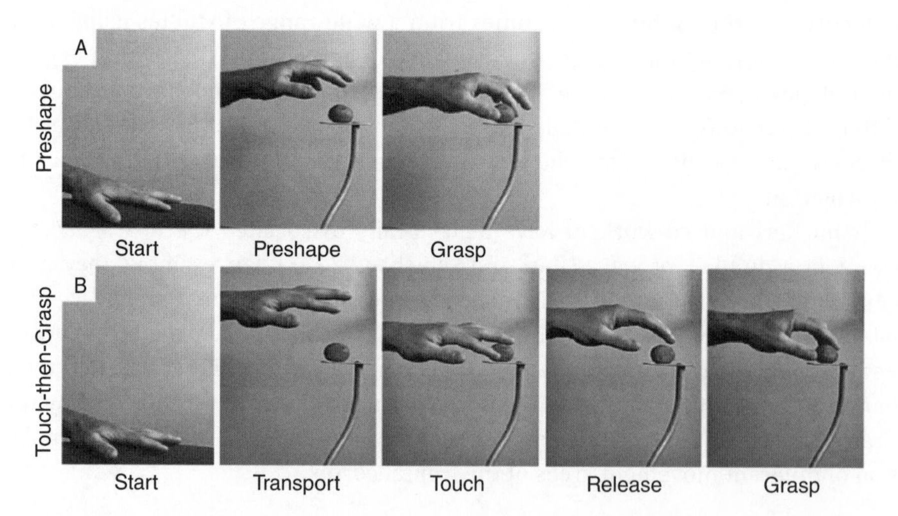

Figure 3.3 When sighted, humans preshape the hand in preparation to grasp the target as they reach towards it, using what is termed a preshaping strategy (A). When vision is occluded, humans use an open and extend hand to reach out and touch the object and only after contact do they shape and close the digits to grasp, using what is termed a Touch-then-Grasp strategy (B). (From Karl and Whishaw, 2013: 208.)

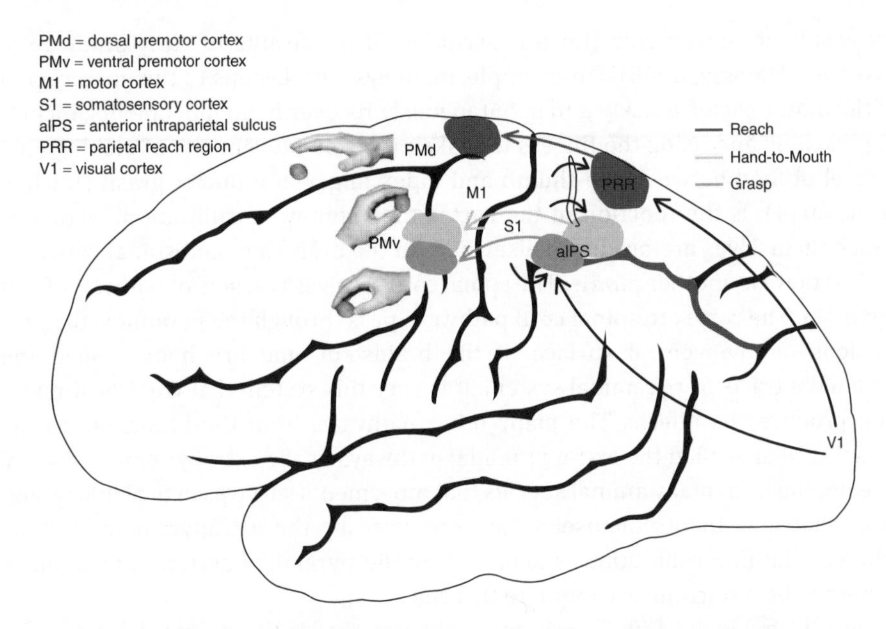

Figure 3.4 Reaching for an object to bring it to the mouth employs three movements: directing the hand to the target, shaping the fingers to grasp, and bringing the hand to the mouth. Separate visuomotor channels mediate a reach and grasp whereas a somatosensory channel brings the food to the mouth. (From Karl and Whishaw, 2013: 208.)

CONCLUSION

Penfield and Boldrey's study provided a remarkable picture of the organization of the motor and somatosensory cortex in humans. Today the interpretation of their findings is quite different from what they had suggested. Rather than having a homuncular organization, the cortex is organized in terms of a small number of postures. These postural regions are linked to other brain regions by pathways that unite sensory events to motor posture. Movements are created by transitions from one posture to another. Just as language is produced by selecting words from the lexicon of language and linking these together in different combinations, motor behavior is produced by selecting postures from the lexicon of postures and transitioning between them.

Nevertheless, this new view still leaves many questions regarding the contributions of motor cortex unanswered. The mystery of movement organization by the motor cortex is not completely solved. Damage to the cortex impairs movement in various ways but it does not abolish movement in many animal species with a well-developed motor cortex. Rats with no neocortex at all display surprisingly complex behavior: they eat and drink, locomote, mate, raise pups, and display most behaviors typical of rats (Whishaw, 1990). Cats and dogs with large cortical lesions can walk about. Nonhuman primates with motor cortex lesions recover hand movements (for a description of the results of early studies see Tyler and Malessa, 2006). For example, monkeys with lesions to the hand region of the motor cortex quickly and spontaneously recover hand movements, including reaching and using the fingers to make precision movements of picking up a morsel of food between the thumb and index finger in a pincer grasp (Darling et al., 2014). Is the function of the cortex then simply to calibrate movements, which themselves are produced elsewhere in the brainstem and spinal cord?

And there is another puzzle. The spinal cord receives two sets of pathways from the brain. The cortex to spinal cord pathways pass through the pyramids, the protrusions on the ventral surface of the brainstem, and are hence called the pyramidal tracts or pyramidal system. It is over this system that cortical stimulation produces its effects. The many other pathways from the brainstem to the spinal cord are called the extrapyramidal pathways or the extrapyramidal system. The finding from many animal species that movements survive cortical injury suggests that movements themselves are produced by the extrapyramidal system, whereas the fine calibration is achieved by the pyramidal system. But humans appear to be a startling exception to this rule.

de Oliveira-Souza (2012) presents evidence that if the pyramidal system is damaged in humans, a profound loss of movement occurs. It matters little whether such damage is caused by a stroke at the level of the brainstem, before pyramidal pathways have begun to branch to their various targets, or by disease, such as by multiple sclerosis (MS), which destroys the myelin, and hence the function, of these fibers. This condition is called "locked-in syndrome", because an individual is left conscious but paralyzed. Eye movements are frequently spared by such damage, allowing an individual to communicate only by blinking. This profound loss of movement in humans is quite unlike the sparing of motor behavior observed in other animals. de Oliveira-Souza suggests that humans are more dependent upon the pyramidal system than are other animals and further suggests that for humans, the extrapyramidal system is vestigial with respect to producing voluntary movement.

There is another finding that suggests that the human pyramidal system is special. Counts of brain and spinal cord cells in different primate species indicate that as the number of cells in the spinal cord increases, there is a proportional increase in the number of cells in the brain, with the exception of motor cortex. The number of cells in the motor cortex increases as the square of the number of cells in the spinal cord (Herculano-Houzel et al., 2016). As a result, for large brain primate species such as apes and humans there is an enormous increase in the number cells in motor cortex that contribute to the pyramidal system. The proposal is, this increase results in the pyramidal system taking over, perhaps by outcompeting, control of movement from the extrapyramidal system.

Thus, we can ask, with respect to the mystery of the motor cortex, is the enrichment of movement so extreme in humans that our movement has become entirely dependent upon our motor cortex? Are the special attributes of humans in skilled hand use and language simply accounted for according to the number of cells in motor cortex? If so, Penfield and Boldrey's homunculus sums it up in an unexpected way; we are human because we have so many motor cortex cells for our hands and mouth.

REFERENCES

Brodmann, K. (1909) *Vergleichende Lokalisationslehre der Grosshirnrinde*. Leipzig: Johann Ambrosius Barth.

Chang, H.T., Ruch, T.C. and Ward, A.A. Jr (1947) Topographical representation of muscles in motor cortex of monkeys. *Journal of Neurophysiology*, 10: 39–56.

Darling, W.G., Morecraft, R.J., Rotella, D.L. et al. (2014) Recovery of precision grasping after motor cortex lesion does not require forced use of the impaired hand in Macaca mulatta. *Experimental Brain Research*, 232: 3929–38.

Descartes, R. (1664) *Traité de l'Homme*. Paris: Angot.

Ferrier, D. (1874) Experiments on the brain of monkeys, No. 1. *Proceedings of the Royal Society of London*, 23: 409–30.

Fritsch, G. and Hitzig, E. (1960 [1870]) On the electrical excitability of the cerebrum. In G. von Bonin (ed.), *The Cerebral Cortex*. Springfield, IL: Charles C. Thomas. pp. 73–96.

Galea, M.P. and Darian-Smith, I. (1994) Multiple corticospinal neuron populations in the macaque monkey are specified by their unique cortical origins, spinal terminations, and connections. *Cerebral Cortex*, 4: 166–94.

Goltz, F. (1888) On the functions of the hemispheres [English translation of Uber die Verrichtungen des Grosshirns]. *Pfuger's Archives*, 42: 419–67. In G.V. Bonin (ed.), *Some Papers on the Cerebral Cortex*. Springfield, IL: Thomas (published 1960). pp. 118–58.

Graziano, M.S.A (2009) *The Intelligent Movement Machine*. New York: Oxford University Press.

Graziano M.S.A. and Afalo T.N. (2007) Mapping behavioral repertoire onto the cortex. *Neuron*, 56: 239–251.

Graziano, M.S.A., Taylor, C.S. and Moore, T. (2002) Complex movements evoked by microstimulation of precentral cortex. *Neuron*, 34: 841–51.

Grunbaum, A. and Sherrington, C. (1901) Observations on the physiology of the cerebral cortex of some of the higher apes (Preliminary communication). *Proceedings of the Royal Society of London*, 69: 206–9.

Herculano-Houzel, S., Kaas, J.H., de Oliveira-Souza, R. (2016) Corticalization of motor control in humans is a consequence of brain scaling in primate evolution. *J Comp Neurol*. 524(3):448–55. doi: 10.1002/cne.23792. Epub 2015 May 12. PMID: 25891512.

Jeannerod, M., Arbib, M.A., Rizzolatt, G. and Sakata, H. (1995) Grasping objects: the cortical mechanisms of visuomotor transformations. *Trends in Neuroscience*, 18: 314–20.

Kaas, J.H., Gharbawie, O.A. and Stepniewaska, I. (2011) The organization and evolution of dorsal stream multisensory motor pathways in primates. *Frontiers in Neuroanatomy*, 5: 1–7.

Karl, J.M. and Whishaw I.Q. (2013) Different evolutionary origins for the reach and the grasp: an explanation for dual visuomotor channels in primate parietofrontal cortex. *Frontiers in Neurology*, 4: 208.

Milner, D. and Goodale, M.A. (2006) *The Visual Brain in Action*. Oxford: Oxford University Press.

de Oliveira-Souza, R. (2015) Damage to the pyramidal tracts is necessary and sufficient for the production of the pyramidal syndrome in man. *Med Hypotheses*. 85(1): 99–110. doi: 10.1016/j.mehy.2015.04.007. Epub 2015 Apr 29. PMID: 25959865.

Penfield, W. and Boldrey, E. (1937) Somatic motor and sensory representation in the cerebral cortex of man as studied by electrical stimulation. *Brain*, 60: 389–443.

Tyler, K.L. and Malessa, R. (2000) The Goltz-Ferrier debates and the triumph of cerebrallocalizationalist theory. *Neurology*, 55: 1015–24.

Ungerleider L.G. and Haxby, J.V. (1994) 'What' and 'where' in the human brain. *Current Opinion in Neurobiology*, 4: 15–165.

Whishaw I.Q. (1990) The decorticate rat. In B. Kolb and R.C. Tees (eds), *The Cerebral Cortex of the Rat*. Cambridge, MA: The MIT Press. pp. 239–67.

Woodworth, R.S. (1899) The accuracy of voluntary movement. *Psychological Reviews*, 3: 1-119.

4 Revisiting Kaas and colleagues: The homunculus: The discovery of multiple representations within the "primary" somatosensory cortex

Leah Krubitzer and Mary Baldwin

There are a few truly momentous occasions in life and most do not occur through any machinations on the part of the individual who enjoys such occasions. Rather, they are happenstance and simply require one to be in the right place at the right time. Such was my experience (Leah Krubitzer) in the 1980s when I found myself lucky enough to be a graduate student in the laboratory of Jon Kaas. A serendipitous twist is to find myself writing this chapter with Mary Baldwin, a fellow student who succeeded me in Jon's laboratory some 20 years later, and today finds herself in the right place at the right time to recount with me this important discovery of our mentor. I began work in Jon's laboratory in 1983, on the heels of his discovery that the sensory neocortex of primates is more complexly organized than previously believed. In the 1970s, the Kaas laboratory utilized multiunit electrophysiological recording techniques in novel ways in visual and somatosensory cortex in non-human primates.

Together with John Allman, Kaas demonstrated that the visual cortex was composed of multiple, topographically organized subdivisions that extended beyond the borders of the traditionally defined primary (V1) and secondary (V2) visual areas (Allman and Kaas, 1971 (MT), 1974 (MTc), 1975 (DM), 1976 (M)). A few years later, in the study that is the highlight of this chapter, Jon discovered that what was once considered a single cortical area, the primary somatosensory area (SI), was actually composed of at least three separate cortical fields with distinct functions (Kaas et al., 1979). This was much more than a point scored by the splitters over the lumpers in their pursuit of defining a functional map of the primate brain. Below we argue that this seminal paper was revolutionary because it demonstrated the power of surveying a large expanse of cortex and the importance of directly relating structure to function, and provided a new perspective on the complexity and number of sensory maps in the primate neocortex.

But before we can address how the field has advanced from this work, it is important to appreciate where we were before these discoveries were made, how these discoveries changed our theoretical framework, and how the use of particular techniques to uncover aspects of cortical organization revolutionized the field.

BEFORE 1979: A BRIEF HISTORY OF CORTICAL CARTOGRAPHY

Appreciating the impact of Jon Kaas's contribution requires that we step back about a century or so. In the late 1800s and early 1900s there were a number of prominent anatomists detailing the cytoarchitecture and myeloarchitecture of the human neocortex, such as Constantin von Economo, Percival Bailey and Gerhardt von Bonin, Oskar and Cécile Vogt, Korbinian Brodmann, Alfred Walter Campbell, and Grafton Elliot-Smith (for reviews see Nieuwenhuya, 2013; Nieuwenhuya et al., 2015; Triarhou, 2007a, b; Kaya et al., 2016). Using Nissl and myelin stains, and in one case unstained, wet tissue and the naked eye (Elliot-Smith, 1907), these scientists pioneered the idea that the neocortex in humans is not a homogeneous structure, but is composed of multiple subdivisions with distinct laminar thicknesses, cell densities, staining and myelination (see Figure 4.1). If the currency of neuroanatomy is others' adoption of one's nomenclature, Brodmann may be the most renowned because his illustrations of the subdivisions of the human neocortex are still found in most textbooks. In fact, Brodmann proposed that these divisions were separate "organs" of the brain that had distinct functions (Brodmann, 1909, cited in Garey, 1994).

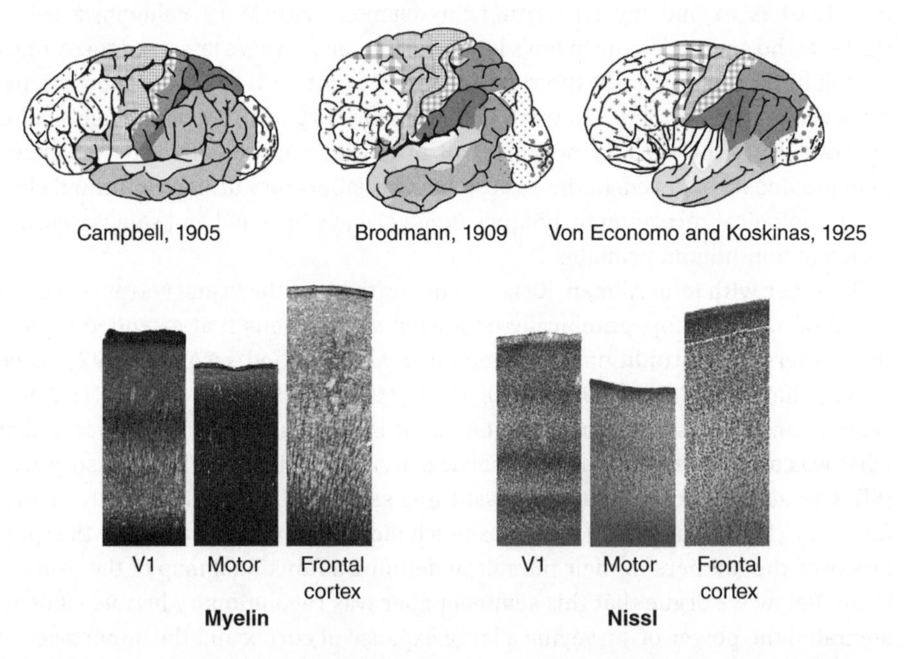

Figure 4.1 Architectonic maps of the neocortex proposed by early anatomists (top row). These maps were constructed by examining sections stained for Nissl substance or myelin (bottom row). Multiple cortical fields were identified in anterior parietal (plaids), posterior parietal (dark gray), occipital (dots), and frontal (crosses). Although the divisions of different anatomists varied, they all agreed that cortex was not homogeneous. Close examination of Nissl and myelin stains reveals clear laminar differences in V1, motor, and frontal cortex.

Despite his work on the human brain, what we consider to be Brodmann's most important contribution to the field goes largely unappreciated: his extensive comparative analysis. He cut and processed tissue from multiple species, including hedgehogs, ground squirrels, rabbits, flying foxes, kinkajous (South American carnivores), lemurs, marmosets, guenons, and humans, and generated a numerical naming system for different cytoarchitectonic fields, some of which were similar across species. Much of his terminology is still commonly used today, including areas 17 and 18 of the occipital lobe, areas 3, 1, and 2 of the parietal lobe, and area 4 of the frontal lobe.

While he and others inferred that these architectonically distinct areas of the neocortex corresponded to functional subdivisions, it was not until the middle of the twentieth century that physiologists such as Clinton Woolsey, Philip Bard, Richard Lende and Wilder Penfield demonstrated that the parietal cortex contained distinct representations of the body surface. These individuals used surface evoked potential techniques to examine where in the neocortex inputs from different portions of the body were represented. Their functional subdivisions roughly corresponded to previously defined architectonic subdivisions by Brodmann and others, although they did not directly relate their functional data to histologically processed tissue. Specifically, a single representation of the contralateral body was found in non-human and human primates in the anterior parietal cortex (see Figure 4.2), a region that roughly corresponded to areas 3, 1 and 2 of Brodmann. This functional field was termed the "primary somatosensory cortex" and abbreviated to SI. Today, arabic numerals have replaced roman numerals in this abbreviation, which is now commonly written as S1. These maps of the body were complete and topographically organized, such that adjacent portions of the skin were represented in adjacent portions of the neocortex. It was during this time that the term "homunculus" became popularized. While Penfield enjoyed much of the notoriety for his studies of the human parietal cortex, it was Clinton Woolsey and colleagues who demonstrated the ubiquity of SI in numerous species including humans.

Lateral to SI, a second representation of the body observed in humans and other mammals was called SII (now commonly referred to as S2) (see Figure 4.2). It should be noted here that during this time the primary auditory area (A1) and the primary visual area (V1) had also been defined in a variety of mammals, roughly corresponding to architectonic divisions of the neocortex (Brodmann's areas 41 and 17 respectively). However, it was not common during this era for physiologists to cut and process the brains from which they recorded. Therefore, a direct relationship between function and structure was rarely well established.

There were several extremely important (and sometimes counterproductive) theoretical concepts that emerged from this era of scientific discovery that still persist today. The first is that when two fields of a given sensory modality were described in evoked potential mapping studies, these were regrettably termed "primary and secondary sensory fields". This, of course, reinforces current theories on hierarchical processing networks in the neocortex, which imply that fields such as the primary somatosensory, auditory and visual areas have ascendency

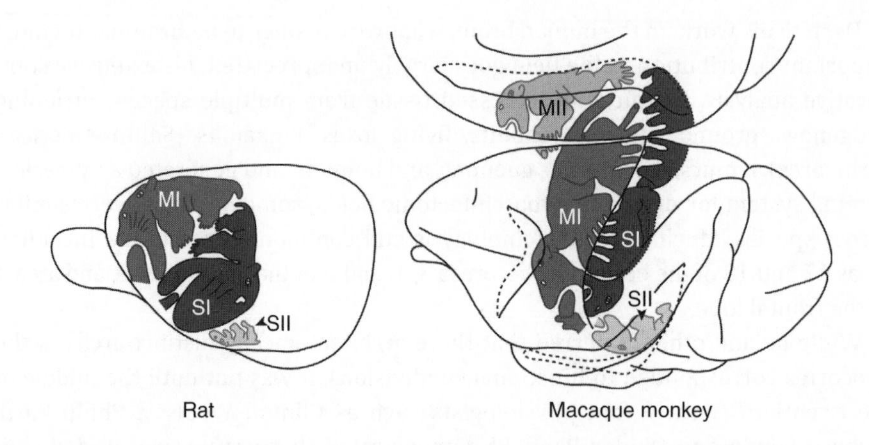

Figure 4.2 Maps of the somatosensory cortex and motor cortex generated from evoked potential studies in anesthetized rats and macaque monkeys. The first (SI) and second (SII) somatosensory areas are depicted as homunculi, or a miniature of the body (normally referring to humans). These maps were originally generated by Clinton Woolsey (see Harlow and Woolsey, 1958).

over secondary areas such as S2 and V2. In addition, this terminology impacted our ideas of cortical field evolution in mammals, perpetuating the idea that primary areas are older, secondary areas are more recently evolved, and other areas (such as V3 and V4) have evolved subsequently and in order. Interestingly, at the time of discovery Woolsey appreciated that the terminology was problematic and stated that these areas were named simply in order of discovery and not ascendancy.

The second theoretical concept that emerged from this era, thanks to the extraordinary comparative studies of Brodmann and later Woolsey, is that mammals share common features of neocortical organization, such as the presence of common anatomical and functional brain areas including S1, S2, V1, V2, and so on. This realization that there were common brain features across species had, and still has, extremely important implications for theories of cortical evolution, and provides a basic framework for assigning homology and understanding the possible cortical organizational scheme that was present in our earliest mammalian ancestors. Of course, since the time of Brodmann's and Woolsey's studies, additional methods have been used to establish the homology of different cortical fields in mammals, but these early works provided the first firm foothold in the field of comparative neuroscience.

The third important consequence of these early comparative studies has been the promotion of the term "association" cortex. In these preparations one or two maps of the sensory epithelium were described. In mammals with small brains, such as rats, the first and second sensory areas (e.g. S1, S2, V1, V2) assumed most of the cortical sheet. In species with larger brains, such as macaque monkeys, these fields were separated by a relatively large expanse of cortex where responses to tactile, auditory or visual stimuli could not be evoked in the anesthetized preparations used to study non-human mammals. This unresponsive cortex consisted of a much larger percentage of the cortical sheet in monkeys and apes than in small-brained mammals, and included the region of cortex between S1/S2 and V2, now known as the posterior parietal cortex. Because stimulation failed to evoke a response, these physiologists assumed this region corresponded to the "association cortex" identified by anatomists at the beginning of the century (e.g. Campbell, 1905; Fleshig, 1905). The association cortex was proposed to be a region where sensory inputs were combined and translated, and perhaps where consciousness was seated (Zeki, 1993). Over time the notion of the posterior parietal cortex (PPC) as the association cortex gradually gave way to an appreciation that the PPC contains a large, complex network of cortical fields, many of which process sensory inputs.

It would be remiss if we did not briefly mention the important contributions by Vernon Mountcastle and his colleagues in the 1950s and '60s. Probably the most significant was his work on the awake behaving monkey, where he directly related neural activity to some aspect of behavior. Discoveries that neurons respond to specific sensory modalities, that particular features of a stimulus such as orientation and direction are coded by specific neurons, and that receptive fields of neurons have a complex organization such as center-surround were also made during this time (see Mountcastle, 1995a, 1995b). Subsequently, he described the cortical column and promoted the notion of a "basic uniformity" of the neocortex at the level of the microcircuit. However, that is a separate story for another day.

1979: THE GAME-CHANGING DISCOVERY OF MULTIPLE REPRESENTATIONS IN THE "PRIMARY" SENSORY CORTEX OF PRIMATES

We have outlined not only the scientific discoveries that occurred before the 1979 Kaas paper but also the theoretical framework prevalent at this time, so that the reader can appreciate why the 1979 study by Kaas, Nelson, Sur, Lin, and Merzenich was so important. Utilizing extracellular electrophysiological recording techniques in a number of New and Old World monkeys under anesthesia, Kaas and colleagues recorded from hundreds of sites within the parietal cortex in each animal while stimulating different body parts. Unlike single-unit studies in which only a few sites within a given monkey cortex are surveyed and stimuli are controlled, in these studies the body was stimulated by handheld probes (a technique termed "hand mapping") lightly tapping or brushing the skin surface for cutaneous stimulation. Deep receptors were stimulated by manipulation of joints or probing muscles. Because neural responses to these manually applied stimuli can be ascertained very rapidly, it is possible to record from hundreds of sites across centimeters of the neocortex. Kaas and his colleagues directly related hundreds of recording sites to cytoarchitectonic boundaries (like those first demonstrated by Brodmann and his contemporaries) to generate functional maps of cortex with architectonic borders.

By sampling so many locations, these studies demonstrated there were multiple representations of the body surface in the parietal cortex rather than only one, each of which was co-extensive with an architectonically distinct field (see Figure 4.3). They found a very detailed and complete representation of the cutaneous body surface in area 3b in which neurons had relatively small receptive fields. Furthermore, a complete mirror reversal representation of the body surface was observed in area 1, and an additional representation of the deep receptors of the muscles and joints was found in area 2. Although they did not characterize a body map in area 3a, they proposed that a separate representation existed there as well, and a complete map of the body in area 3a has been subsequently described in a variety of mammals (see Figure 4.3C; also see Krubitzer et al., 2004, for a review). Thus, rather than a single map or homunculus of the body that spanned four cytoarchitectonic fields (see Figure 4.3A), Kaas and colleagues discovered two topographically organized representations corresponding to areas 3b and 1, and one roughly topographic representation of deep receptors that co-registered with area 2 (see Figure 4.3B).

This discovery that the anterior parietal cortex was much more complexly organized than was previously believed was the game changer, because for the first time fundamental questions could be asked about the hierarchy of information processing, about homology across species, and about the evolutionary genesis of cortical field formation (i.e. how new cortical fields were added). For example, how can we determine which areas emerged early in mammalian evolution and which areas evolved more recently if more than a single sensory area is present in some mammals versus others, as is the case for the classical macaque monkey SI (now known to contain areas 3a, 3b/S1, 1 and 2) versus rodent SI? Further, of those more recently emerging areas, which ones are or are not shared across extant species, and which areas are specializations to a particular clade, or even species? Finally, as new areas emerge in the course of evolution, to what degree is their contribution to the network dependent on the processing that occurs in already-existing areas? It seems highly unlikely that fields are added into a network in a strictly sequential or hierarchical fashion (despite the unfortunate nomenclature of V1, V2, V3, V4, etc.), but instead emerge between existing areas within the network.

In addition to changing the theoretical framework of how we think about the organization and function of the somatosensory cortex, this study was an important demonstration of a methodological approach that complemented other techniques. For example, Kaas underscored the importance of combining different techniques to subdivide the neocortex accurately. Unlike most previous studies, he directly related his electrophysiological recording data with his

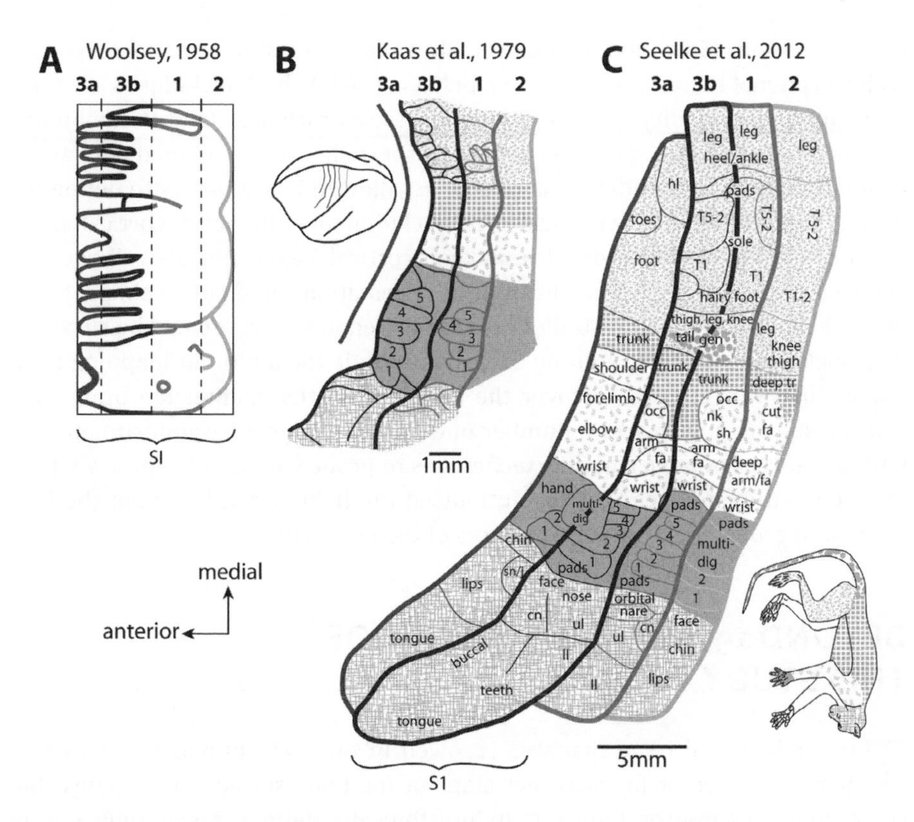

Figure 4.3 Changing concepts of the organization of somatosensory cortex over time. Early evoked potential maps of Woolsey (A) suggest that a single representation of the body was co-extensive with four cytoarchitectonic fields (3a, 3b, 1, and 2). The 1979 study by Kaas and colleagues discovered that cytoarchitectonic areas 3b and 1 each contained a separate and complete representation of the body (B). Subsequent studies summarized by Seelke et al. (2012) demonstrate that detailed maps of the body exist in each of the four cytoarchitectonic fields, at least in macaque monkeys (C).

cytoarchitectonic data. In subsequent papers and reviews he argued convincingly that the best way to subdivide the neocortex was to use multiple criteria, including an examination of the connection patterns of a presumptive cortical field (Kaas, 1982). In addition, the use of multiunit electrophysiological recording techniques made surveying a huge swath of cortex possible, and allowed one to pose questions about the overall cortical organization. While a number of important discoveries in parietal and posterior parietal cortex in the same era were made using single-unit electrophysiological techniques (see Mountcastle 1995a, b, for a review), the types of results yielded by the two techniques were synergistic rather than mutually exclusive. Studies of single neurons in awake animals are limited in terms of the amount of cortex that can be surveyed, but they allow one to determine the response properties of neurons in much greater detail and with more precision. Moreover, it is extremely valuable to study neural responses while an animal is conscious and even behaving. While both techniques are critical to our understanding of cortical function, the extracellular multiunit mapping method pioneered by Kaas and colleagues, especially when combined with architectonic analysis and studies of connections, allowed investigators to define the overall organization of the cortical sheet including the number of cortical areas, the topographic organization of individual cortical fields, and the relative size and location of cortical fields. This type of information could not be elucidated in awake behaving single unit studies because not enough cortex could be surveyed in a single animal. This is still the case today. While the multiunit mapping techniques that were used to uncover the mysteries of the mammalian brain have showed staying power over a number of decades, it is perhaps unfortunate that with the advent of powerful new techniques to probe cortical networks we tend to throw out techniques that revolutionized the field and still remain the best method to probe organizational features of the neocortex.

BEYOND 1979: THE INFLUENCE OF THIS STUDY TODAY

The fundamental issue that was resolved in the 1979 publication was that there are three or four distinct maps of the body surface in primates that correspond to three or four cytoarchitectonically defined areas, only one of which is homologous to S1 (area 3b) as defined in non-primate mammals. Unfortunately, many scientists who work on the anterior parietal cortex in cats and monkeys still refer to these multiple architectonic fields as S1, a bone of contention that Kaas himself appreciated and vocalized in a compelling review entitled "What, if anything, is S1?" (Kaas, 1983). This is more than a semantic issue. The inappropriate use of the term S1 promotes a misunderstanding of both cortical circuitry and homology.

In a number of ways the seminal paper by Kaas and colleagues (1979) has laid the foundations for much of what we know about cortical organization and function today. First, this study re-emphasized that there is often a good correspondence between structure and function. That is, the Kaas multiunit mapping studies demonstrate that the multiple anatomically identified cortical areas in the anterior parietal cortex correspond with topographically organized somatosensory cortical fields. Further, multiple somatosensory fields not only exist in primates but are also found in squirrels, rats, flying foxes, tree shrews, and a variety of other species (see Figure 4.4).

Second, the 1979 study was pivotal in the discovery of cortical plasticity. In a series of studies published in subsequent years by Jon Kaas and Mike Merzenich, using the same multiunit extracellular techniques as those used in the 1979 study, they demonstrated that there were large alterations in cortical maps in S1 (3b) in adults that emerged as a consequence of alterations in the periphery, such as a

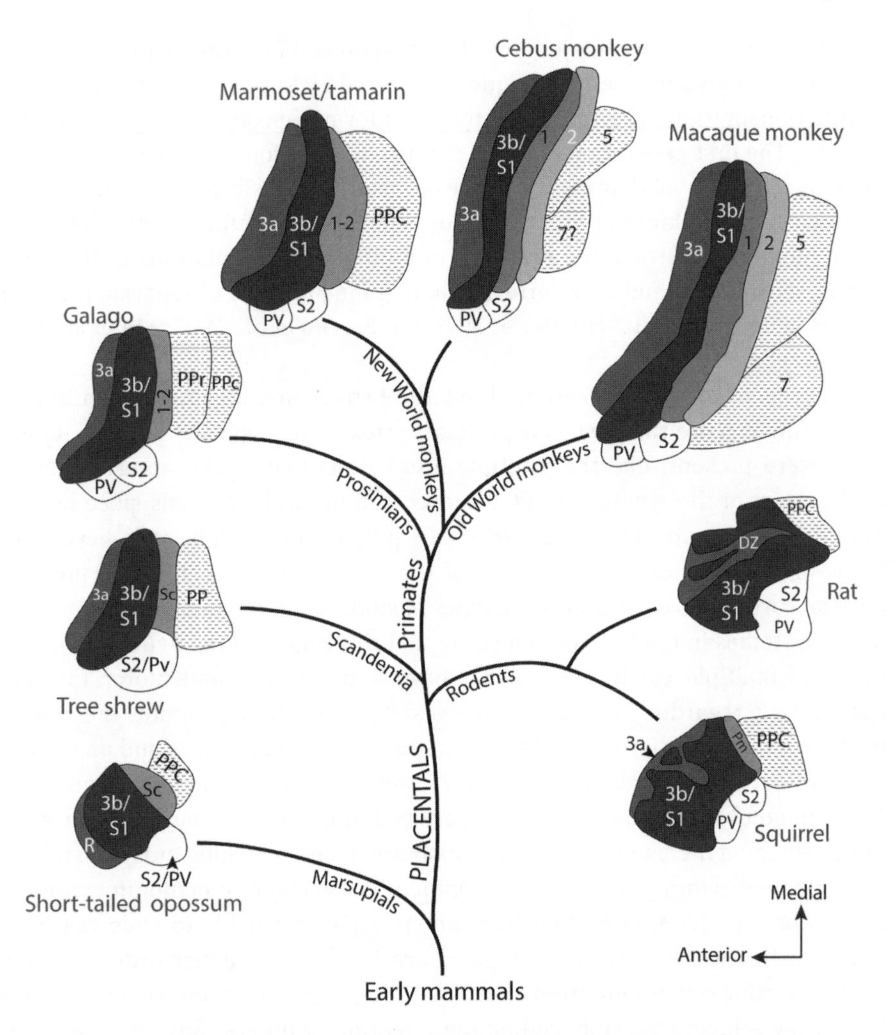

Figure 4.4 A cladogram showing the phylogenetic relationship of different species of mammals and the organization of their parietal cortex. Modern comparative analysis utilizes multiple techniques (e.g. electrophysiological and architectonic) to subdivide the neocortex. This figure demonstrates that all species examined have a primary somatosensory area (S1) and a second sensory area (S2). Recent studies demonstrate that a field rostral to S1 (termed area 3a or R) has been identified across groups as well as a posterior parietal cortex. However, the posterior parietal cortex is greatly expanded in primates. To date only primates have four separate anterior parietal fields (3a, 3b, 1 and 2), only one of which (area 3b) should be considered as S1.

nerve section or a loss of digits and limbs due to injury (Merzenich et al., 1983). This led us to re-examine the idea of strict critical periods during development. It also led to an explosion of experiments in both labs and spawned a new generation of investigators who studied adult plasticity, including Gregg Recanzone,

Hubert Dinse, Tim Pons, Jim Jenkins, Randy Nudo and Preston Garraghty, to name but a few, all of whom used techniques pioneered by Kaas and Merzenich.

Additionally, the 1979 study was the basis for non-invasive imaging studies in humans. The first generation of functional magnetic resonance imaging studies in humans sought to validate the existence of areas previously described in monkeys in which extracellular multiunit techniques were used to define a field. These studies in humans confirmed the presence of multiple representations in the visual, somatosensory and auditory cortex, including the presence of separate representations in 3a, 3b, 1 and 2 (Binder et al., 1994; Sereno et al., 1995; Lin et al., 1996; Disbrow et al., 1999).

Subsequent to the publication of the 1979 study, investigators began to more thoroughly explore the posterior parietal cortex to see whether distinct subdivisions were present, like the multiple areas described in the anterior parietal cortex. Most of the studies in PPC are executed in awake animals since neurons here do not respond well in anesthetized preparations. While it is beyond the scope of this chapter to consider all of the single-unit studies on the posterior parietal cortex, it is clear that these types of studies in monkeys, and non-invasive imaging studies in humans, demonstrate that this region of the neocortex is composed of multiple cortical fields that have both sensory and motor functions (particularly regarding the hands and eyes). Specifically, in primates the anterior portion of the posterior parietal cortex in the intraparietal sulcus and on the inferior parietal lobule (including areas 5, 7a and 7b of Brodmann and PF and PG of von Economo) is dominated by areas involved in intentional reaching and grasping, preshaping the hand to match a visual target, and matching grasp postures to shape and behavioral context. These regions are also involved in monitoring tactile and proprioceptive information from anterior parietal fields to code reach and grasp kinematics. Finally, these regions are involved in higher-order functions such as parsing self motion from object motion, generating an internal representation of the self, and possibly coding the intention of others (Snyder et al., 1997; Fogassi et al., 2005; Bisley and Goldberg, 2010).

What is probably the most important concept to have emerged in recent years is that the parietal cortex, including both the anterior and posterior parietal cortex, is actually involved in motor control, and not strictly in sensory processing (Gharbawie et al., 2011a, 2011b; Stepniewska et al., 2014). Further, we now know that neurons in somatosensory areas in the parietal cortex and lateral sulcus are modulated by attention, and even by stimulation from other modalities such as vision (Haggard et al., 2007; Burton et al., 2008). Thus, function is distributed across cortical networks that are made up of individual nodes or cortical fields, rather than being strictly localized to a particular cortical field. This means processing is not strictly hierarchical, that there is not a one-to-one correspondence between cortical field and function, and that differential activation of nodes within a network can generate a variety of complex behaviors. It is our hope that the next generation of scientists will seek to understand how characteristics such as perception, intention and consciousness emerge from such networks and the multiple cortical fields that compose them.

REFERENCES

Allman, J.M. and Kaas, J.H. (1971) A representation of the visual field in the caudal third of the middle temporal gyrus of the owl monkey (Aotus trivirgatus). *Brain Research*, 31: 85–105.

Allman, J.M. and Kaas, J.H. (1974) A crescent-shaped cortical visual area surrounding the middle temporal area (MT) in the owl monkey (Aotus trivirgatus). *Brain Research*, 81: 199–213.

Allman, J.M. and Kaas, J.H. (1975) The dorsomedial cortical visual area: a third tier area in the occipital lobe of the owl monkey (Aotus trivirgatus). *Brain Research*, 100: 473–87.

Allman, J.M. and Kaas, J.H. (1976) Representation of the visual field on the medial wall of the occipital-parietal cortex in the owl monkey. *Science*, 191: 572–5.

Binder, J.R., Rao, S.M., Hammeke, T.A. et al. (1994) Functional magnetic resonance imaging of human auditory cortex. *Annals Neurology*, 35: 662–72.

Bisley, J.W. and Goldberg, M.E. (2010) Attention, intention, and priority in the parietal lobe. *Annual Review of Neuroscience*, 33: 1–21.

Burton, H., Sinclair, R.J. and McLaren, D.G. (2008) Cortical network for vibrotactile attention: a fMRI study. *Human Brain Mapping*, 29: 207–21.

Campbell, A.W. (1905) *Histological Studies on the Localization of Cerebral Function*. Cambridge: Cambridge University Press.

DeYoe, E.A., Carman, G.J., Bandettini, P. et al. (1996) Mapping striate and extrastriate visual areas in human cerebral cortex. *Proceedings of the National Academy of Sciences, USA*, 93: 2382–6.

Disbrow, E., Roberts, T. and Krubitzer, L. (1999) The use of fMRI for determining the topographic organization of cortical fields in human and nonhuman primates. *Brain Research*, 829: 167–73.

Elliot-Smith, G. (1907) A new topographical survey of the human cerebral cortex, being an account of the distribution of the anatomically distinct cortical areas and their relationship to the cerebral sulci. *Journal of Anatomy and Physiology*, 41: 237–54.

Fleshig, P. (1905) *Gehirnphysiologie und Willenstheorien*. In Fifth International Psychology Congress. pp. 73–89. (Translated by G. von Bonin, 1960, in *Some Papers on the Cerebral Cortex*. Springfield, IL: CC Thomas.)

Fogassi, L., Ferrari, P.F., Gesierich, B. et al. (2005) Parietal lobe: from action organization to intention understanding. *Science*, 308: 662–7.

Garey, L.J. (1994) *Brodmann's Localization in the Cerebral Cortex: The Principles of Comparative Localization in the Cerebral Cortex Based on Cytoarchitectonics*. (By Dr K. Brodmann, translated by Laurence J. Garey, Smith-Gordon Company Limited, London.)

Gharbawie, O.A., Stepniewska, I. and Kaas, J.H. (2011a) Cortical connections of functional zones in posterior parietal cortex and frontal cortex motor regions in new world monkeys. *Cerebral Cortex*, 21: 1981–2002.

Gharbawie, O.A., Stepniewska, I., Qi, H. and Kaas, J.H. (2011b) Multiple parietal-frontal pathways mediate grasping in macaque monkeys. *Journal of Neuroscience*, 31: 11660–77.

Haggard, P., Christakou, A. and Serino, A. (2007) Viewing the body modulates tactile receptive fields. *Experimental Brain Research*, 180: 187–93.

Kaas, J.H. (1982) The segregation of function in the nervous system: Why do the sensory systems have so many subdivisions? *Contributions to Sensory Physiology,* 7: 88–117.

Kaas, J.H. (1983) What, if anything, is SI? Organization of first somatosensory area of cortex. *Physiological Reviews,* 63: 206–31.

Kaas, J.H., Nelson, R.J., Sur, M. Lin, C.S. and Merzenich, M.M. (1979). Multiple representations of the body within the primary somatosensory cortex of primates. *Science,* 204: 521–3.

Kaya, Y., Uysal, H., Akkoyunlu, G. and Sarikcioglu, L. (2016) Constantin von Economo (1876–1931) and his legacy to neuroscience. *Child's Nervous System* 32(2): 217–20.

Krubitzer, L., Huffman, K.J., Disbrow, E. and Recanzone, G. (2004) Organization of area 3a in macaque monkeys: contributions to the cortical phenotype. *Journal of Comparative Neurology,* 471: 97–111.

Lin, W., Kuppusamy, K., Haacke, E.M. and Burton, H. (1996) Functional MRI in human somatosensory cortex activated by touching textured surfaces. *Journal of Magnetic Resonance Imaging,* 6: 565–72.

Merzenich, M.M., Kaas, J.H., Wall, J. et al. (1983) Topographic reorganization of somatosensory cortical areas 3b and 1 in adult monkeys following restricted deafferentation. *Neuroscience,* 8: 33–55.

Mountcastle, V.B. (1995a) The evolution of ideas concerning the function of the neocortex. *Cerebral Cortex,* 5: 289–95.

Mountcastle, V.B. (1995b) The parietal system and some higher brain functions. *Cerebral Cortex*, 5: 377–90.

Nieuwenhuya, R. (2013) The myeloarchitectonic studies on the human cerebral cortex of the Vogt-Vogt school, and their significance for the interpretation of functional neuroimaging data. *Brain Structure and Function,* 218: 303–52.

Nieuwenhuya, R., R., Broere, C.A., and Cerliani, L. (2015) A new myeloarachitectonic map of the human neocortex based on data from the Vogt-Vogt school. *Brain Structure and Func*tion, 220: 2551–73.

Seelke, A.M., Padberg, J.J., Disbrow, E. et al. (2012) Topographic maps within Broadman's Area 5 of macaque monkeys. *Cerebral Cortex,* 22: 1834–50.

Sereno, M.I., Dale, A.M., Reppas, J.B. et al. (1995) Borders of multiple visual areas in humans revealed by functional magnetic resonance imaging. *Science,* 268: 889–93.

Snyder, L.H., Batista, A.P. and Ansersen, R.A. (1997) Coding of intention in the posterior parietal cortex. *Nature*, 386: 167–70.

Stepniewska, I., Gharbawie, O.A., Burish, M.J., and Kaas, J.H. (2014) Effects of muscimol inactivations of functional domains in motor, premotor, and posterior parietal cortex on complex movements evoked by electrical stimulation. *Journal of Neurophysiology*, 111: 1100–19.

Triarhou, L.C. (2007a) The Economo-Koskinas atlas revisited: Cytoarchitectonics and functional context. *Stereotactic and Functional Neurosurgery* 85: 195–203.

Triarhou, L.C. (2007b) A proposed number system for the 107 cortical areas of Economo and Koskinas, and Brodmann area correlations. *Stereotactic and Functional Neurosurgery*, 85: 204–15.

Woolsey, C.N. (1958) Organization of somatic sensory and motor areas of the cerebral cortex. In H.F. Harlow and C.N. Woolsey (eds), *Biological and Biochemical Basis of Behavior*. Madison, WI: University of Wisconsin Press. pp. 63–81.

Zeki, S. (1993) The visual association cortex. *Current Opinion in Neurobiology* 3: 155–9.

5 Revisiting Ungerleider and Mishkin: Two cortical visual systems

Jason W. Flindall and Claudia L.R. Gonzalez

"The most striking feature," wrote Nicholas Humphrey and Larry Weiskrantz in 1967, "is the evidence for spatial localization by the de-striate monkey." This conclusion was reached after examining the visual behaviour of two monkeys who were lacking their entire visual cortex (Humphrey and Weiskrantz, 1967). In this article, the authors described the animals' remarkable ability to accurately reach for and grasp objects despite the complete removal of the striate cortex (primary visual cortex; V1). The monkeys appeared blind in a number of visual discrimination tasks (e.g. for object features, or distance discrimination), but were able to successfully orient and guide the arm and hand towards a moving object (the same object that the monkeys could not otherwise identify). Humphrey and Weiskrantz suggested that in the absence of the striate cortex, the "residual vision" was likely mediated by the superior colliculus (part of the tectum, a structure in the midbrain responsible for visual-oriented behaviour).

Two years later, Gerald Schneider published an article describing the effects of lesions to either the superior colliculus or the striate cortex on the visual behaviour of hamsters (Schneider, 1969). Animals were trained to attend and orient to the location in which a sunflower seed would be handed to them (visual orienting task). They were also trained to discriminate patterns: black versus white cards; horizontal versus vertical stripes; or a speckled pattern versus diagonal stripes (visual discrimination task). Removal of the superior colliculus resulted in severe impairments in the visual orienting task, but preserved ability in the visual discrimination task. Animals with visual cortex lesions showed the opposite behaviour, i.e. no deficits in the visual orienting task, but an inability to discriminate between patterns even after hundreds of training trials. Schneider interpreted these results as "a dissociation between mechanisms for two types of visuomotor control . . . one mechanism is concerned with the locating of objects . . . the other mechanism is concerned with the specific identification of objects."

Thirteen years after Schneider published his findings, Leslie Ungerleider and Mortimer Mishkin presented behavioural, electrophysiological, and anatomical evidence for two distinct *cortical* visual systems in the rhesus monkey (Ungerleider and Mishkin, 1982). In this chapter, it is argued that although a dissociation between two types of vision (one for spatial localization and the other for object identification) had been made before, this was related to subcortical and cortical systems (see Schneider, 1969, for example). In contrast, Ungerleider and Mishkin demonstrated that a system originating in the striate cortex and projecting (a) dorsally to the posterior parietal cortex, and (b) ventrally to the inferior temporal cortex was responsible for mediating spatial perception (locating *where* an object is) and object perception (identifying *what* an object is), respectively. Evidence for this dissociation came from a series of studies showing the specific deficits displayed by monkeys with lesions to one or the other pathway. Animals were trained in two tasks: a landmark discrimination task, and a visual discrimination task (see Figure 5.1). In the landmark discrimination task, animals were presented with two food wells – one on the right, the other on the left. The monkey was rewarded when choosing the one closer to a striped cylinder (i.e. the landmark). This task assessed the animal's ability to discriminate which food well was going to contain the food based on the position (i.e. the 'where') of the landmark. In the visual discrimination task the food wells were marked with a place card (either a positive sign or the outline of a square). To locate the reward, the monkey had to learn to differentiate between the two cards based on their outward appearance (i.e. the 'what'). When lesions were made to the posterior parietal cortex, animals were severely impaired on the landmark discrimination task. This visuospatial deficit was in sharp contrast to the monkey's preserved ability in the visual discrimination task. Damage to the inferior temporal cortex, on the other hand, compromised visual discrimination performance, but spared visuospatial function. This evidence demonstrated that two distinct cortical systems were responsible for (a) the spatial localization of objects (where) and (b) the perception of the object's properties (what).

The proposed theory of distinct "where" and "what" pathways revolutionized the field of visual neuroscience. Hundreds of scientific articles aimed at characterizing the two systems were published in the following years. Most notably, an article by Melvyn Goodale and David Milner presented evidence of the two separate visual systems in humans. In "Separate visual pathways for perception and action" (1992), Goodale and Milner concurred with the view of a "what" pathway, but they proposed a revision of the "where" pathway to a "how" pathway. They argued that "how" "captures more appropriately the functional dichotomy between the ventral and dorsal projections" (1992). The emphasis, according to them, is on the output requirements, on a visual system that allows the necessary sensorimotor transformations to guide action. In this article they describe the behaviour of one patient (DF; see below for more details) who suffered damage to the ventral stream. "Despite her profound inability to recognize the size, shape, and orientation of visual objects," reported Goodale and Milner, "DF showed strikingly accurate guidance of hand and finger movements directed at the very same objects." Goodale and Milner's "two visual streams theory" (TVST) has been highly influential in our understanding of how the human visual system is organized. Below is a review of the evidence supporting the TVST and the recent revisions/additions that have been made to the theory.

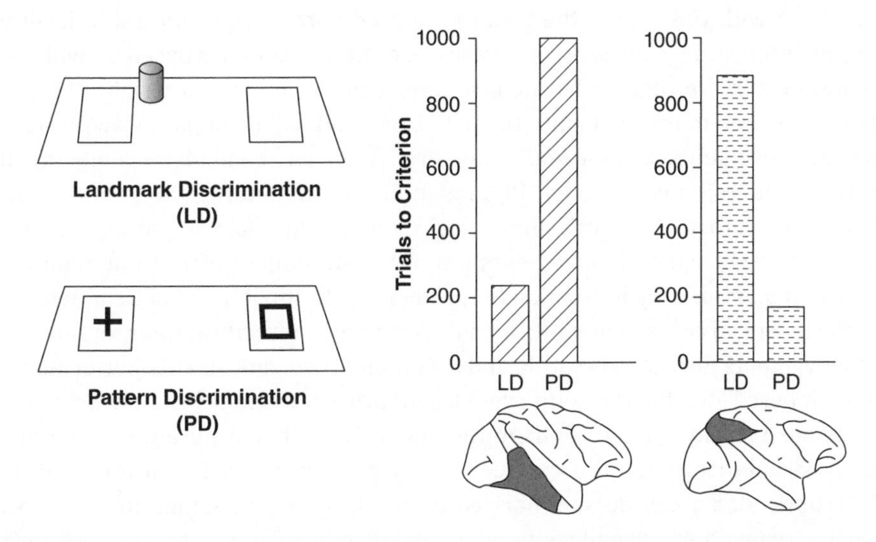

Figure 5.1 Landmark discrimination (LD) and pattern discrimination (PD) tasks used by Ungerleider and Mishkin. Selective deficits in PD arise following temporal cortex lesions, while LD is selectively impaired following parietal cortex lesions. (Adapted from Ungerleider and Mishkin, 1982.)

Source: Ingle et al. (1982) Modified figures from "Two Cortical Visual Systems," © 1982 Massachusetts Institute of Technology, by permission of The MIT Press.

TWO VISUAL STREAMS THEORY

Perhaps the most compelling evidence for TVST comes from case studies on those with neurological disorders; specifically, the contrasting symptoms of optic ataxia and visual form agnosia. Optic ataxia, sometimes called dysmetria, is a disorder characterized by misreaching. Patients with optic ataxia often have normal vision, and no weakness or spasticity in their limbs, and yet are unable to accurately reach-to-touch or grasp target objects in the visual periphery. Perenin and Vighetto (1988) describe the behaviour of six patients (PP, PD, AV, JR, JP, and CB) who suffered from optic ataxia. This is not a visual disorder – these six patients, for example, all had normal or nearly normal visual acuity. Nor is it a general motor disorder – these patients were able to reach to touch locations on their own bodies without difficulty, and would reach for objects in the central visual field as would you or I. However, when asked to reach for a target in the peripheral visual field, and even though they had a full, unobstructed view of both their moving limb and the target, they were unable to smoothly bring one to the other. Optic ataxia usually manifests following a unilateral lesion to the posterior parietal cortex (PPC). Up to 96% of movements into the contralesional hemifield (i.e. the field of vision opposite the side of the lesion) are degraded, especially when those movements are performed with the contralesional hand. Meanwhile, movements into the *ipsilesional* hemifield (i.e. the field of vision coinciding with the side of the lesion) are no different from those of a healthy control. Interestingly, the level of severity of optic ataxia varies based on whether the lesion is in the right or left hemisphere. If the lesion is in the right PPC (e.g. CB and JP), then movements into the left visual field will be impaired (with right-handed movements usually more impaired than left-handed movements). If, however, the lesion is to the left PPC (as in PP, PD, AV, and JR), then not only are movements into the contralesional (right) hemifield again impaired for both hands, but right-handed movements into the ipsilesional (left) visual hemifield are *also* impaired. Only left-handed movements into the left visual hemifield are unaffected. In all cases, if patients are allowed to look directly at their target, then the movements proceed smoothly and accurately. Also, movements performed to a remembered location (i.e. with eyes closed) are less degraded than when movements are visually guided. Thus, optic ataxia is neither a purely motor nor a purely visual disorder, but a confluence of visuomotor deficits. In terms of the TVST, optic ataxia can be summarized as resulting from damage to the dorsal visual stream; only visually guided reaching movements into contralesional space are severely degraded. Visual form agnosia, on the other hand, is nearly the exact opposite of optic ataxia in terms of symptoms; patients with this disorder are unable to recognize or describe objects by vision alone, and yet are often fully capable of interaction with those objects through prehension.

Visual form agnosia is a rare condition, with very few patients described in the literature. Visual form agnosia may be caused by an ischemic stroke (patient JS; Karnath et al., 2009), but more often is the result of acute carbon dioxide poisoning. Patients HC, MDS, RC, and DF developed visual form agnosia from critical CO_2 intoxication events (for a review, see Heider, 2000). While concentrated CO_2 exposure can cause widespread and diffuse damage to white and grey matter, the occipitotemporal cortex is particularly susceptible to CO_2 poisoning, lying as it does on a so-called "watershed" region at the junction of areas supplied by two major cerebral arteries. DF – the most famous and extensively described patient in the visual form agnosia literature – suffered acute CO_2 intoxication while showering in a poorly ventilated bathroom. When she awoke from unconsciousness, DF found that her vision had been severely degraded, to the point where she could no longer recognize shapes, textures, faces, writing, or projected images. In short, she had lost the ability to "see" objects. Despite this visual loss, DF was paradoxically still able to interact with objects in a remarkably normal fashion. When researchers would present DF with two rectangular blocks, each with different dimensions but identical surface areas, she was unable to report which block was wider or longer. Whether she was reporting size verbally or by manual estimation (i.e. by moving her finger and thumb apart to the perceived width of the block), DF fared no better than would be predicted by guessing.

However, when asked to reach out and grasp the blocks placed in front of her, not only did she reach straight toward the block, she also accurately scaled her grip aperture in flight to the size of the target. In healthy people, maximum grip aperture (MGA; the widest distance between your fingertips as you reach to grasp an object) scales with target size; you naturally open your hand wider when grasping a large object, like an apple, than you do when you're grasping a small object, like a grape. DF unconsciously scaled her MGA to the target size, despite being unable to verbally report, recognize, or estimate in any conscious way the size of the target placed before her.

DF's blindsight abilities extended beyond grasping stationary rectangles. When presented with a rotatable slot cut into a board, she was unable to report the angle of the slot's rotation, but was able to quickly and accurately "post a letter" into that slot. When given smooth, irregular, pebble-like shapes, DF not only accurately scaled her grasp, but she also chose the most appropriate points at which to grasp the target such that she would support its centre of mass between her opposing digits.

JS, another patient with visual form agnosia, shares DF's blindsight abilities (Karnath et al., 2009). Whereas DF suffered from CO_2 poisoning, JS's disability arose following an ischemic stroke. Thus, the diffuse damage associated with acute CO_2 intoxication was not present alongside JS's injury, allowing researchers to localize the exact region responsible for his symptoms. JS's stroke destroyed medial aspects of the ventral occipitotemporal cortex bilaterally; specifically, the fusiform and lingual gyri, as well as the isthmus of the posterior cingulate gyrus. These structures are thought to be critical for the function of the ventral visual stream; indeed, MRI data from DF (along with other patients described in the literature, including one Subject 006) suggest that the medial aspect of the occipitotemporal cortex is critical for form and contour processing. Without these structures, one is unable to form conscious visual representations of objects.

Despite their remarkable retention of visually guided function, neither DF nor JS are completely without fault when it comes to prehension. Both show considerably more variability in their hand shapes than healthy controls during grasping. DF can distinguish between high-contrast luminance differences (as would be required for grasping irregularly shaped pebbles), but cannot distinguish edges if they are based on texture differences (Goodale et al., 1994; Westwood and Goodale, 2011). Her abilities also depend on binocular vision; if asked to close one eye, DF's performance on a prehension task suffers dramatically. If a brief delay is introduced between DF "seeing" the target and acting upon it, her ability likewise deteriorates. These findings are in total agreement with the TVST, which posits that (1) complex edge detection requires ventral stream processing, (2) the dorsal stream flows to the parietal cortex, where the majority of neurons have binocular receptive fields, and (3) the dorsal stream does not store visual information to inform motor actions, instead using only up-to-the-second visual information for guidance.

Because the injuries suffered by the patients described above are never perfectly, cleanly delineated in the brain, critics will often say that their cases alone are not enough to prove the division of labour predicted by the TVST. After all, their symptoms could conceivably be related to the diffuse damage affecting other brain regions. However, the relative separation of the vision for action and vision for perception streams in healthy, neurologically intact brains can be revealed via size contrast illusions. Size contrast illusions, by definition, are illusions in which an object is perceived as larger or smaller than its actual size when viewed with respect to other features or objects in the scene (see Figure 5.2). In other words, because of the *allocentric* comparisons we make between an object and its surroundings, the vision-for-perception system is fooled into misinterpreting the dimensions of a focused object. The vision-for-action system, however – relying as it does on *egocentric*, effector-based location information and retinotopically-based size information – is resistant to such illusions. Thus, even though someone might *perceive* a target in an illusion array as larger (or smaller) than the same object viewed alone, they will shape their hand accurately (as measured by the MGA) to the *true* size of the target when reaching-to-grasp, regardless of the presence of an illusory array.

Consider two identical circles, as those presented in the Ebbinghaus-Titchener illusion (see Figure 5.2, left). The circle on the left, surrounded by comparatively smaller circles, appears larger than the circle on the right, which is surrounded by comparatively larger circles. When asked to grasp such equally sized (but perceptually different) targets, most people will scale their grip apertures the same way for both, in direct opposition to conscious perception (Aglioti et al., 1995). In other words, the ventral stream sees the illusion, and reports two different circles; the dorsal stream sees the truth, and produces grip scaling information appropriately. We may observe this behaviour and infer a functional separation between the vision-for-action and vision-for-perception streams. Like DF's blindsight abilities, the dorsal stream's resistance to an illusion relies on online visual feedback: if a delay is introduced between seeing the illusion and grasping it, then a person's MGA will scale instead to the *perceived* size of the target. Again, this change in grip-aperture scaling may be understood in terms of the dorsal stream's inability to hold in memory visual information. Because delayed grasps are based on a memory of the visual scene, they are therefore susceptible to the ventral stream's inaccurate perception/interpretation of the array. Critics of illusion studies point to the illusion's effect on delayed grasps and interpret it as evidence against the TVST, saying that since perception affects action, the two streams must not be as functionally independent as posited (Schenk and McIntosh, 2010). However, Goodale and Milner never stated that the two visual streams were absolutely independent of one another; armchair logic alone refutes a theory of absolute division. After all, a particularly well-made children's doll may have a life-like size, weight, and appearance, yet a doll will elicit different grasping behaviours than will a real live baby. As far as the dorsal stream is concerned the doll and the baby are identical, yet the vision-for-perception stream will (thankfully) influence the care we show when picking up an infant.

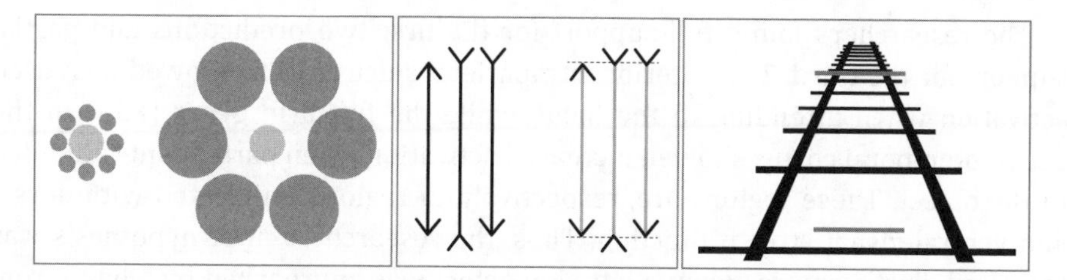

Figure 5.2 Size-contrast illusions commonly used in prehension studies. Left: Ebbinghaus-Titchener illusion. The grey centre circles are equal in size, but the circle on the left appears larger due to the contrasting surrounding circles. Centre: Muller-Lyer illusion. The length of a line segment is made to appear longer or shorter by the addition of arrows at either end. The array on the right shows that all three segments are identical in length. Right: Ponzo illusion. The illusion uses implied perspective to change the perceived length of two identically-sized line segments, shown here in grey.

Evidence for the TVST has not been limited to lesion and behavioural studies. While methodological considerations make reach-to-grasp fMRI studies difficult to perform (see Culham et al., 2003), neuroimaging has nevertheless been used to show the functional division between the dorsal and ventral streams. Shmuelof and Zohary (2005) conducted a study wherein participants were shown videos of simple objects (cups, jars, scissors, etc.) located to one side of a fixation point. These videos also featured a hand, approaching the object from the other side of the screen, grasping the object in a variety of different ways. While watching these videos, participants were instructed to either (a) covertly name the object, or (b) count the number of fingers with which the hand grasped the target. In this manner the viewer's attention was drawn to either the object or the action. Cortical activity was monitored via fMRI while participants performed these tasks. The researchers began with three predictions. First, that participants' dorsal and ventral streams would show differential activation based on the task – that is, the dorsal stream would show increased activation when participants attended to the hand (in the counting task), and the ventral stream would show increased activation when participants attended the object (in the naming task). Second, because the videos showed an object on one side with a moving hand on the other side, they predicted that, due to the crossed nature of the visual system, the dorsal and ventral streams would show increased activation on the side contralateral to the stimuli to which they would predominantly respond. That is, if a cup were shown on the left side of the screen with a hand reaching in from the right, then the right-ventral stream would be more active than the left-ventral stream, and the dorsal stream would be more active in the left hemisphere. Third, knowing that the fMRI response is reduced following multiple presentations of the same stimulus, they predicted that the dorsal stream would show adaptation with repeated presentations of the same type of grasp, regardless of the object being grasped, while the ventral stream would show adaptation to repeated presentations of the same object, regardless of the type of grasp being performed.

The researchers found full support for the first two predictions and partial support for the third. The anterior intraparietal sulcus (aIPS) showed increased activation when attending to the hand, while the fusiform gyrus (FuG) in the occipitotemporal cortex showed increased activation when participants attended to the object. These regions are, respectively, in regions implicated with dorsal and ventral visual stream function. Thus, the researchers' first hypothesis was validated. Furthermore, when a left-side object was approached by a hand from the right, the right FuG and the left aIPS were more active than their left and right counterparts respectively. The reverse was true when a right-side object was approached from the left, in full agreement with the researcher's second hypothesis. Finally, regarding adaptation, the FuG showed reduced fMRI response following multiple presentations of the same target object, but no reduced response following repeated grasp types. The aIPS, in contrast, showed a reduced response to both repeated hand configurations *and* repeated object presentations. Thus, the human ventral stream responded to object parameters in the visual scene, whereas the dorsal stream responded to both hand configurations and objects with which the hand was to interact. This last result was presumably due to the graspable nature of the objects presented. More recent studies have shown that objects elicit responses based on the actions that they afford. In other words, a cup would activate both ventral and dorsal streams because its identity would not be dissociated from the action of grasping it.

When considering the TVST one should not think of the dorsal and ventral streams as isolated and independent, but rather as imbricated circuits, each one specialized in outputting information to be used for ultimately different purposes. In a seminal study, Cavina-Pratesi et al. (2007) used a robust fMRI paradigm and specially designed equipment to dissociate the visual processing of objects for grasping versus processing for recognition. In the first part of this study, participants were asked to either reach-to-touch or reach-to-grasp 3D objects, while in the second part of the study they viewed 2D pictures of objects or scrambled versions of these same pictures. By contrasting the difference in brain activation during these tasks (touch vs. grasp, and real vs. scrambled), researchers were able to identify the regions involved in object processing for different tasks. During the active tasks, regions in the dorsal stream (specifically, the aIPS in both hemispheres, and the junction of the intraparietal sulcus and postcentral sulcus in the left hemisphere) showed heightened activity. In particular, the aIPS was active when viewing 3D objects, but less so during the touching task than during the grasping task. That is, this aIPS appears to be particularly important for control of grasping actions, when the hand must be appropriately shaped to the target. During the passive viewing task, the occipitotemporal cortex – specifically the lateral occipital complex (LOC) – showed increased bilateral activation when the images were intact compared to when they were scrambled. That is, the LOC is very active during object recognition, and less active during passive viewing of pixels. The aIPS showed no increased activation during the viewing task, while the LOC showed no increased activation during the reaching and grasping tasks. Again, in healthy participants, researchers have shown dissociation between the functions of dorsal and ventral visual streams.

Some neuroimaging studies not only show posterior parietal activation for prehensile actions, but also differential activation of subregions within the dorsal stream depending on the motor action being performed. In fact, there is evidence of sub-pathways within the dorsal stream, subserving reaching and grasping actions independently. The dorsomedial pathway – including the superior parieto-occipital cortex (SPOC) and dorsal premotor cortex (PMd) – is thought to control reaching and online correction of movement. In contrast, the dorsolateral pathway, including aIPS and ventral premotor (PMv) areas, subserves hand movements (i.e. grasping) and action understanding. This pathway's strong connections to ventral stream areas, along with its location next to the inferior parietal lobule (IPL, which is implicated in tool use and gesture production), probably assist greatly in its role for action understanding. The dorsal stream also contains subregions specialized for controlling eye movements (parietal eye fields; PEF) and alternate effector actions (e.g. potential foot or eye movements, if a hand movement is planned), as well as regions specialized for determining the direction of a movement (superior parietal lobule; SPL), regardless of that movement's effector. Finally, advanced neuroimaging can differentiate not only between reaching and grasping, but also between different types of grasping (power vs. precision grasps, for example). Altogether, these recent studies suggest that the parietal cortex is incredibly complex, housing effector-based, direction-based and task-based networks, all of which serve the production of action. The ventral stream is no less complex; fMRI has identified subregions within the occipitotemporal complex that code for objects (e.g. LOC), faces (fusiform face area; FFA), bodies (extrastriate body area; EBA), places (parahippocampal place area; PPA), tools ($LOTC_{TOOL}$), and hands ($LOTC_{HANDS}$). As might be expected, these last two areas show stronger connectivity with dorsal stream grasping regions than do the others. Taken together, modern neuroimaging studies suggest that the human visual system is more intricate than we once believed, and we have only begun to scratch the surface in understanding the nature of its complexities.

REFERENCES

Aglioti, S., Desouza, J. and Goodale, M. (1995) Size-contrast illusions deceive the eye but not the hand. *Current Biology*, 5: 679–85.

Cavina-Pratesi, C., Goodale, M. and Culham, J.C. (2007) FMRI reveals a dissociation between grasping and perceiving the size of real 3D objects. *PLoS ONE*, 2: e424.

Culham, J.C., Danckert, S.L., De Souza, J.F. et al. (2003) Visually guided grasping produces fMRI activation in dorsal but not ventral stream brain areas. *Experimental Brain Research*, 153: 180–9.

Goodale, M., Jakobson, L., Milner, A. et al. (1994) The nature and limits of orientation and pattern processing supporting visuomotor control in a visual form agnosic. *Journal of Cognitive Neuroscience*, 6: 46–56.

Goodale, M. and Milner, A. (1992) Separate visual pathways for perception and action. *Trends in Neuroscience*, 15.

Heider, B. (2000) Visual form agnosia: neural mechanisms and anatomical foundations. *Neurocase*, 6: 1–12.

Humphrey, N.K. and Weiskrantz, L. (1967) Vision in monkeys after removal of the striate cortex. *Nature*, 215: 595–7.

Karnath, H.-O., Rüter, J., Mandler, A. and Himmelbach, M. (2009) The anatomy of object recognition—visual form agnosia caused by medial occipitotemporal stroke. *Journal of Neuroscience*, 29: 5854–62.

Perenin, M. and Vighetto, A. (1988) Optic ataxia: a specific disruption in visuomotor mechanisms. *Brain*, 111: 643–74.

Schenk, T. and Mcintosh, R.D. (2010) Do we have independent visual streams for perception and action? *Cognitive Neuroscience*, 1: 52–62.

Schneider, G.E. (1969) Two visual systems. *Science*, 163: 895–902.

Shmuelof, L. and Zohary, E. (2005) Dissociation between ventral and dorsal fMRI activation during object and action recognition. *Neuron*, 47: 457–70.

Ungerleider, L. and Mishkin, M. (1982) Two cortical visual systems. In D. Ingle, M. Goodale and R. Mansfield (eds), *Analysis of Visual Behaviour*. Cambridge, MA: MIT Press. pp. 549–86.

Westwood, D.A. and Goodale, M. (2011) Converging evidence for diverging pathways: neuropsychology and psychophysics tell the same story. *Vision Research*, 51: 804–11.

6 Revisiting Sperry: What the split brain tells us

Michael C. Corballis

In 1961, Roger W. Sperry published a paper in *Science* entitled "Cerebral organization and behavior" that heralded an astonishing era in behavioral neuroscience. The focus of the article was the splitting of the brain through cutting the corpus callosum, by far the largest and most important of the commissures connecting the two sides. The effect of this so-called split-brain operation was to have profound influences on our understanding of the nature of consciousness and on the different ways in which the two sides of the brain operate—influences that extended beyond neuroscience to such diverse disciplines as anthropology, art, literature, philosophy, psychology, and even business. Twenty years after publication of the *Science* article, Sperry belatedly received the Nobel Prize in Physiology or Medicine "for his discoveries concerning the functional specialization of the cerebral hemispheres."

The 1961 article summarized the work that Sperry and his colleagues had carried out on split-brained cats and monkeys. By splitting the optic chiasm as well as the corpus callosum, the researchers were able to present visual information to one side of the brain by presenting it to just one eye (see Figure 6.1). Thus a visual input to the left eye would be projected to the left brain, and input to the right eye only to the right brain. The results revealed a remarkable disconnection. If the split-brain animal learned to make a visual discrimination, such as responding to a cross but not to a circle, when viewing with one eye, it was then unable to make that discrimination when tested with the other eye. The memory for the discrimination was therefore contained entirely within the trained hemisphere of the brain, and inaccessible to the other. In the animal with the corpus callosum intact, in contrast, the discrimination learned with one eye would be available to the other, since the corpus callosum allows transfer of learning from one side of the brain to the other.

Each side of the brain could even be taught conflicting discriminations, with one eye and therefore one side of the brain learning to respond to a cross but not to a circle, and the other eye–brain combination learning the opposite. In monkeys,

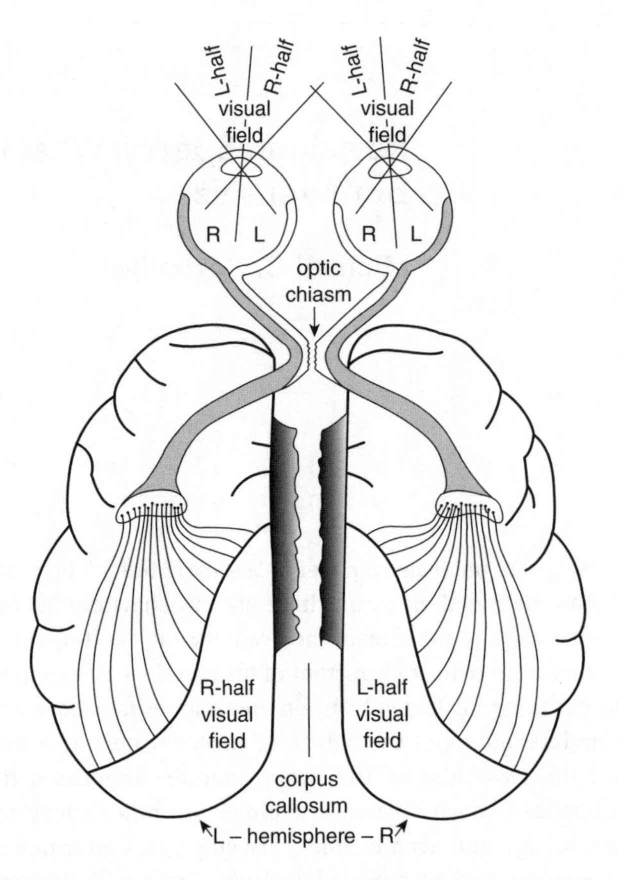

Figure 6.1 Schematic diagram showing how section of the optic chiasm and corpus callosum restricts visual flow to the opposite cerebral hemisphere (adapted from Sperry, 1961)

use of polarized filters enabled the researchers to present conflicting information to the two eyes simultaneously, and the animal learned both discriminations without any apparent conflict. When later given a free choice, the animals would alternate periodically between the two discriminations, as though each side of the brain would take control for a while before yielding to the other. As Sperry put it, "The split-brain cat or monkey is thus in many respects an animal with two separate brains that may be used either together or in alternation." (1961: 133)

HUMAN STUDIES

But what of the human brain? The idea of splitting the human brain had long been of philosophical interest, especially as a test of Cartesian dualism. Gustav Fechner, the nineteenth-century experimental psychologist, and William McDougall, a prominent British psychologist of the early twentieth century, both wondered what would happen to consciousness if the brain were split. Indeed

McDougall is said to have tried to persuade the esteemed physiologist, Charles S. Sherrington, to sever his—McDougall's—corpus callosum if he should become incurably ill, thereby separating the two sides of his brain. McDougall was a convinced dualist who believed his mind would remain intact if his brain were divided, but his personal request was never fulfilled.

The surgeon Joseph E. Bogen had taken note of Sperry's split-brain findings, and considered the possibility of sectioning the forebrain commissures, including the corpus callosum, for the relief of intractable epilepsy in humans. This was not without precedent, since William van Wagenen, a surgeon in Rochester NY, had already pioneered the operation in the 1940s, and psychological examination of his patients had revealed many of the effects that were later to attract widespread interest (Akelaitis, 1941a, b, 1942, 1945; Akelaitis et al., 1942). At the time, though, the operation had not been deemed successful in treating epilepsy, but Bogen and his colleague Philip J. Vogel nevertheless elected to pursue the operation in a series of patients in Los Angeles.[1] With the help of his PhD student Michael S. Gazzaniga, Sperry then set out to adapt the procedures used in the animal research for the study of split-brained humans.

The visual technique used in monkeys and cats was no longer feasible, since it was impractical (and of no medical benefit) to sever the optic chiasm as well as the corpus callosum in human patients. Instead, Sperry and Gazzaniga took advantage of the fact that, with the optic chiasm intact, each visual hemifield of each eye projects only to the opposite hemisphere. That is, visual patterns projected to the right of the point of fixation were projected to the left hemisphere, and those to the left of fixation to the right hemisphere. Visual inputs were projected quickly, typically in less than 200 milliseconds, so the participants did not have time to move their eyes away from fixation and so allow input to switch from one hemisphere to the other.

As in the split-brained cats and monkeys, each cerebral hemisphere was shown to process visual and tactile information independent of the other. In a lecture delivered in Stockholm when he received the Nobel Prize, and later published in *Science*, Sperry summarized as follows:

> Each disconnected hemisphere behaved as if it were not conscious of cognitive events in the partner hemisphere . . . Each brain half, in other words, seemed to have its own largely separate cognitive domain with its own private perceptual, learning, and memory experiences, all of which were seemingly oblivious to corresponding events in the other hemisphere. (1982: 1224)

Although the independence of the disconnected hemisphere was of considerable psychological and philosophical importance, the aspects of the split-brain

[1]Bogen and Vogel sectioned all or nearly all of the forebrain commissures, but in later surgery accomplished elsewhere only the corpus callosum was severed. Psychological testing has revealed no consistent differences between these groups, indicating that sectioning the corpus callosum, by far the largest and most important of the commissures, was responsible for the psychological effects described in this chapter.

work that attracted most attention were the functional differences between the hemispheres—and indeed it was these that won Sperry his prize. The patients proved unable to read words or name pictures of objects flashed to the left visual hemifield, since in these cases the information was relayed to the right cerebral hemisphere, revealing that this hemisphere does not possess the power of speech. These tasks were readily accomplished when the information was relayed to the language-dominant left hemisphere (Gazzaniga et al., 1965, 1967; Sperry, 1982).

The split-brain studies also revealed the right brain to be superior at a number of nonverbal operations, perhaps balancing the left brain's verbal dominance. In Sperry's words:

> Examples include reading faces, fitting designs into larger matrices, judging whole circle size from a small arc, discriminating and recalling nondescript shapes, making mental spatial transformations, discriminating musical chords, sorting block sizes and shapes into categories, perceiving wholes from a collection of parts, and the intuitive perception and apprehension of geometric principles. (1982: 1225)

Rather surprisingly, the mute right hemisphere of the split brain proved capable of understanding language. The patients could generally identify objects or words projected to the right hemisphere by pointing to the corresponding name or object respectively, and were generally capable of following instructions as to how to respond to information presented to the right hemisphere. Zaidel (1976) developed a contact-lens apparatus that enabled prolonged viewing while restricting visual input to a single hemisphere, and found that the isolated right hemispheres of two adult split-brained patients had comprehension scores equivalent to those of the average 16 year-old and 11 year-old respectively. These findings ran counter to evidence that left-hemisphere damage can result in profound deficits in both the production and comprehension of language. Gazzaniga (1983) has suggested that the patients sampled by Zaidel were not representative of the totality of operated patients and overestimated normal right-hemisphere function. Sperry (1982) himself suggested that it was the evidence from unilateral brain damage that was misleading, arguing that right-hemisphere function is suppressed when the left hemisphere is damaged.

The degree of right-hemisphere comprehension remains a somewhat controversial issue, but it does seem clear that left-hemisphere dominance is much more pronounced for the production of language than for its comprehension, whether in speech or writing.

THE DUAL BRAIN

One lasting legacy of the split-brain research has been the notion of the dual brain, with the two sides regarded as complementary opposites. Bogen (1969) wrote of the left brain as *propositional* and the right brain as *appositional*, relating the dichotomy to age-old distinctions such as the Confucian concepts of

Yin and *Yang*, the Buddhist concepts of *buddhi* and *manas*, or Lévi-Strauss's distinction between the *positive* and the *mystic*. The dual brain was pursued by the psychologist Robert E. Ornstein in his best-selling (1972) book *The Psychology of Consciousness*, with the left hemisphere portrayed as linear, rational, analytic, and fundamentally Western in its style of thought, and the right hemisphere as divergent, intuitive, holistic, emotional, and fundamentally Eastern. The duality may have been exaggerated, even mythologized, by the tumultuous events of the 1960s and early 1970s, with the protests over US involvement in the Vietnam War, and the emergence of a drug culture. The left hemisphere came to be associated with the military-industrial establishment and the right with the creative, peace-loving East. In the slogan "make love not war," the left hemisphere no doubt stood for war and the right for love and peace (Corballis, 1980).

Yet the left brain/right brain cult has persisted well beyond those times, and well beyond the domain of neuroscience, with calls for greater attention to be given to the creative, emotional right hemisphere in activities as diverse as art, education, history, literature, and even business. And even the staid *American Heritage ® Dictionary of the English Language* (2008) offers the following definitions:

> **Left-brained** *adj*: 1. Having the left brain dominant. 2. Of or relating to the thought processes, such as logic and calculation, generally associated with the left brain. 3. Of or relating to a person whose behavior is dominated by logic, analytical thinking and verbal communication, rather than emotion and creativity.
>
> **Right-brained** *adj*: 1. Having the right brain dominant. 2. Of or relating to the thought processes involved in creativity and imagination, generally associated with the right brain. 3. Of or relating to a person whose behavior is dominated by emotion, creativity, intuition, nonverbal communication and global reasoning rather than logic and analysis.[2]

Another recent example is Iain McGilchrist's acclaimed (2009) book *The Master and His Emissary*, which portrays the history of Western civilization as shaped by the counter-forces of the left and right brains. McGilchrist reverses the traditional notion of the left hemisphere as the dominant or major hemisphere; in his book it's the right brain that's the master and the left brain the emissary. Indeed, at the time of writing a Google search for "right brain" elicited around 98 million hits, while "left brain" produced only around 55 million.

Within neuroscience itself, the duality of the brain has not worn so well. Anatomically and physiologically, the two sides of the brain are much more alike than different. To be sure, the left hemisphere is dominant for speech in most people, and the right more specialized for spatial processing and processing facial emotion (e.g. Badzakova-Trajkov et al., 2010), but there is little evidence for

[2]I explained to my 5 year-old granddaughter that the left brain is responsible for speech, and asked her what she thought the right brain does. "It just lazes about," she said.

right-hemispheric dominance for creativity; rather, both hemispheres are activated when people undertake creative tasks (e.g. Badzakova-Trajkov et al., 2011; Ellamil et al., 2012; Moore et al., 2009). Moreover, factor analysis of asymmetries in resting-state activity in the brain suggest that there are at least four independent dimensions of asymmetry, not just one, with two favoring the left and two favoring the right (Liu et al., 2009). The asymmetry of the brain can no longer be viewed as a dichotomy, let alone a single one.

HOW SPLIT IS THE SPLIT BRAIN?

Aside from the evidence showing that the two hemispheres of the split brain were differently specialized, evidence from cross-hemispheric matching seemed to confirm the functional independence of the separated hemispheres. Split-brained patients cannot judge whether pairs of digits, letters, colours, or line drawings of faces projected to opposite visual hemifields are the same or different (Johnson, 1984), and this seems to extend even to the matching of simple objects differing only in such elementary features as luminance and stimulus size (Corballis and Corballis, 2001). Yet everyday observations of split-brain patients seemed to suggest a degree of perceptual unity.

Sperry himself observed that lower-order vegetative functions remained connected, and that there did seem to be sharing of emotion between the two sides of the split brain, "apparently through crossed fibers in the undivided brain stem" (1982: 1225). In everyday life, moreover, split-brain patients seem surprisingly normal, displaying what Bogen (1993) called "social ordinariness." Sperry too noted that, after sufficient post-operative recovery, "a person with complete section of the forebrain commissures would usually go undetected as a rule in a casual first meeting or conversation or even through an entire routine medical exam" (1982: 1223). One split-brain patient drove his truck for many years on the roads and highways of New Hampshire without serious incident.

From the mid-1980s, experimental evidence began to emerge showing greater integration between the disconnected hemispheres than was implied either by the matching experiments or the experiments on cerebral specialization. One split-brained patient, for instance, was able to perceive apparent motion across the visual midline. If a disk appeared in one visual hemifield followed shortly by a disk in the other, he saw a single disk apparently jump from one side to the other (Ramachandran et al., 1986)—this is the well-known phi phenomenon. This result seemed clearly to reflect interhemispheric integration, since there is no apparent motion within each visual hemifield. Gazzaniga (1987) suggested that the patient may have simply inferred motion through his attention being drawn to the disk that appeared second, and deduced that it must have jumped from the other side, but this possibility was effectively ruled out in later studies (Naikar and Corballis, 1996).

Spatial attention also crosses between the disconnected hemispheres. Holtzman (1984) showed that split-brained patients could direct an eye movement to a specific location in one visual hemifield on the basis of a locational probe flashed briefly in the other. Location was specified relative to a 2 × 2 matrix of Xs, the probe being a circular shape briefly surrounding one of the Xs. The subjects could quickly direct their gaze to the equivalent position in a matrix in the other hemifield, even though the two matrices were on different sides. But there were limits. The patients could not direct eye movements across hemifields if the probe in one hemifield depicted a shape rather than a location. For example, if the probe was a cross, the subjects were unable to move their eyes to the cross in a 2 × 2 matrix that also comprised a circle, a square, and a triangle.

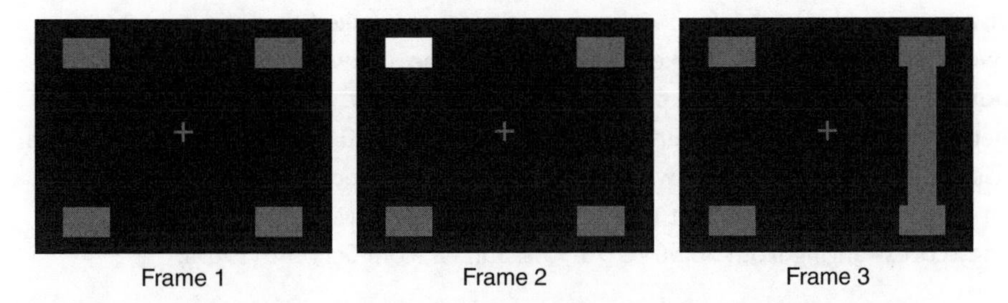

Figure 6.2 A typical sequence in an experiment showing transfer of visual attention between visual fields. *Frame 1* shows the background display, in *Frame 2* one rectangle is briefly illuminated, in *Frame 3* a line connects two rectangles in the opposite visual hemifield. Normal subjects, three acallosals, and one of two cases with surgical section of the corpus callosum see the line as though spreading in a direction away from the illuminated rectangle, in this case, downward (adapted from Corballis et al., 2004).

The transfer of spatial attention is also illustrated by a variant of the line-motion illusion (Hikosaka et al., 1993). Subjects are shown an array of four rectangles, two in each visual hemifield. One hemifield is briefly illuminated, and then a vertical line joins two of the rectangles. If one of these rectangles is illuminated, the line appears to spread from that rectangle to the other (i.e. downward if the top rectangle is illuminated, upward if the bottom rectangle is illuminated), probably because attention is attracted by the illumination and processing is speeded in that vicinity. This illusion, though, can occur between visual hemifields, such that if a rectangle is illuminated in one hemifield it can induce perceived motion in the other (see Figure 6.2). This was true of neurologically normal participants, and with three individuals with agenesis of the corpus callosum and one of two patients with section of the corpus callosum (Corballis et al., 2004). (The other split-brained patient had shown only weak effects when the illusion was induced within hemifields).

Split-brained patients can even integrate simple visual properties between the disconnected hemispheres. Sergent (1987) showed that two split-brained patients could accurately decide whether two sloping lines in opposite visual half-fields

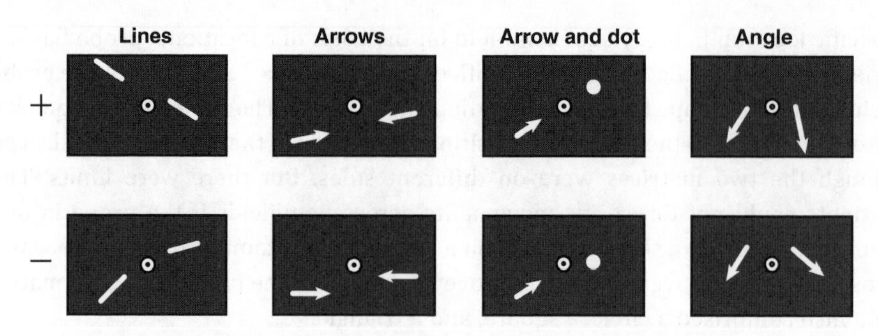

Figure 6.3 Samples of stimuli used by Sergent to test perceptual unity between visual hemifields. From left to right, the patients were asked to decide: (1) Are the two lines straight (+) or broken (–)? (2) Would the arrows meet head-on (+) or not (–)? (3) Does the arrow point to the dot (+) or not (–)? (4) Is the angle between the two lines smaller (+) or greater (–) than 90 degrees? The stimuli were flashed for 150 milliseconds while the patients maintained fixation on the central circle. One split-brain patient (L.B.) scored over 93% on all tasks, the other (N.G.) had scores ranging from 76.56 to 98-95%. Stimuli from Sergent (1987).

were aligned or not, whether inward pointing arrows would or would not meet head on, whether an arrow on one side did or did not point to a dot on the other side, or whether two arrows if connected would form an angle of greater or less than 90 degrees (see Figure 6.3). These findings invited skepticism,[3] but in an independent study one of the split-brain patients tested by Sergent proved as adept as neurologically normal controls in judging alignment across the visual half-fields (Corballis and Trudel, 1993).

These various findings might be understood in terms of what has been called the second visual system (Bogen, 1990), a subcortical system involving projection to the superior colliculi, and thence to the pulvinar nuclei of the thalamus (see Corballis, 1995, for a more detailed discussion). This system has also been implicated in the phenomenon known as "blindsight," in which patients with damage to the primary visual cortex can respond to visual stimuli without consciously "seeing" them (Weiskrantz et al., 1974). This system allows interhemispheric transfer via the tectal commissure or thalamic commissure, both of which remain intact in the split-brain patients. Discussing her findings, Sergent wrote:

> This subcortical coordination of hemisphere activity may thus underlie the behavioural integration displayed by commissurotomized patients in their daily activities, allowing them to relate different parts of the visual field and to maintain a unity of purpose in their action. (1987: 1389)

[3]Justine Sergent tragically took her own life in May 1994, following allegations that she had fabricated data—a charge she vigorously denied. A few of Sergent's split-brain experiments were methodologically flawed and indeed failed replication (Corballis, 1994; Seymour et al., 1994), but those described here appear to have been valid, and indeed informative and insightful.

This system may indeed explain why split-brained patients appear to navigate normally in the visual world, and why one of them could even drive without incident.

The limitations of the second visual system may apply primarily to color and shape which are processed cortically. This is why split-brained patients cannot match shapes or colors between hemifields, and why they could not direct attention on the basis of visual shape. Judging sameness and difference per se may also be beyond the capacity of the second visual system, which may be why a split-brained patient could not judge whether stimuli differing only in size or luminance were the same or different when shown in opposite hemifields (Corballis and Corballis, 2001). The second visual system seems to have lower temporal resolution than the cortical system. Although a split-brained patient could judge the apparent motion of a dot across the visual midline, and could tell whether two dots appeared simultaneously or in succession when succession was relatively slow (Naikar and Corballis, 1996), he proved unable to tell whether the dots were simultaneous or successive when these were presented more rapidly. Neurologically normal controls, in contrast, were unaffected by the increase in speed (Corballis, 1996).

The second visual system probably evolved earlier in evolution than the cortical system, and is sufficient for the basic visual requirements of navigation, perception of movement, and attention to events across the visual world. It is unaffected by section of the forebrain commissures. It was later supplemented by a cortical system providing for the perception of color and shape, and for enhanced temporal resolution—as required, for example, for the perception of speech. One difficulty is that work on blindsight suggests that the second visual system operates without awareness. Perhaps the fact that the cerebral cortices themselves remain intact in the split brain, but not in cases of blindsight, is sufficient to enable a degree of awareness. Split-brain studies may help unravel some of the difficulties and controversies over the nature of the second visual system.

CONCLUSION

Sadly, though, the split-brain era is largely over, at least in the case of humans. In modern times, with the development of more effective drugs, callosotomy as a treatment for epilepsy is largely restricted to extreme cases, greatly restricting the scope of psychological testing. The early cases were often of normal or above-normal intelligence, and gave a fairly accurate picture of what would happen if the normal brain were split—a picture that might well have satisfied the curiosity of William McDougall. Increasingly, too, partial callosotomy is preferred to full callosotomy, but does not reveal the dramatic effects of full cortical disconnection.

Nevertheless the split-brain patients, now mostly of the past, had a profound effect not only on neuroscience but also on philosophy and culture. They showed in dramatic fashion that the mind does indeed depend on the brain, to the point that some philosophers, spearheaded by Patricia Churchland (1986), began to write of the mind-brain as an entity. Curiously Sperry himself, who had done so much to demonstrate the mechanistic nature of the brain, seemed to argue for a more holistic view of consciousness, although he claimed not to be a dualist. In his Nobel address he wrote of the split-brain studies as follows:

> The key development is a switch from prior noncausal, parallelist views to a new causal, or "interactionist" interpretation that ascribes to inner experience an integral causal control role in brain function and behavior. In effect, and without resorting to dualism, the mental forces of the conscious mind are restored to the brain of objective science from which they had long been excluded on materialist-behaviorist principles. (Sperry, 1982: 1226)

Early work on the split brain showed how aspects of consciousness could be split, to the point that each side of the brain could accommodate different, and even conflicting, ideas. Later work began to show aspects in which the mental processing in the divided brain was integrated. This was not an affirmation of dualism, but instead showed that aspects of the conscious perceptual world could remain integrated through a subcortical system in spite of the separation of cortical structure. This aspect of perception has still not been fully explored.

REFERENCES

Akelaitis, A.J. (1941a) Psychobiological studies following section of the corpus callosum: a preliminary report. *American Journal of Psychiatry*, 97: 1147–57.

Akelaitis, A.J. (1941b) Studies on the corpus callosum: II. The higher visual functions in each homonymous field following complete section of the corpus callosum. *Archives of Neurology & Psychiatry (Chicago)*, 45: 788–96.

Akelaitis, A.J. (1942) Studies on the corpus callosum: V. Homonymous defects for color, object and letter recognition (homonymous hemiamblyopia) before and after section of the corpus callosum. *Archives of Neurology & Psychiatry (Chicago)*, 48: 108–118.

Akelaitis, A.J. (1945) Studies on the corpus callosum: IV. Diagnostic dyspraxia in epileptics following partial and complete section of the corpus callosum. *American Journal of Psychiatry*, 101: 594–9.

Akelaitis, A.J., Risteen, W.A., Herren, R. and van Wagenen, W.P. (1942) Studies on the corpus callosum: III. A contribution to the study of dyspraxia in epileptics following partial and complete section of the corpus callosum. *Archives of Neurology & Psychiatry (Chicago)*, 47: 971–1008.

American Heritage Dictionary of the English Language, 4th edition (2008). New York: Houghton Mifflin.

Badzakova-Trajkov, G., Häberling, I.S. and Corballis, M.C. (2011) Magical ideation, creativity, handedness, and cerebral asymmetries: a combined behavioural and fMRI study. *Neuropsychologia*, 40: 2896–903. http://doi:10.1016/j.neuropsychologia.2011.06.016

Badzakova-Trajkov, G., Häberling, I.S., Roberts, R.P. and Corballis, M.C. (2010) Cerebral asymmetries: complementary and independent processes. *PLoS ONE*, 5(3): e9682. http://doi:10.1371/journal.pone.0009682

Bogen, J.E. (1969) The other side of the brain: II. An appositional mind. *Bulletin of the Los Angeles Neurological Society*, 34: 135–62.

Bogen, J.E. (1990) Partial hemispheric independence with the neocommissures intact. In C. Trevarthen (ed.), *Brain Circuits and Functions of the Mind: Essays in Honor of R.W. Sperry*. Cambridge: Cambridge University Press. pp. 211–30.

Bogen, J.E. (1993) The callosal syndromes. In K.M. Heilman and E. Valenstein (eds), *Clinical Neuropsychology*, 3rd edition. New York: Oxford University Press. pp. 337–407.

Churchland, P.M. (1986) *Neurophilosophy: Toward a Unified Science of the Mind-Brain*. Cambridge, MA: The MIT Press.

Corballis, M.C. (1980) Laterality and myth. *American Psychologist*, 35: 254–65.

Corballis, M.C. (1995) Visual integration in the split brain. *Neuropsychologia*, 33: 937–59.

Corballis, M.C. (1996) Hemispheric interactions in temporal judgments about spatially separated stimuli. *Neuropsychology*, 10: 42–50.

Corballis, M.C., Barnett, K.J., Fabri, M. et al. (2004) Hemispheric integration and differences in perception of a line-motion illusion in the divided brain. *Neuropsychologia*, 42: 1852–7.

Corballis, M.C. and Corballis, P.M. (2001) Interhemispheric visual matching in the split-brain. *Neuropsychologia*, 39: 1395–400.

Corballis, M.C. and Trudel, C. I. (1993) The role of the forebrain commissures in interhemispheric integration. *Neuropsychology*, 7: 306–24.

Ellamil, M., Dobson, C., Beeman, M. and Christoff, K. (2012) Evaluative and generative modes of thought during the creative process. *NeuroImage*, 59: 1783–94.

Gazzaniga, M.S. (1983) Right hemisphere language following brain bisection: a twenty-year perspective. *American Psychologist*, 38: 525–37.

Gazzaniga, M.S. (1987) Perceptual and attentional processes following callosal section in humans. *Neuropsychologia*, 25: 119–33.

Gazzaniga, M.S., Bogen, J.E. and Sperry, R.W. (1965) Observations of visual perception after disconnexion of the cerebral hemispheres in man. *Brain*, 88: 221–30.

Gazzaniga, M.S., Bogen, J.E. and Sperry, R.W. (1967) Dyspraxia following division of the cerebral hemispheres. *Archives of Neurology*, 16: 606–612.

Hikosaka, O., Miyauchi, S. and Shimojo, S. (1993) Visual attention revealed by an illusion of motion. *Neuroscience Research*, 18: 11–8.

Holtzman, J.D. (1984) Interactions between cortical and subcortical visual areas: evidence from human commissurotomy patients. *Vision Research*, 24: 801–13.

Johnson, L.E. (1984) Bilateral cross-integration by human forebrain commissurotomy subjects. *Neuropsychologia*, 22: 167–75.

Liu, H., Stufflebeam, S.M., Sepulcrea, J. et al. (2009) Evidence from intrinsic activity that asymmetry of the human brain is controlled by multiple factors. *Proceedings of the National Academy of Sciences*, 106: 20499–503.

McGilchrist, I. (2009) *The Master and His Emissary: The Divided Brain and the Making of the Western World*. New Haven, CT: Yale University Press.

Moore, D.W., Bhadelia, R.A., Billings, R.L. et al. (2009) Hemispheric connectivity and the visual–spatial divergent-thinking component of creativity. *Brain and Cognition*, 70: 267–72.

Naikar, N. and Corballis, M.C. (1996) Perception of apparent motion across the retinal midline following commissurotomy. *Neuropsychologia*, 34: 297–309.

Ornstein, R.E. (1972) *The Psychology of Consciousness*. San Francisco, CA: Freeman.

Ramachandran, V. S., Cronin-Golomb, A. and Myers, J. J. (1986) Perception of apparent motion by commissurotomy patients. *Nature,* 320: 358–9.

Sergent, J. (1987) A new look at the human split brain. *Brain*, 110: 1375–92.

Seymour, S.E., Reuter-Lorenz, P.A. and Gazzaniga, M.S. (1994) The disconnection syndrome: basic findings reaffirmed. *Brain*, 117: 105–15.

Sperry, R.W. (1961) Cerebral organization and behavior. *Science*, 133: 1749–57.

Sperry, R.W. (1969) A modified concept of consciousness. *Psychological Review*, 76: 532–36.

Sperry, R.W. (1982) Some effects of disconnecting the cerebral hemisphere. *Science*, 217: 1223–7.

Weiskrantz, L., Warrington, E.K., Sanders, M.D. and Marshall, J. (1974) Visual capacity in the hemianopic field following a restricted occipital ablation. *Brain*, 97: 709–28.

Zaidel, E. (1976) Auditory vocabulary of the right hemisphere following brain bisection or hemidecortication. *Cortex*, 12: 191–211.

PART 2

Cortical Functions

7 Revisiting Hebb: The organization of behavior

Richard E. Brown

Donald Olding Hebb's book *The Organization of Behavior* (1949a) introduced the concepts of synaptic plasticity and cell assemblies to account for the neural events underlying behaviour, providing a theory of the neurophysiological basis of behaviour and revolutionizing psychology. Hebb's ideas, as presented in this book and other writings, have influenced all areas of psychology and neuroscience. In this chapter I summarize his early research and the writing of *The Organization of Behavior*. I then examine how the book was received at the time, its impact on psychology in the 1950s, and its lasting influence on psychology and neuroscience in the twenty-first century.

INTRODUCTION

Hebb's book, *The Organization of Behavior*, published in 1949, introduced the concepts of synaptic change and cell assemblies to account for the neural events underlying behaviour, and revolutionized psychology by establishing a biological basis for psychological phenomena. Hebb later wrote a unique and original introductory textbook in psychology (Hebb, 1958), which was, in many ways, a follow-up to *The Organization of Behavior*. After he retired, he wrote his third book, *Essay on Mind* (1980a). This chapter reviews Hebb's early research leading up to the writing of *The Organization of Behavior*, and the influence of *The Organization of Behavior* on psychology and neuroscience from 1949 to the present day.

HEBB'S EARLY RESEARCH CAREER

Donald Olding Hebb (see Figure 7.1) was born on 22 July 1904, in Chester, Nova Scotia, and graduated with his BA in English and Philosophy from

Figure 7.1 Donald O. Hebb at Dalhousie University in 1979, aged 75. (Photograph taken by Richard Brown.)

Dalhousie University in 1925. He obtained a teaching certificate from the Provincial Normal College in Truro, Nova Scotia, and taught school in both Chester, Nova Scotia, and Montreal, Quebec (Hebb, 1980b). He obtained his MA in psychology as a part-time student at McGill University under Professor Chester Kellogg while he taught school during the day. As a teacher, Hebb was interested in changing the school environment to improve learning and school performance in children (Hebb, 1930). At McGill, he studied Sherrington's *Integrative Activity of the Nervous System* and Pavlov's *Conditioned Reflexes*, and wrote a theoretical MA thesis (Hebb, 1932) which contained his first thoughts on the nature of synaptic activity during conditioning (Brown and Milner, 2003). Hebb began his PhD research on field orientation in rats (Hebb, 1938) at the University of Chicago with Karl Lashley, and when Lashley moved to Harvard, Hebb completed his research at Harvard. He received his PhD from Harvard in 1936 for his thesis on visual abilities of rats reared in the dark (Hebb, 1937). Lashley (Bruce, 1996;

Dewsbury, 2002) had a profound influence on Hebb's thinking about the neurobiological bases of behaviour.

In 1937, Hebb began to study the psychological effects of temporal lobe and frontal lobe surgery on the patients of Wilder Penfield at the Montreal Neurological Institute (MNI), and this work initiated the scientific study of human neuropsychology. His most complete study (Hebb and Penfield, 1940) was on patient KM, who had had a frontal lobotomy. His experiences in testing patients at the MNI led to a number of ideas about the nature of intelligence and how it should be tested (Hebb, 1939a; Hebb and Morton, 1944). With N.W. Morton of the McGill Psychology Department, Hebb developed two new intelligence tests, the verbal Adult Comprehension test and the non-verbal Picture Anomaly test (Hebb and Morton, 1943). During this time he published his review of the functional effects of frontal lobe lesions in humans (Hebb, 1945a), and his observations that lesions of different brain areas produced different cognitive impairments (Hebb, 1939b). He also published his theories that the age at which a brain injury occurred was important in determining its effects on intelligence, and that intelligence was composed of two components: a fixed or innate component and a variable component that could be influenced by environmental experiences (Hebb, 1942).

In 1939 Hebb became a Lecturer in Experimental Psychology at Queens University in Kingston, Ontario, where he and his student, Kenneth Williams, designed a variable path maze for testing intelligence in rats (Hebb and

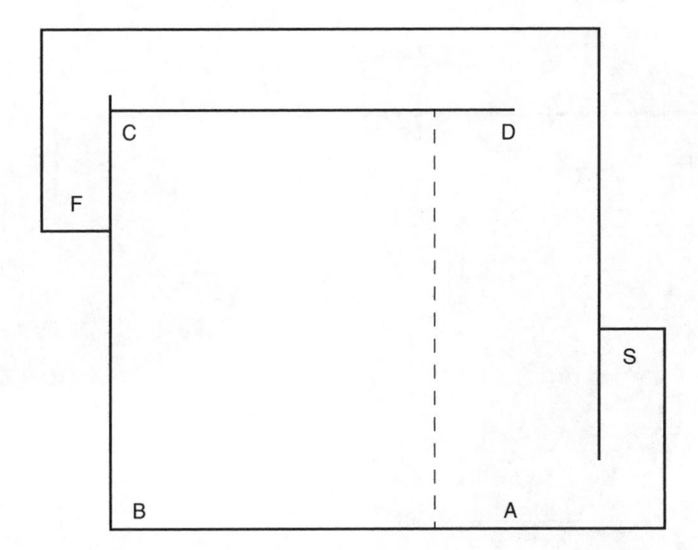

Figure 7.2 An example of one of the maze problems used for testing intelligence in rats in the Hebb–Williams maze. S, start box; F, food in the goal box; A, B, C, D points in the maze. The line CD represents the barrier used in this test problem. In other problems, barriers placed in different locations present the rat with more difficult problems. Crossing the dotted line constitutes an error. (*Source*: Hebb, 1949a/2002: 137)

Williams, 1946). This Hebb–Williams maze (see Figure 7.2) has since been used in a plethora of studies of comparative learning in animals and humans (Shore et al., 2001). In 1942 Hebb moved to the Yale Primate Laboratories at Orange Park, Florida, and worked there until 1947. He studied fear and anger (see Figure 7.3) in chimpanzees (Hebb, 1945b, 1949b) and related this research to human emotionality (Hebb, 1946, 1947a). In addition to his primate research, Hebb published one of the first studies on the behaviour of dolphins (McBride and Hebb, 1948) and continued his work on the development of rat intelligence. To determine the effects of early experience on learning, Hebb reared rats as pets at home and showed that enriched experience during development resulted in improved maze learning in adulthood. Although these results were only published as an abstract (Hebb, 1947b), they formed the basis of studies on the effects of environmental enrichment on behaviour and neural development (Krech et al., 1962). The study of environmental enrichment became one of the most important areas of developmental psychology, and continues to influence research (van Praag et al., 2000). Recently, environmental enrichment has been used as a new strategy for neuro-rehabilitation (Bondi et al., 2014).

Figure 7.3 Objects which stimulated fear responses when presented to adult chimpanzees. Left, a plaster of Paris cast from a death mask of an adult chimpanzee; right, a clay model of an infant chimpanzee's head, nearly life size. (*Source*: Hebb, 1972: 204)

WRITING *THE ORGANIZATION OF BEHAVIOR*

During his years in Florida, Hebb completed the first five chapters of the manuscript of *The Organization of Behavior,* in which he outlined a new way of understanding behaviour in terms of brain function. In a 1959 paper Hebb (1959a) explained how his theory came into being. He stated that his work on the effects of brain lesions at different ages on intelligence (Hebb, 1942) led him to conclude that "intelligence itself, and not merely the ability to do well on intelligence tests must be a product of experience" (1980b: 292). He then began to think of learning, perception, modes of thought, and intelligence in terms of neural mechanisms. He pointed out a number of stumbling blocks that he had to circumvent in order to develop his neuropsychological theory. These included the inter-relatedness of psychological concepts; the assumption that behaviour is under sensory control; and the lack of neurophysiological data. Hebb's most difficult problem in the development of his theory was to separate neural activity from sensory input. The solution to this problem was provided by the work of Lorente de No (1938), who suggested that lasting structural changes occur to reinforce and perpetuate the temporary, dynamic facilitation of neural activity by a sensory stimulus. Hebb integrated this idea of a reverberatory circuit into his theory of synaptic change and the cell assembly, and made the assumption that there was a dual mechanism of the memory trace, i.e. a "dynamic" plus a "structural engram" (see Figure 7.4). This led him to postulate that:

> When an axon of cell A is near enough to excite cell B and repeatedly or persistently takes part in firing it, some growth process or metabolic change takes place in one or both cells such that A's efficiency, as one of the cells firing B, is increased. (Hebb, 1949a: 62)

As noted by Tim Bliss, "Hebb's postulate must be one of the most quoted sentences in the literature of neuroscience" (2003: 621). The concept of reverberatory neural circuits continues to be used to study neural networks underlying memory (Tegnér et al., 2002; Johnson et al., 2009).

The second influence on Hebb's ideas in the preparation of *The Organization of Behavior* was the work of von Senden (1932) on the development of visual ability in congenitally blind people, after surgical operations to remove cataracts had enabled them to see for the first time. Hebb used the case studies in this book, and the work of Riesen (1947) on monkeys reared in the dark at the Yerkes Primate Laboratory, to argue that learning played an important role in perception. Although Wertheimer (1951) disagreed with Hebb's interpretation of von Senden's results, they were crucial for Hebb's development of the concept of the cell assembly and the idea that perceptual organization occurs through a process of learning.

Hebb sent the first five chapters of *The Organization of Behavior* to a number of his colleagues, including E.G. Boring, Henry Nissen and Karl Lashley, for comments, and during the summer of 1947 he taught a graduate seminar at Harvard using the manuscript of his book. One of the students in this class was Mark Rosenzweig (1998), who has written that:

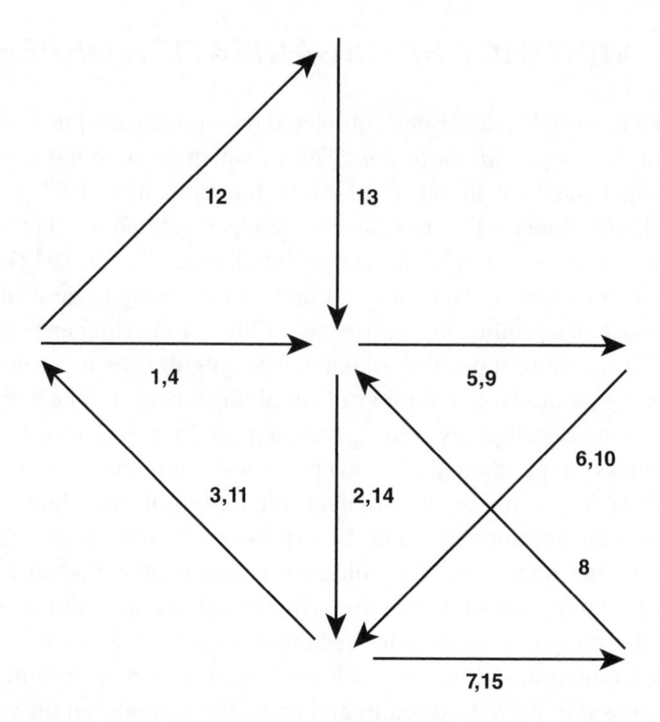

Figure 7.4 Hebb's theoretical model of the cell assembly. Arrows represent a simple "assembly" of neural pathways or open multiple chains firing according to the numbers on each (the pathway "1,4" fires first and fourth, and so on), illustrating the possibility of an "alternating" reverberation which would not extinguish as readily as that in a simple closed circuit. This alternating reverberation formed the phase sequence. (*Source*: Hebb, 1949a/2002: 73)

> I took a graduate seminar with Donald O. Hebb at Harvard in the summer of 1947 where the text was a mimeographed version of Hebb's influential book *The Organization of Behavior* which appeared in print in 1949. (I wish I had had the foresight to save that 1947 version.) Hebb's creative suggestions revitalized theorizing and research on learning and memory, and I benefitted directly from them and from further contacts with him.

After the summer at Harvard, Hebb became Professor of Psychology at McGill University, where he completed the book, re-writing the chapters critiqued by his colleagues, and adding new chapters on emotion, motivation, and intelligence. In 1949, John Wiley & Sons published *The Organization of Behavior.*

In this book, Hebb examined the issues concerning psychologists in the 1940s and showed how these problems might be dealt with using a set of neurophysiological postulates. He proposed two concepts, synaptic plasticity and cell assemblies, which became central tenants in neuroscience (Spatz, 1996) and remain central to neuroscience today (Wallace and Kerr, 2010; Favero et al., 2014). In this book, Hebb integrated the research and theories of the most prominent psychologists of the time (Lashley, Kohler, Tolman and Hull) through a common neurophysiological process. He also reinterpreted his own previous research on Penfield's neurosurgery patients, intelligence and how it develops, animal models of intelligence, and his work on emotions and fear in chimpanzees and humans, and integrated these into his theory. He took "mental" processes, such as attention, and related them to neurophysiological activity. He was critical of the Pavlovian stimulus–response (S-R) association model, on which Hull's learning theory was based, and put emphasis on Tolman's stimulus–stimulus (S-S) associations. He was also critical of the Gestalt school's field theory explanation of generalization in the visual system because it could not explain how recognition of the field was acquired. He was critical of behaviourist theory because, although it could explain learning, it was vague as to how a pattern falling on different receptors could reach the same learned recognition structure.

The neurochemical synapse was not formally recognized until Eccles published his seminal paper in 1954 (Eccles et al., 1954). Before this, it was believed that synapses in the CNS were electrical. Thus, in 1945, when Hebb was developing his ideas about the neural basis of behaviour, chemical transmission in the CNS was unknown and he did not discuss neurotransmitters. He envisioned the brain as a series of electrical circuits, and changes in the "bio-electric fields" were thought to underlie learning. Hebb developed his idea of "cell assemblies" from Lorente de No's (1938) theory of recurrent (reverberating) nerve circuits. Eccles later developed the idea of inhibitory synapses (Brooks and Eccles, 1947) but Hebb did not incorporate the concept of inhibition into his theory; it had to wait for Milner's (1957) revision. Hebb thus employed the neuroanatomical and neurophysiological knowledge of the day to develop his ideas of synaptic change and cell assemblies, which were linked by patterns of neural activity that he called phase sequences. He envisioned phase sequences as neural representations of images and concepts, and the use of phase sequences allowed Hebb to define consciousness in neuropsychological terms (1949a: 45), which he later developed into a theory of mind (Hebb, 1959b). Hebb realized that his theory would need revision in light of new discoveries, but the fact that his ideas on synaptic plasticity (Favero et al., 2014), cell assemblies (Lansner, 2009; Wallace and Kerr, 2010) and phase sequences (Almeida-Filho et al., 2014) continue to stimulate new research and discussion today is a tribute to his theory (see Figure 7.5).

REVIEWS OF *THE ORGANIZATION OF BEHAVIOR*

The published reviews of *The Organization of Behavior* were uniformly positive, even when they were critical. Manford H. Kuhn (1950) stated that "this book will probably come to be regarded as a landmark in psychological theory." W. J. Brogden (1950) said "The neural theory is admittedly gross, and probably impossible to test, but its presentation results in provocative discussion." Fred Attneave (1950) stated that "I believe *The Organization of Behavior* to be the most important contribution to psychological theory in recent years." In a lengthy review, Leeper (1950) stated that "There are so many respects in which Hebb's

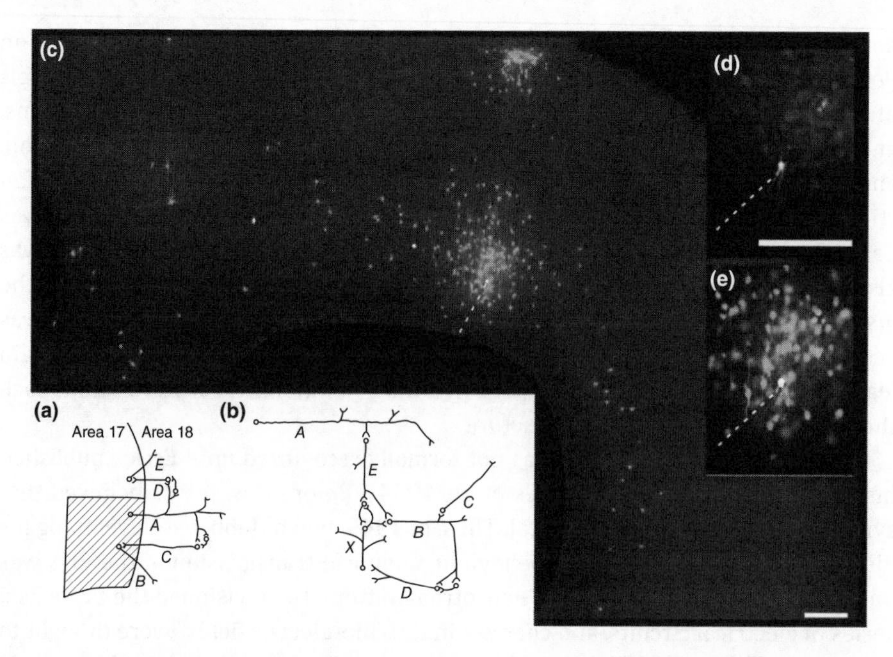

Figure 7.5 Hebb's illustration of the concepts of synaptic plasticity and the formation of a cell assembly. (a) A hypothetical group of neurons in Area 17 and 18 of the visual cortex, illustrating Hebb's concept of how repeated stimuli in the visual receptive field area corresponding to the shaded area may lead to a recurrence of firing patterns between these neurons, with the consequence that the connections AC and CB are strengthened. (b) Illustration of another hypothetical situation in which A, B and C are neurons in visual Area 18, and are all strongly activated by a particular stimulus. The remaining neurons D, E and X form direct or indirect connections with the initial three neurons. Hebb proposed that the appropriate stimuli would result in strengthening of numerous connections including AE, BC, BD, eventually resulting in an increase in the probability of coordinated activity among the various neuronal pairs. (c) Identification of synaptically coupled neurons using monosynaptically restricted transsynaptic tracing with a deletion-mutant rabies virus. The mutant virus lacks a gene encoding a glycoprotein essential to the viruses' ability to infect synaptically coupled neurons. Additionally, infection of a neuron with the mutant rabies virus requires expression in the target neuron of a specific receptor protein not normally present in mammalian neurons. In an initial step, a subpopulation of cortical neurons were transfected with genes encoding: first, the viral receptor protein, thus allowing infection with the modified rabies virus; second, the normal rabies virus glycoprotein that allows the virus to infect neurons presynaptically connected to the infected cell; and third, dsRed for identification. The mutant virus expresses green fluorescent protein (GFP), thus neurons presynaptic to the initially infected neurons appear green. Initially infected neurons express both dsRed and GFP, and appear yellow. (d and e) dsRed expression alone and transsynaptically labelled neurons expressing GFP around the neuron identified at the end of the dashed white line in (c). Scale bars 200 mm. (*Source*: Wallace and Kerr, 2010)

book is so high in quality and is so delightfully written that it will have an assured status in psychology."

HOW *THE ORGANIZATION OF BEHAVIOR* INFLUENCED PSYCHOLOGY AND NEUROSCIENCE IN THE SECOND HALF OF THE TWENTIETH CENTURY

Hebb's book lived up to reviewers' predictions and became one of the most important contributions to psychology in the twentieth century. Almost every book on psychology in the 1950s and 1960s discussed Hebb's theories and *The Organization of Behavior* became one of the most cited books in psychology and neuroscience. Five of the seven volumes of Sigmund Koch's series *Psychology: A Study of a Science* (1959–1963), containing chapters on perception, learning, motivation, ethology, psychoanalytic theory, social interaction, personality and biological psychology, refer to Hebb's theory. In his chapter on physiological psychology, Davis had a special section on "Hebb's theory" in which he said that "By far the most thoroughgoing and comprehensive system of physiological psychology in several decades is D.O. Hebb's *Organization of Behavior*" (Davis, 1962: 267). Hebb's ideas were also prominent in chapters of major textbooks, including Osgood's *Method and Theory in Experimental Psychology* (1953); S.S. Stevens's *Handbook of Experimental Psychology* (1951); and Wolman's *Handbook of General Psychology* (1973). Hilgard wrote a two-page summary of *The Organization of Behavior* in his book on *Theories of Learning,* stating that "Because of its influence in renewing interest in neurophysiological speculation Hebb's (1949) theory deserves special attention" (Hilgard, 1956: 454). Allport (1955) devoted his entire Chapter 7, entitled "The association approach, cell assembly and phase sequence," to a discussion of Hebb's ideas on perception. Hebb's theories also permeate other books of the time, including Harlow and Woolsey's *Biological and Biochemical Bases of Behavior* (1965) and Peter Milner's *Physiological Psychology* (1970).

Once Hebb had completed *The Organization of Behavior,* he saw his previous research in a new way and extended his old experiments in light of his new theories. Hebb was not an author on many of these papers, thus to understand how his theories stimulated new research one must look to his students. Rabinovitch and Rosvold (1951) developed a standardized procedure for the Hebb–Williams maze and tested rats with cortical damage. Other students used the Hebb–Williams maze to test the effects of environmental experience and lesions (Smith, 1959), the effects of blindness and early rearing experience (Hymovitch, 1952), and the effects of environmental enrichment (Forgays and Forgays, 1952) on learning and memory. Other students extended the studies on the effects of lesions on intelligence that Hebb began with Penfield. Rosvold and Mishkin (1950) examined the effects of prefrontal lobotomy on intelligence as Hebb had done in 1937 (Hebb and Penfield, 1940; Hebb, 1945a); D.G. Forgays (1952) studied the development of cognitive dysfunction after surgery; and Brenda Milner

(1954) studied the intellectual functions of the temporal lobes as Hebb had done in 1939 (Hebb, 1939b). Thompson and Heron continued Hebb's studies of fear and emotionality using pure-bred Scottish terriers, which were tested in studies of early rearing experience as Hebb had done with rats reared at home (Thompson and Heron, 1954). Melzack and Thompson continued Hebb's studies of emotional behaviour and the development of social behaviour (Melzack, 1954; Melzack and Thompson, 1956), while Olds and Milner (1954) discovered the pleasure centres in the brain in Hebb's lab (see Figure 7.6), and his work on sensory deprivation (see Figure 7.7; also Bexton et al., 1954) became rather infamous (Brown, 2007b).

Figure 7.6 Olds and Milner's first demonstration of reward circuits in the brain. The rat has been implanted with electrodes, connected with a light flexible wire so he can move about freely. If the experimenters give a brain stimulation in one corner of the maze, the rat will persist in coming back to it, just as if he were hungry and found food there. Note the maze is an early version of the Hebb–Williams maze shown in Figure 7.2. (*Source*: Hebb, 1972: 183; also published in Milner, 1989: 63)

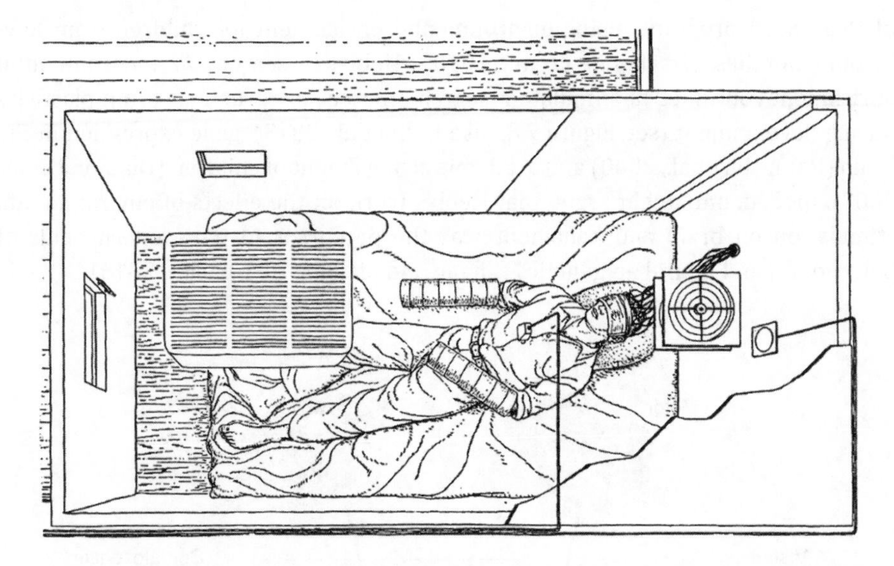

Figure 7.7 A drawing of a subject in a sensory deprivation experiment in Hebb's laboratory. Cuffs were worn to prevent somesthetic perception by the hands; the plastic shield over the eyes admitted light but prevented pattern vision. The subject had a foam rubber U-shaped cushion covering his ears; here it has been removed so that EEG tracings can be taken. An air-conditioner is shown on the ceiling, upper left, and the microphone by which the subject could report his experiences is hanging just above his chest. (*Source*: Heron, 1957: 212)

Following the publication of *The Organization of Behavior*, Hebb became a major theorist in psychology. He reviewed the field of animal and physiological psychology for the first *Annual Review of Psychology* (Hebb, 1950) and discussed the problem of separating genetic and environmental components of behaviour in his presentation to the British Association for the Study of Behaviour (Hebb, 1953a). In the *Handbook of Social Psychology*, Hebb and Thompson (1954) outlined the social significance of animal research for human behaviour as an example of the biological basis of behaviour. In this chapter Hebb and Thompson argued that emotionality and psychopathology result from abnormalities in cell assemblies. In other reviews, Hebb presented evidence for the relevance of his neurological approach to the theory of motivation (Hebb, 1955a), personality (Hebb, 1951), levels of adjustment at maturity (Hebb, 1955b), human thought (Hebb, 1953b), consciousness and introspection (Hebb, 1954) and the theory of mind (Hebb, 1959b).

Hebb's ideas founded two areas of research: the effects of environmental enrichment and impoverishment in early development (Hunt, 1979) and the field of sensory deprivation (Solomon et al., 1961). Hebb's work also stimulated research on the effects of early experience on problem solving, neuroanatomy and neurochemistry (Mohammed et al., 2002), and his ideas helped to facilitate the development of Head Start programs using environmental enrichment for children from low-income families (Hackman et al., 2010). Hebb's research on environmental enrichment continues to stimulate research on areas as diverse as brain plasticity during development (see Figure 7.8; also Kolb et al., 2013), gene expression in the brain (Rampson et al., 2000) and academic achievement of children (Diamond et al., 2007). Indeed, one might argue that Hebb's work on the effects of environmental stimulation on brain and behaviour was the precursor to the modern fields of behavioural and neural epigenetics (Zhang and Meaney, 2010; Isles, 2015).

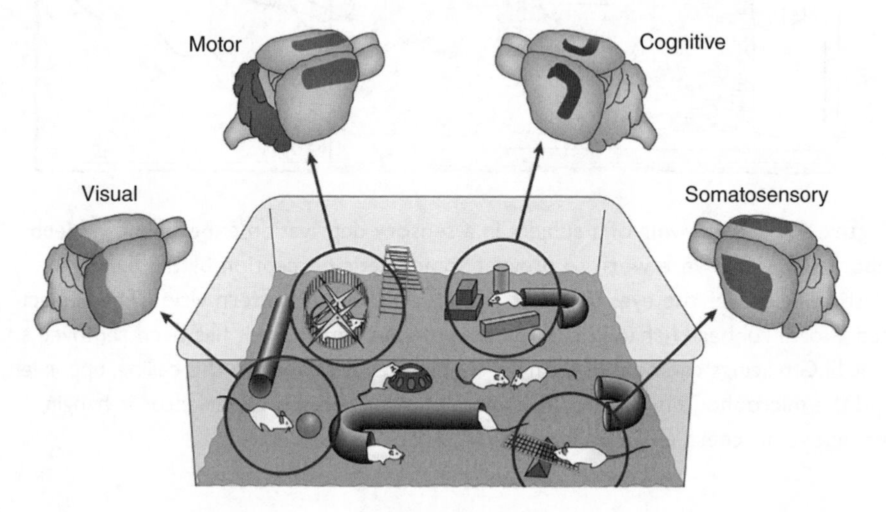

Figure 7.8 Environmental enrichment and the effects of enhanced sensory, cognitive and motor stimulation on different brain areas. Enrichment can promote neuronal activation, signalling and plasticity throughout various brain regions. Enhanced sensory stimulation, including increased somatosensory and visual input, activates the somatosensory (red) and visual (orange) cortices. Increased cognitive stimulation – for example, the encoding of information relating to spatial maps, object recognition, novelty and modulation of attention – is likely to activate the hippocampus (blue) and other cortical areas. In addition, enhanced motor activity, such as naturalistic exploratory movements (including fine motor skills that differ radically from wheel running alone), stimulates areas such as the motor cortex and cerebellum (green). (*Source*: Nithianantharajah and Hannan, 2006)

THE INFLUENCE OF HEBB IN THE TWENTY-FIRST CENTURY

Hebb died on 20 August 1985, following complications from surgery. Numerous obituaries were published which described his legacy in every area of psychology and neuroscience (Harnad, 1985; Fentress, 1987; Klein, 1999).

Modern neuropsychology is based on Hebb's work with Penfield; the study of environmental effects on development derives from Hebb's pet rats reared at home in an enriched environment; and computer models of the brain are based on Hebb's ideas of the synapse and cell assembly. Long-term potentiation (Bliss and Lomo, 1973) is the experimental analysis of Hebbian synaptic plasticity, and the work of Hubel and Wiesel on neural plasticity of visual development (Stent, 1973) derives from the first five chapters of *The Organization of Behavior*. There are very few areas of neuroscience today that have not been influenced by Hebb's work (Jusczyk and Klein, 1980/2010) and the field of computational neuroscience is largely based on his ideas (Sommer and Wennekers, 2003).

On the 50th anniversary of its publication, there were almost as many reviews of *The Organization of Behavior* as when it was originally published (Martinez and Glickman, 1994; Nicolelis et al., 1997; Sejnowski, 1999) and its impact has been compared with that of Darwin's *Origin of Species* (Adams et al., 1998). Because *The Organization of Behavior* was out of print, Peter Milner and I had it republished in 2002 with a Foreword about the importance of the book and a bibliography of Hebb's work (Hebb, 1949a). This reissue has also been reviewed (Tees, 2003). Hebb's ideas of synaptic plasticity and cell assemblies have now become fundamental concepts in psychology and neuroscience (Spatz, 1996; Kolb, 2003; Sejnowski, 2003), and Posner and Rothbart (2004) have argued that Hebb's ideas provide the basis for an integration of the disparate sub-fields of psychology. We have examined the origins of *The Organization of Behavior* (Brown and Milner, 2003), and Cooper (2005) provides a history and commentary on the Hebb synapse and learning rule. Jeff Hawkins used Hebb's ideas in his theory of machine intelligence (Hawkins, 2004), and modern theories of biological cybernetics use Hebb's theories of cell assemblies to understand the brain–mind relationships underlying cognition (Pulvermüller et al., 2014; Li, Liu and Tsien, 2016).

WHAT DOES THE FUTURE HOLD FOR HEBB'S IDEAS?

Based on the current literature, Hebb's influence on research in psychology and neuroscience is greater than ever. Current PUBMED topics under the title "Hebb" (July 2015) give "The Hebb repetition effect", "Hebbian phase sequences", "Non-Hebbian spike-timing-dependent plasticity", "Hebbian learning" and "A neo-Hebbian framework for episodic memory". Other PUBMED topics include terms such as hebbian plasticity, hebbian learning, hebbian theory, hebb williams maze, hebbian synapse, non-hebbian and many others. There are at least seven areas in which Hebb's ideas will stimulate future research: (1) neural and behavioural epigenetics; (2) the long-term effects of environment on health and lifetime achievement; (3) studies of memory at all levels; (4) aging; (5) computer modelling of the brain; (6) robotics; and (7) neurorehabilitation. Finally, Hebb will become a focus for studies on the history of psychology and neuroscience (Brown, 2007a; Ghassemzadeh et al., 2013).

Here is a brief look into my crystal ball.

(1) *Neural and behavioural epigenetics* Epigenetics is the study of how environmental stimuli affect gene expression and thus brain and behaviour. Hebb's ideas on gene–environment interaction were precursors to this field and epigenetic approaches are being used to study Hebbian learning and neural plasticity (Boyce and Kobor, 2015). As research into neural and behavioural epigenetics develops, there will be a need to understand exactly how environmental stimuli cause short- and long-term changes in the brain and behaviour. If neural plasticity operates through Hebb synapses, cell assemblies and phase sequences, then the question will be how epigenetic processes such as DNA methylation and histone modification are involved in Hebbian brain mechanisms (Rampson et al., 2000; Boyce and Kobor, 2015; Guzman-Karlsson et al., 2014).

(2) *The long-term effects of environment on lifetime health and achievement* Hebb's ideas on environmental enrichment and isolation are embedded in the study of how early life events, poverty and other environmental stressors have long-term effects on brain and behaviour (see Figure 7.9), and his ideas will be important in understanding how these environmental events have long-term effects on the brain and health over the lifespan (Luby et al., 2012; Isles, 2015). For example, studies on how early life stressors which activate the Hypothalamic-Pituitary-Adrenal system modulate neural development will examine changes in Hebbian synapses and cell assemblies over the lifespan (Huang, 2014) and the development of clinical disorders will focus on synaptic changes within cell assemblies (Licznerski and Duman, 2013).

(3) *Studies of memory at all levels from molecular biology to human cognition* Hebb's ideas continue to stimulate new research on the physiological mechanisms of learning and memory (Schrader et al., 2002; Lisman et al., 2011; Johansen et al., 2014), learning and development (Munakata and Pfaffly, 2004), memory span (Oberauer et al., 2015), decision making (Wang, 2012) and language learning (see Figure 7.10; also Wennekers et al., 2006). The recent discoveries that glial cells show synaptic plasticity (Haydon and Nedergaard, 2015; Sampedro-Piquero et al., 2014) will initiate research to determine whether this is Hebbian or non-Hebbian in nature.

(4) *Aging* The Hebb–Williams maze (Kobayashi et al., 2002) and the Hebb repetition effect (Turcotte et al., 2005) are experimental tools used to study aging. But for understanding the mechanisms of age-related changes in neural plasticity and cognitive abilities, it will be important to understand the effects of aging on Hebbian synapses and cell assemblies (Burke and Barnes, 2006; Dasgupta, 2014; Li et al., 2016).

(5) *Computer modeling of the brain* Computer models of the brain use Hebb learning rules and cell assemblies (Wennekers, 2007) to build neural networks (see Figure 7.11) based on spike-timing-dependent plasticity (Markham et al., 2011). The role of Hebbian theory in neural network modelling will continue to

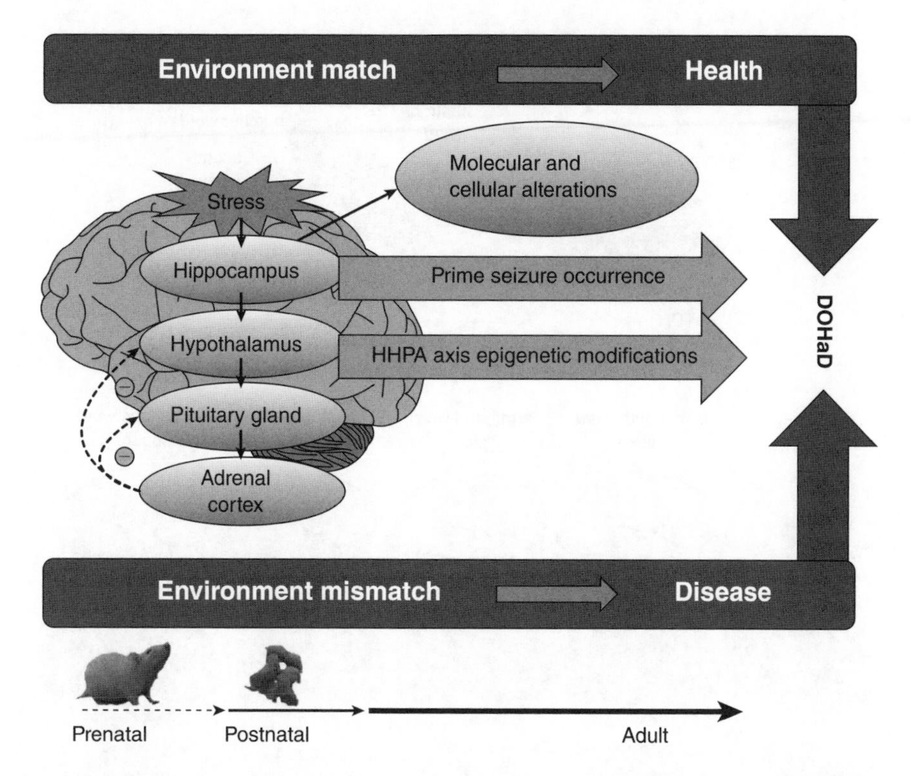

Figure 7.9 A conceptual diagram of how prenatal and postnatal stress act on the hippocampal hypothalamic-pituitary-adrenal (HHPA) axis to lead to neuropsychiatric disorders through epigenetic modifications – the so-called developmental origins of the health and disease (DOHaD) model, as presently understood. If the early programming environment matches the later adult environment, the adults are healthy. If a mismatch occurs, the adults are more likely to have diseases. This figure also shows the increased seizure propensity in the context of early-life stress. (*Source*: Huang, L.-T., 2014)

grow in importance (Buzsáki, 2010; Fotouhi et al., 2015). One day soon there may be Hebbian computers and artificial brains.

(6) *Robotics* and (7) *Neurorehabilitation* Hebbian theories of environmental enrichment are now used in neuro-rehabilitation (Bondi et al., 2014), and with the use of Hebbian learning rules and cell assemblies in robotics (Wang et al., 2009; Calderon et al., 2013) it is now possible for Hebbian learning rules to control brain–robot interfaces in rehabilitation (see Figure 7.12; also Gomez-Rodriguez et al., 2011; Takeuchi and Izumi, 2015).

Thus it is clear that Hebb's ideas have grown in importance in psychology and neuroscience in the twenty-first century and will continue to grow and influence research in the future. In 1971 B.F. Skinner wrote *Beyond Freedom and Dignity*, a book in which he saw the use of his conditioning methods for modifying behaviour used to build a new society. In fact, it appears that without trying to do so, Hebb's methods for understanding the neural basis of psychological functions are, in fact, being used to build a new society.

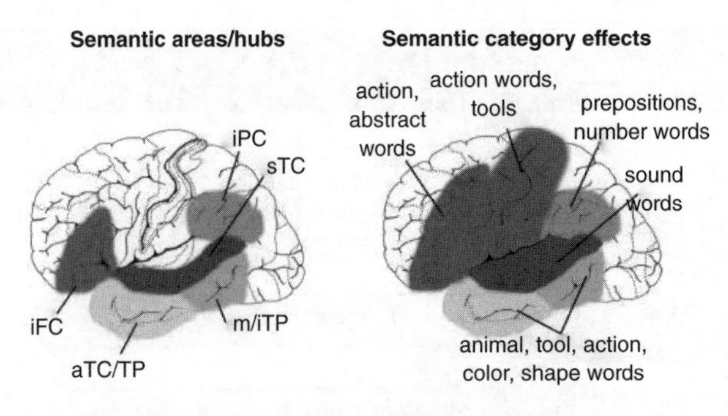

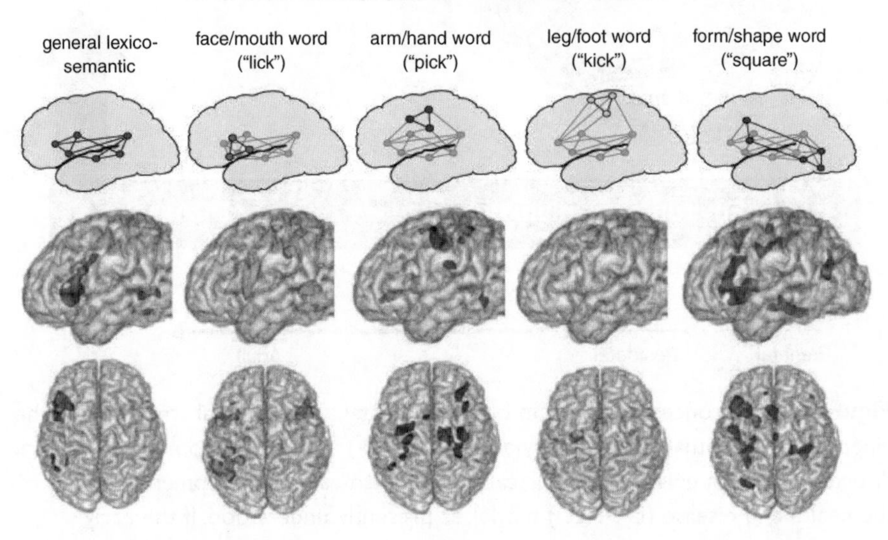

Figure 7.10 The concept of Hebb cell assemblies used to understand brain circuits underlying semantic processing. Top left panel: areas of particular importance for general semantic processing as proposed in the literature; iFC, inferior frontal cortex; iPC, inferior parietal cortex; sTC, superior temporal cortex; m/iTC, middle/inferior temporal cortex; aTC, anterior temporal cortex; TP, temporal pole. Top right panel: cortical areas where semantic category specificity was reported in neuropsychological patient studies and neuroimaging research – for word categories semantically related to actions (for example, "grasp"), numbers ("seven"), space (prepositions, e.g. "under"), sound ("bell"), colour ("green"), shape ("square"), animals ("cat"), tools ("knife") and abstract entities ("love", "beauty"). Middle panel: model of general lexico-semantic circuits shared by all word types (leftmost graph) and category-specific circuits for four different semantic word types (from left to right: face-related, arm-related and leg-related action words, form-related word). Bottom panel: brain activation for the same types of words as revealed by fMRI experiments and cluster analysis. (*Source*: Pulvermüller et al., 2014)

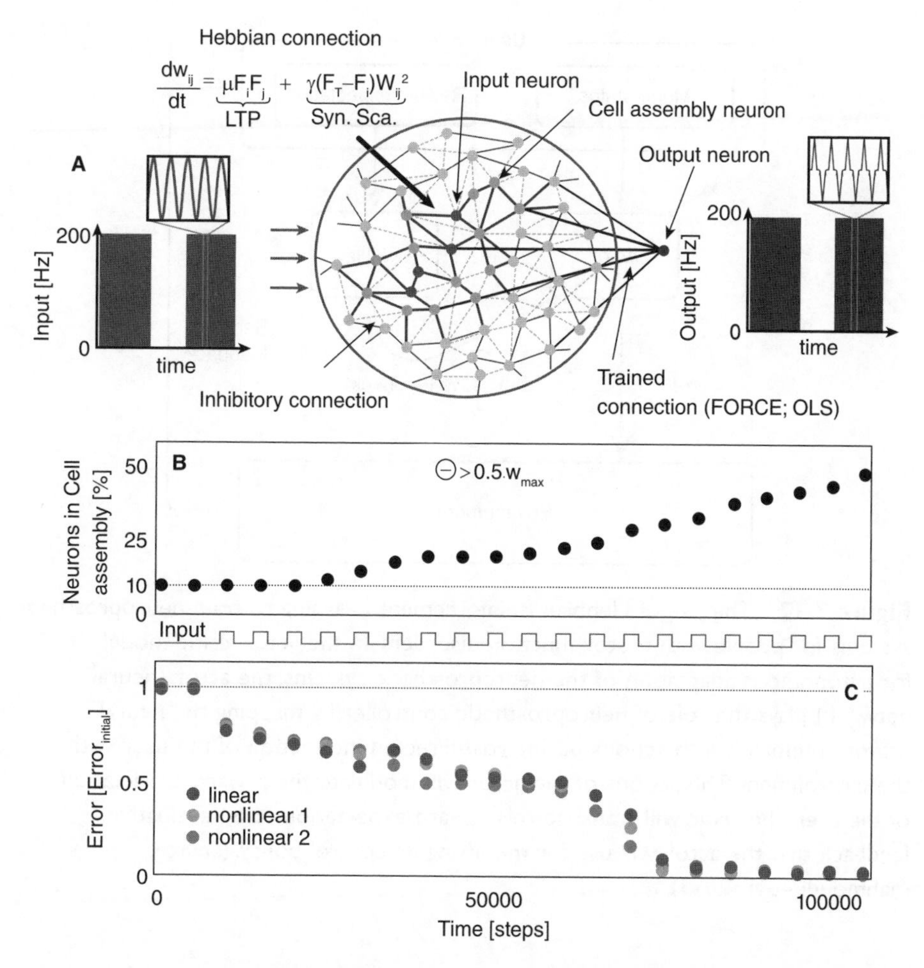

Figure 7.11 Changes in the cell assembly during learning: cell assembly size and computational performance are correlated. (A) An input is delivered to several neurons (red disks) in a random neuronal network with very weak but plastic excitatory (black lines) and constant inhibitory connections (green dashed lines). The interaction of long-term potentiation (LTP) and synaptic scaling (Syn. Sca.) enables the formation of a Hebbian cell assembly (red and orange disks) by increasing synaptic efficacies (thicker lines) when repeating the input several times. Note this network is not topographically organized. The neighbourhood ordering shown here is for graphical reasons only. Output neurons (blue disk) are connected (blue dashed) to the full network and trained in a conventional way to create the desired output. Here we used a single output neuron, but several can be connected without additional constraints. (B) With more learning trials the cell assembly grows and integrates more neurons. (*Source*: Tetzlaff et al., 2015)

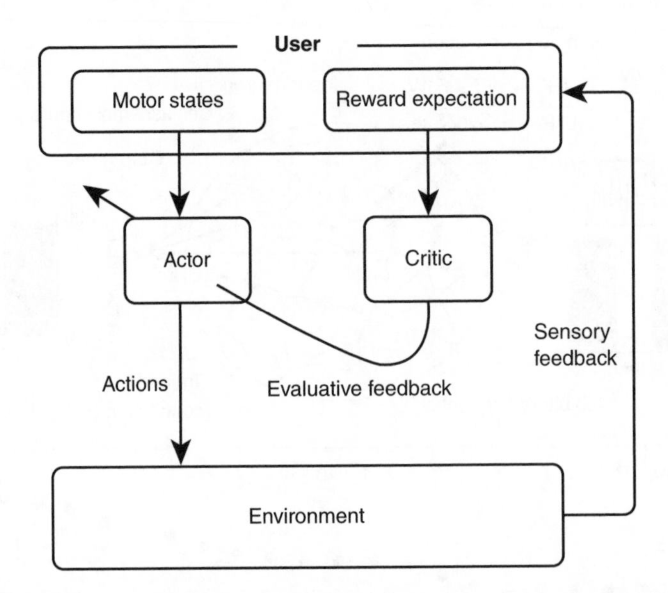

Figure 7.12 The use of Hebbian Reinforcement Learning to train neuroprosthetic systems for people with motor function disorders. In the actor–critic model for autonomous adaptation of the neuroprosthetic systems, the actor (neural network) plays the role of neuroprosthetic controller by mapping the neural motor commands into actions during goal-directed interaction of the user with the environment. The actions of the actor will modulate the reward expectation of the user. The critic will translate this reward expectation to an evaluative feedback that the actor will use for modifying its control policy. (*Source*: Mahmoudi et al., 2013)

ACKNOWLEDGEMENTS

I would like to thank the McGill University archives, the University of Chicago, Harvard University and Queen's University, Kingston, Ontario, for access to Hebb's files, and to Mary Ellen Hebb and Jane Hebb Paul for access to family photographs and letters.

REFERENCES

Adams, P. (1998) Hebb and Darwin. *Journal of Theoretical Biology,* 195: 419–38.

Allport, F.H. (1955) *Theories of Perception and the Concept of Structure.* New York: John Wiley and Sons.

Almeida-Filho, D.G., Lopes-dos-Santos, V., Vasconcelos, N.A., Miranda, J.G., Tort, A.B. and Ribeiro, S. (2014) An investigation of Hebbian phase sequences as assembly graphs. *Frontiers in Neural Circuits,* 8(8): 34. doi: 10.3389/fncir.2014.00034.

Attneave, F. (1950) The organization of behavior: a neuropsychological theory. *American Journal of Psychology*, 63: 633–5.

Bexton, W.H., Heron, W. and Scott, R.H. (1954) Effects of decreased variation in the sensory environment. *Canadian Journal of Psychology,* 8: 70–6.

Bliss, T.V.P. (2003) A journey from neocortex to hippocampus. *Philosophical Transactions of the Royal Society B: Biological Sciences,* 358: 621–3.

Bliss, T.V.P. and Lomo, T. (1973) Long-lasting potentiation of synaptic transmission in the dentate area of the anaesthetized rabbit following stimulation of the perforant path. *Journal of Physiology*, 232: 331–56.

Bondi, C.O., Klitsch, K.C., Leary, J.B. and Kline, A.E. (2014) Environmental enrichment as a viable neurorehabilitation strategy for experimental traumatic brain injury. *Journal of Neurotrauma*, 31: 873–88.

Boyce, W.T. and Kobor, M.S. (2015) Development and the epigenome: the 'synapse' of gene-environment interplay. *Developmental Science,* 18: 1–23.

Brogden, W.J. (1950) Organization of behavior. *The Scientific Monthly*, 71: 283–4.

Brooks, C.M. and Eccles, J.C. (1947) An electrical hypothesis of central inhibition. *Nature*, 159: 760–4.

Brown, R.E. (2007a) The life and work of Donald Olding Hebb: Canada's greatest psychologist. *The Proceedings of the Nova Scotian Institute of Science*, 44: 1–25.

Brown, R.E. (2007b) Alfred McCoy, Hebb, the CIA and torture. *Journal of the History of the Behavioural Sciences*, 43: 205–13.

Brown, R.E. and Milner, P.M. (2003) The legacy of Donald O. Hebb: more than the Hebb Synapse. *Nature Reviews Neuroscience,* 4: 1013–19.

Bruce, D. (1996) Lashley, Hebb, connections, and criticisms. *Canadian Psychology*, 37: 129–36.

Burke, S.N. and Barnes, C.A. (2006) Neural plasticity in the ageing brain. *Nature Reviews Neuroscience,* 7: 30–40.

Buzsáki, G. (2010) Neural syntax: cell assemblies, synapsembles, and readers. *Neuron*, 68: 362–85.

Calderon, D., Baidyk, T. and Kussul, E. (2013) Hebbian ensemble neural network for robot movement control. *Optical Memory and Neural Networks*, 22: 166–83. doi:10.3103/S1060992X13030028.

Cooper, S.J. (2005) Donald O. Hebb's synapse and learning rule: a history and commentary. *Neuroscience and Biobehavioral Reviews,* 28: 851–74.

Dasgupta, S. (2014) Cognitive aging as interplay between Hebbian learning and criticality. Cornell University Library. https://arxiv.org/abs/1402.0836. [Concise version of MSc thesis, Neural Models of the Ageing Brain, University of Edinburgh, 2010. 64 pages, 20 figures.]

Davis, R.C. (1962) Experiment and theory in physiological psychology. In S. Koch (ed.), *Psychology: A Study of a Science*. New York: McGraw-Hill. Volume 4, 242–79.

Dewsbury, D.A. (2002) The Chicago Five: A family group of integrative psychobiologists. *History of Psychology,* 5: 16–37.

Diamond, A., Barnett, W.S., Thomas, J. and Munro, S. (2007) Preschool program improves cognitive control. *Science*, 318: 1387–8.

Eccles, J.C., Fatt, P. and Koketsu, K. (1954) Cholinergic and inhibitory synapses in a pathway from motor-axon collaterals to motoneurones. *Journal of Physiology* (London), 126: 524–62.

Favero, M., Cangiano, A. and Busetto, G. (2014) Hebb-based rules of neural plasticity: are they ubiquitously important for the refinement of synaptic connections in development? *Neuroscientist*, 20: 8–14.

Fentress, J.C. (1987) D. O. Hebb and the developmental organization of behavior. *Developmental Psychobiology*, 20: 103–9.

Forgays, D.G. (1952) Reversible disturbances of function in man following cortical insult. *Journal of Comparative & Physiological Psychology*, 45: 209–15.

Forgays, D.G. and Forgays, J.W. (1952) The nature of the effect of free-environmental experience in the rat. *Journal of Comparative & Physiological Psychology*, 45: 322–8.

Fotouhi, M., Heidari, M. and Sharifitabar, M. (2015) Continuous neural network with windowed Hebbian learning. *Biological Cybernetics*, 109: 321. doi:10.1007/s00422-015-0645-7.

Ghassemzadeh, H., Posner, M.I. and Rothbart, M.K. (2013) Contributions of Hebb and Vygotsky to an integrated science of mind. *Journal of the History of Neuroscience*, 22: 292–306.

Gomez-Rodriguez, M., Grosse-Wentrup, M., Hill, J., Gharabaghi, A., Scholkopf, B. and Peters, J. (2011) Towards brain-robot interfaces in stroke rehabilitation. *IEEE International Conference on Rehabilitation Robotics*, 2011: 5975385. doi: 10.1109/ICORR.2011.5975385.

Guzman-Karlsson, M.C., Meadows, J.P., Gavin, C.F., Hablitz, J.J. and Sweatt, J.D. (2014) Transcriptional and epigenetic regulation of Hebbian and non-Hebbian plasticity. *Neuropharmacology*, 80: 3–17.

Hackman, D.A., Farah, M.J. and Meaney, M.J. (2010) Socioeconomic status and the brain: mechanistic insights from human and animal research. *Nature Reviews Neuroscience*, 11: 651–9.

Harlow, H.F. and Woolsey, C.N. (eds) (1965) *Biological and Biochemical Bases of Behavior*. Madison: University of Wisconsin Press.

Harnad, S. (1985) D.O. Hebb: father of cognitive psychobiology: 1904–1985. *Behavioural and Brain Sciences*, 8: opp. 529.

Hawkins, J. (2004) *On Intelligence*. New York: Henry Holt & Co.

Haydon, P.G. and Nedergaard, M. (2015) How do astrocytes participate in neural plasticity? *Cold Spring Harbor Perspectives in Biology*, 2015; 7: a020438.

Hebb, D.O. (1930) Elementary school methods. *Teacher's Magazine*, 12: 23–6.

Hebb, D.O. (1932) Conditioned and Unconditioned Reflexes and Inhibition. Unpublished MA Thesis, McGill University, Montreal, Quebec.

Hebb, D.O. (1937) The innate organization of visual activity: I. Perception of figures by rats reared in total darkness. *Journal of Genetic Psychology*, 51: 101–26.

Hebb, D.O. (1938) Studies of the organization of behavior. I. Behavior of the rat in a field orientation. *Journal of Comparative Psychology*, 25: 333–53.

Hebb, D.O. (1939a) Intelligence in man after large removals of cerebral tissue: report of four left frontal lobe cases. *Journal of General Psychology*, 21: 73–87.

Hebb, D.O. (1939b) Intelligence in man after large removals of cerebral tissue: defects following right temporal lobectomy. *Journal of General Psychology*, 21: 437–46.

Hebb, D.O. (1942) The effects of early and late brain injury upon test scores, and the nature of normal adult intelligence. *Proceedings of the American Philosophical Society*, 85: 275–92.

Hebb, D.O. (1945a) Man's frontal lobes: a critical review. *Archives of Neurology and Psychiatry,* 54: 10–24.

Hebb, D.O. (1945b) The forms and conditions of chimpanzee anger. *Bulletin of the Canadian Psychological Association,* 5: 32–5.

Hebb, D.O. (1946) On the nature of fear. *Psychological Review,* 53: 259–76.

Hebb, D.O. (1947a) Spontaneous neurosis in chimpanzees: theoretical relations with clinical and experimental phenomena. *Psychosomatic Medicine,* 9: 3–19.

Hebb, D.O. (1947b) The effects of early experience on problem solving at maturity. *American Psychologist*, 2: 306–7.

Hebb, D.O. (1949a) *The Organization of Behavior: A Neuropsychological Theory*. New York: Wiley. (Reprinted 2002 by Lawrence Erlbaum Associates, Mahwah, New Jersey.)

Hebb, D.O. (1949b) Temperament in chimpanzees: I. Method of analysis. *Journal of Comparative & Physiological Psychology,* 42: 192–206.

Hebb, D.O. (1950) Animal and physiological psychology. *Annual Review of Psychology,* 1: 173–88.

Hebb, D.O. (1951) The role of neurological ideas in psychology. *Journal of Personality,* 20: 39–55.

Hebb, D.O. (1953a) Heredity and environment in mammalian behaviour. *British Journal of Animal Behaviour,* 1: 43–7.

Hebb, D.O. (1953b) On human thought. *Canadian Journal of Psychology,* 7: 99–110.

Hebb, D.O. (1954) The problem of consciousness and introspection. In J.F. Delafresnaye (ed.), *Brain Mechanisms of Consciousness.* Springfield, IL: Charles C. Thomas. pp. 402–21.

Hebb, D.O. (1955a) Drives and the C.N.S. (conceptual nervous system). *Psychological Review,* 62: 243–54.

Hebb, D.O. (1955b) The mammal and his environment. *American Journal of Psychiatry,* 111: 826–31.

Hebb, D.O. (1958) *A Textbook of Psychology*. Philadelphia: Saunders.

Hebb, D.O. (1959a) A neuropsychological theory. In S. Koch (ed.), *Psychology: A Study of a Science* (Vol. 1). New York: McGraw Hill. pp. 622–43.

Hebb, D.O. (1959b) Intelligence, brain function and the theory of mind. *Brain*, 82: 260–75.

Hebb, D.O. (1980a) *Essay on Mind*. Hillsdale, NJ: Erlbaum.

Hebb, D.O. (1980b) D.O. Hebb. In G. Lindzey (ed.), *A History of Psychology in Autobiography,* Vol. VII. San Francisco, CA: W.H. Freeman. pp. 273–309.

Hebb, D.O. and Morton, N.W. (1943) The McGill adult comprehension examination: Verbal situation and picture anomaly series. *Journal of Educational Psychology,* 34: 16–25.

Hebb, D.O. and Morton, N.W. (1944) Note on the measurement of adult intelligence. *Journal of General Psychology,* 30: 217–23.

Hebb, D.O. and Penfield, W. (1940) Human behavior after extensive bilateral removal from the frontal lobes. *Archives of Neurology and Psychiatry,* 44, 421–38.

Hebb, D.O. and Thompson, W.R. (1954) The social significance of animal studies. In G. Lindzey (ed.), *Handbook of Social Psychology*, Vol.1. Cambridge, MA: Addison-Wesley. pp. 532–61.

Hebb, D.O. and Williams, K.A. (1946) Method of rating animal intelligence. *Journal of General Psychology,* 34: 59–65.

Hilgard, E.R. (1956) *Theories of Learning*, 2nd edition. New York: Appleton-Century-Crofts.

Huang, L.T. (2014) Early-life stress impacts the developing hippocampus and primes seizure occurrence: cellular, molecular, and epigenetic mechanisms. *Frontiers in Molecular Neuroscience,* 10; 7: 8. doi: 10.3389/fnmol.2014.00008.

Hunt, J. McV. (1979) Psychological development: early experience. *Annual Review of Psychology,* 30: 103–43.

Hymovitch, B. (1952) The effects of experimental variations on problem solving in the rat. *Journal of Comparative & Physiological Psychology,* 45: 313–21.

Isles, A.R. (2015) Neural and behavioral epigenetics: what it is, and what is hype. *Genes Brain and Behavior,* 14: 64–72. doi: 10.1111/gbb.12184.

Johansen, J.P., Diaz-Mataix, L., Hamanaka, H., Ozawa, T., Ycu, E., Koivumaa, J., Kumar, A., Hou, M., Deisseroth, K., Boyden, E.S. and LeDoux, J.E. (2014) Hebbian and neuro-modulatory mechanisms interact to trigger associative memory formation. *Proceedings of the National Academy of Sciences USA*, 111(51): E5584–92. doi: 10.1073/pnas.1421304111.

Johnson, L.R., Ledoux, J.E. and Doyère, V. (2009) Hebbian reverberations in emotional memory micro circuits. *Frontiers in Neuroscience*, 3(2):198–205. doi: 10.3389/neuro.01.027.2009.

Jusczyk, P.W. and Klein, R.M. (eds) (1980/2010) *The Nature of Thought: Essays in Honor of D.O. Hebb.* Hillsdale, NJ: Lawrence Erlbaum.

Klein, R.M. (1999) The Hebb legacy. *Canadian Journal of Experimental Psychology,* 53: 1–3.

Kobayashi, S., Ohashi, Y. and Ando, S. (2002) Effects of enriched environments with different durations and starting times on learning capacity during aging in rats assessed by a refined procedure of the Hebb-Williams maze task. *Journal of Neuroscience Research,* 70: 340–6.

Koch, S. (ed.) (1959–1963) *Psychology: A Study of a Science.* New York: McGraw-Hill. Volumes 1–5.

Kolb, B. (2003) The impact of the Hebbian Learning Rule on research in behavioural neuroscience. *Canadian Psychology*, 44: 14–16.

Kolb, B., Mychasiuk, R., Muhammad, A. and Gibb, R. (2013) Brain plasticity in the developing brain. *Progress in Brain Research,* 207: 35–64.

Krech, D., Rosenzweig, M.R. and Bennett, E.L. (1962) Relations between chemistry and problem-solving among rats raised in enriched and impoverished environments. *Journal of Comparative & Physiological Psychology,* 55: 801–7.

Kuhn, M.H. (1950) Hebb, D.O. Organization of Behavior: A Neuropsychological Theory. *The Annals of the American. Academy of Political and Social Science,* 217: 216–17.

Lansner, A. (2009) Associative memory models: from the cell-assembly theory to bio-physically detailed cortex simulations. *Trends in Neuroscience,* 32: 178–86.

Leeper, R. (1950) The Organization of Behavior: A Neuropsychological Theory (review). *Journal of Abnormal and Social Psychology,* 45: 768–75.

Li, M., Liu, J., and Tsien, J.Z. (2016) Theory of connectivity: Nature and nurture of cell assemblies and cognitive computation. *Frontiers in Neural Circuits*, 10: 34. doi: 10.3389/fncir.2016.00034.

Licznerski, P. and Duman, R.S. (2013) Remodeling of axo-spinous synapses in the pathophysiology and treatment of depression. *Neuroscience,* 251: 33–50.

Lisman, J., Grace, A.A. and Duzel, E. (2011) A neoHebbian framework for episodic memory: role of dopamine-dependent late LTP. *Trends in Neuroscience,* 34: 536–47.

Lorente de No, R. (1938) Synaptic stimulation of motoneurons as a local process. *Journal of Neurophysiology,* 1: 195–206.

Luby, J.L., Barch, D.M., Belden, A., Gaffrey, M.S., Tillman, R., Babb, C., Nishino, T., Suzuki, H. and Botterton, K.N. (2012) Maternal support in early childhood predicts larger hippocampal volumes at school age. *Proceedings of the National Academy of Sciences USA,* 109: 2854–9.

Mahmoudi, B., Pohlmeyer, E.A., Prins, N.W., Geng, S., and Sanchez, J.C. (2013) Towards autonomous neuroprosthetic control using Hebbian reinforcement learning. *Journal of Neural Engineering,* 10 (2013): 066005. doi:10.1088/1741-2560/10/6/066005.

Markram, H., Gerstner, W. and Sjöström, P.J. (2011) A history of spike-timing-dependent plasticity. *Frontiers in Synaptic Neuroscience*, 3: 4. doi: 10.3389/fnsyn.2011.00004.

Martinez, J.L., Jr. and Glickman, S.E. (1994) Hebb revisited: perception, plasticity, and the Hebb Synapse. *Contemporary Psychology,* 39: 1018–20.

McBride, A.F. and Hebb, D.O. (1948) Behavior of the captive bottle-nose dolphin, *Tursiops truncatus. Journal of Comparative & Physiological Psychology,* 41: 111–23.

Melzack, R. (1954) The genesis of emotional behavior: an experimental study of the dog. *Journal of Comparative and Physiological Psychology*, 47: 166–8.

Melzack, R. and Thompson, W.R. (1956) Effects of early experience on social behaviour. *Canadian Journal of Psychology,* 10: 82–90.

Milner, B. (1954) Intellectual function of the temporal lobes. *Psychological Bulletin,* 51: 42–62.

Milner, P. (1970) *Physiological Psychology*. New York: Holt, Rinehart and Winston.

Milner, P.M. (1957) The cell assembly: Mark II. *Psychological Review,* 64: 242–52.

Mohammed, A.H., Zhu, S.W., Darmopil, S., Hjerling-Leffler, J., Ernfors, P., Winblad, B., Diamond, M.C., Eriksson, P.S. and Bogdanovic, N. (2002) Environmental enrichment and the brain. *Progress in Brain Research,* 138: 109–33.

Munakata, Y. and Pfaffly, J. (2004) Hebbian learning and development. *Developmental Science,* 7: 141–8.

Nicolelis, M.A., Fanselow, E.E. and Ghazanfar, A.A. (1997) Hebb's dream: the resurgence of cell assemblies. *Neuron*, 19: 219–21.

Oberauer, K., Jones, T. and Lewandowsky, S. (2015) The Hebb repetition effect in simple and complex memory span. *Memory & Cognition,* 43: 852–65.

Olds, J. and Milner, P. (1954) Positive reinforcement produced by electrical stimulation of septal area and other regions of rat brain. *Journal of Comparative and Physiological Psychology,* 47: 419–27.

Osgood, C.E. (1953) *Method and Theory in Experimental Psychology.* New York: Oxford University Press.

Posner, M.I. and Rothbart, M.K. (2004) Hebb's neural networks support the integration of psychological science. *Canadian Psychology*, 45: 265–78.

Pulvermüller, F., Garagnani, M. and Wennekers, T. (2014) Thinking in circuits: toward neurobiological explanation in cognitive neuroscience. *Biological Cybernetics,* 108: 573–93.

Rabinovitch, M.S. and Rosvold, H.E. (1951) A closed-field intelligence test for rats. *Canadian Journal of Psychology,* 5: 122–8.

Rampon, C., Jiang, C.H., Dong, H., Tang, Y.P., Lockhart, D.J., Schultz, P.G., Tsien, J.Z. and Hu, Y. (2000) Effects of environmental enrichment on gene expression in the brain. *Proceedings of the National Academy of Sciences USA,* 97: 12880–4.

Riesen, A.H. (1947) The development of visual perception in man and chimpanzee. *Science,* 106: 107–8.

Rosenzweig, M.R. (1998) Some historical background of topics in this conference. *Neurobiology of Learning and Memory,* 70: 3–13.

Rosvold, H.E. and Mishkin, M. (1950) Evaluation of the effects of prefrontal lobotomy on intelligence. *Canadian Journal of Psychology,* 3: 122–7.

Sampedro-Piquero, P., De Bartolo, P., Petrosini, L., Zancada-Menendez, C., Arias, J.L. and Begega, A. (2014) Astrocytic plasticity as a possible mediator of the cognitive improvements after environmental enrichment in aged rats. *Neurobiology of Learning and Memory,* 114: 16–25.

Schrader, L.A., Anderson, A.E., Varga, A.W., Levy, M. and Sweatt, J.D. (2002) The other half of Hebb: K+ channels and the regulation of neuronal excitability in the hippocampus. *Molecular Neurobiology,* 25: 51–66.

Sejnowski, T.J. (1999) The book of Hebb. *Neuron,* 24: 773–6.

Sejnowski, T.J. (2003) The once and future Hebb synapse. *Canadian Psychology,* 44: 17–20.

Shore, D.I., Stanford, L., MacInnes, J.W., Klein, R.M. and Brown, R.E. (2001) Of Mice and Men: Virtual Hebb-Williams mazes permit comparison of spatial learning across species. *Cognitive, Affective and Behavioral Neuroscience,* 1: 83–9.

Smith, C.J. (1959) Mass action and early environment in the rat. *Journal of Comparative & Physiological Psychology,* 52: 154–6.

Solomon, P., Kubzansky, P.E., Leidermann, P.H., Mendelson, J.H., Trumbell, R. and Wexler, D. (eds) (1961) *Sensory Deprivation*. Cambridge, MA: Harvard University Press.

Sommer, F.T. and Wennekers, T. (2003) Models of distributed associative memory networks in the brain. *Theory in Biosciences,* 122: 55–69.

Spatz, H.C. (1996) Hebb's concept of synaptic plasticity and neuronal cell assemblies. *Behavioural Brain Research,* 78: 3–7.

Stent, G.S. (1973) A physiological mechanism for Hebb's postulate of learning. *Proceedings of the National Academy of Sciences USA,* 70: 997–1001.

Stevens, S.S. (1951) *Handbook of Experimental Psychology*. New York: John Wiley and Sons.

Takeuchi, N. and Izumi, S. (2015) Combinations of stroke neurorehabilitation to facilitate motor recovery: perspectives on Hebbian plasticity and homeostatic metaplasticity. *Frontiers in Human Neuroscience,* 9: 349. doi: 10.3389/fnhum.2015.00349

Tees, R.C. (2003) The organization of behavior: a neuropsychological theory. *Canadian Psychology,* 44: 74–6.

Tegnér, J., Compte, A. and Wang, X.J. (2002) The dynamical stability of reverberatory neural circuits. *Biological Cybernetics,* 87: 471–81.

Tetzlaff, C., Dasgupta, S., Kulvicius,T. and Wörgötter, F. (2015) The use of Hebbian cell assemblies for nonlinear computation. *Scientific Reports,* 5: 12866. doi: 10.1038/srep12866.

Thompson, W.R. and Heron, W. (1954) The effects of restricting early experience on the problem-solving capacity of dogs. *Canadian Journal of Psychology*, 8: 17–31.

Turcotte, J., Gagnon, S. and Poirier, M. (2005) The effect of old age on the learning of supraspan sequences. *Psychology and Aging*, 20: 251–60.

van Praag, H., Kempermann, G. and Gage, F.H. (2000) Neural consequences of environmental enrichment. *Nature Reviews Neuroscience*, 1: 191–8.

von Senden, M. (1932) *Raum-und Gestaltauffassung bei Operierten Blindgebornen von und nach der Operation.* Liepzig: J.A. Barth. (Translated as von Senden, M. (1960) *Space and Sight: The Perception of Space and Shape in the Congenitally Blind Before and After Operation*. London: Methuen & Co. English translation by Peter Heath, with Appendices by A. H. Riesen, G.J. Warnock and J.Z. Young, and a select bibliography of books and papers citing this work.)

Wallace, D.J. and Kerr, J.N. (2010) Chasing the cell assembly. *Current Opinion in Neurobiology*, 20: 296–305.

Wang, X.J. (2012) Neural dynamics and circuit mechanisms of decision-making. *Current Opinion in Neurobiology*, 22: 1039–46.

Wang, Y., Wu, T., Orchard, G., Dudek, P., Rucci, M. and Shi, B.E. (2009) Hebbian learning of visually directed reaching by a robot arm. *IEEE Biomedical Circuits and Systems, Beijing, China, November 2009*. http://aplab.bu.edu/assets/download/PDFs/proceedings/Wang-biocas09.pdf.

Wennekers, T. (2007) A cell assembly model for complex behaviour. *Neurocomputing*, 70: 1988–92.

Wennekers, T., Garagnani, M. and Pulvermüller, F. (2006) Language models based on Hebbian cell assemblies. *Journal of Physiology, Paris*, 100: 16–30.

Wertheimer, M. (1951) Hebb and Senden on the role of learning in perception. *American Journal of Psychology*, 64: 133–7.

Wolman, B.B. (ed.) (1973) *Handbook of General Psychology*. Englewood Cliffs, NJ: Prentice-Hall.

Zhang, T.Y. and Meaney, M.J. (2010) Epigenetics and the environmental regulation of the genome and its function. *Annual Review of Psychology*, 61: 439–66.

8 Revisiting Scoville and Milner: Loss of recent memory after bilateral hippocampal lesions

Robert J. Sutherland

Scoville and Milner (1957) provided post-operative clinical descriptions of 10 patients with hippocampus and hippocampal gyrus damage of varying extent. One of these, Case 1, became the most famous patient in neuroscience, Henry Gustave Molaison (b. 1926 – d. 2008), known for most of his life in publications as HM.

The authors' intention was to emphasize the consistent connection between bilateral damage to the hippocampal complex with an isolated, persisting, severe memory deficit. This connection has proven to be one of the most fertile, both in inspiring research and theoretical speculation, in the history of neuroscience. As is illustrated in Figure 8.1, Brenda Milner certainly knew how to get the hay down to where the goats can get it. A glance at this figure highlights the ongoing importance of the study of the role of the hippocampus in memory.

Although there have been descriptions of many patients with temporal lobe damage and memory deficits, HM remains central with respect to our understanding of the role of the temporal lobes in memory. Nevertheless, the report on his deficit is also the stimulus for many other approaches to investigating the role of the brain in memory as well, and parenthetically, for a reconsideration of neurosurgical approaches to the treatment of brain disease.

THE HIPPOCAMPUS AND MEMORY

Milner began studying memory deficit after surgical excisions of the medial temporal lobe in 1951 (Milner, 1977) at the Montreal Neurological Institute with Wilder Penfield, the most innovative and creative neurosurgeon of his generation. Within a series of patients who received unilateral temporal lobectomy to treat severe epilepsy, two patients were unique in displaying profound deficits. First, they were impaired in creating long-term memories from new experiences (anterograde amnesia). Second, not all kinds of memory were equally affected. Third, they displayed loss of memories already established from several months to

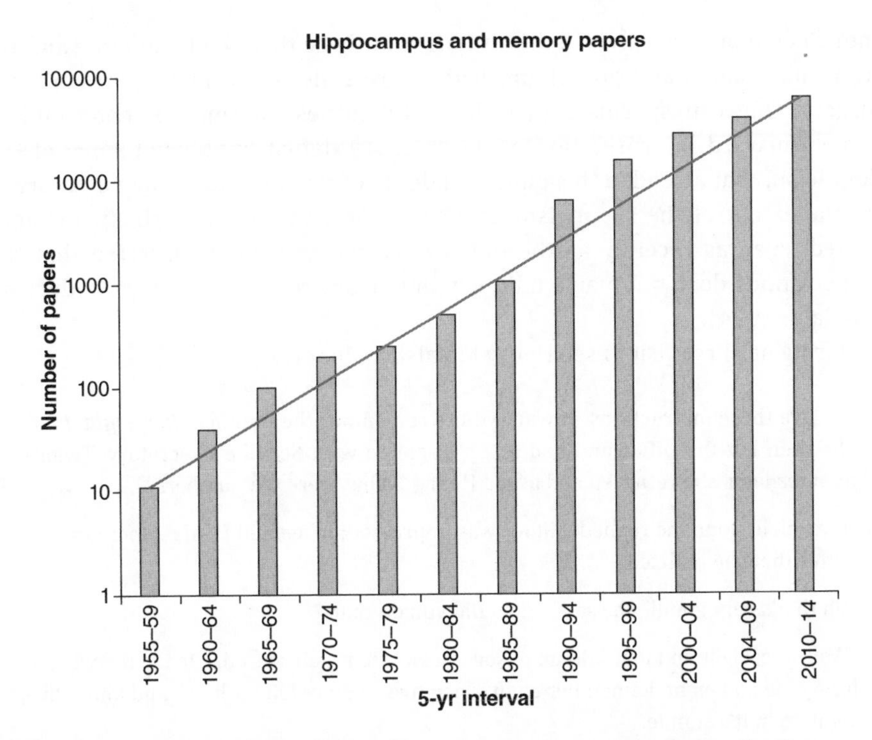

Figure 8.1 Number of journal articles on hippocampus and memory by 5-year intervals from the interval containing Scoville and Milner's paper to present. Results from Web of Science (search terms for topic: hippocamp* AND (memory OR amnes*), conducted 8 November 2015). The linearity in a semilog plot indicates that the follow-up papers have been increasing exponentially since the publication interval. In the previous 5-year interval there were no papers on this topic (data not shown).

years before surgery (retrograde amnesia). Each of these features of memory became focal points for subsequent investigations of the brain substrates of memory, as we will describe below.

In an initial report on these two patients, Milner and Penfield (1955) proposed that the memory deficit was due to bilateral hippocampal damage. They hypothesized that both patients had a pre-existing and unsuspected hippocampal lesion in the temporal lobe opposite the temporal lobectomies. Prompted by that report, William Scoville, a neurosurgeon working in Boston, contacted Penfield concerning some of his own surgical cases. For several years Scoville had been performing experimental psychosurgery on psychotic patients in Boston. In nine of these cases the surgeries included bilateral medial temporal removals, which had no reliable effect on psychosis or personality. One case was Henry Molaison, a young man with uncontrolled and severe epilepsy. As a result, Milner traveled to Boston to perform psychological assessments of this series of patients, resulting in this landmark 1957 paper.

It is important to acknowledge difficulties in formal testing of the psychotic patients. Three of these could not complete testing. There was apparently enough evidence to convince the authors that these three had moderate anterograde amnesia. In addition, Henry had severe epilepsy for many years prior to surgery. One can speculate that these difficulties, to some extent clouding straightforward interpretation of the memory deficit, motivated some of the skepticism that attended this paper. Students of memory now may not appreciate the extent of the skepticism about Scoville and Milner's principal claims. Indeed, even as recently as the mid-1990s neuroscientists asserted that the hippocampus does not play a uniquely important role in memory (Vanderwolf and Cain, 1994).

During her first visit to see Henry Molaison, Milner:

> . . . gave these instructions: "I want you to remember the numbers *five, eight, four.*" She then left the office and had a cup of coffee with Scoville's secretary. Twenty minutes later, she returned and asked Henry, "What were the numbers?"
>
> "Five, eight, four," he replied. Milner was impressed; it seemed Henry's memory was better that she realized.
>
> "Oh that's very good!" she said. "How did you do that?"
>
> "Well, five, eight and four add up to seventeen." Henry answered. "Divide by two, you have nine and eight. Remember eight. Then five – you're left with five and four – five, eight, four. It's simple."
>
> "Well that's very good. And do you remember my name?"
>
> "No, I'm sorry. My trouble is my memory."
>
> "I'm Dr. Milner, and I come from Montreal."
>
> "Oh, Montreal, Canada," Henry said. "I was in Canada once – I went to Toronto."
>
> "Oh. Do you still remember the number?"
>
> "Number? Was there a number?" (Corkin, 2013)

Much of what Milner's work with bilateral medial temporal lobectomy patients contributed to our understanding of memory and brain is illustrated in this first conversation between HM and Milner. She registered surprise when Henry correctly recalled the number after such a long delay, as her previous work with patients in Montreal suggested that memories rapidly disappeared after even short delays. Henry was hopeless when asked to recall her name, but he came up with a series of arithmetic steps that kept the numbers in his mind as long as he was continuing these operations. When those operations were interrupted by Milner the numbers were quickly gone.

MULTIPLE MEMORY SYSTEMS

Milner made the important observation that after bilateral medial temporal lobe damage, intelligence, even in formal IQ testing, was preserved.

The preserved aspects of intelligence include *working memory*, the active state of memory for new or old information that is being manipulated or rehearsed in some way. This clearly showed that memory in the brain was not a unitary construct, that there were *multiple memory systems*. Milner and her students soon learned that medial temporal lobe amnesics could learn and retain many skills or habits as well as control participants, despite being unable to recall the events in the learning episodes.

Milner inferred that the hippocampal region must be necessary for the process (*memory consolidation*) that enables recent memories to be recalled even when they have not been activated for a long time. It is equally clear that retrieval of memories from the remote past does not require the hippocampal region. This is evident from Henry's knowledge of the fact that he had taken a trip to Toronto, Canada, and his knowledge of the fact that Montreal was in Canada. Many investigators have subsequently speculated as to why this kind of knowledge is independent of the hippocampal region. Does it represent retrieval from a fact memory system (*semantic memory*) and/or does this knowledge represent the retrieval of memories from a sufficiently remote past (*spared remote memories*)? Milner often noted the consistent presence of severe retrograde amnesia in medial temporal lobe patients, but it varied in duration from patient to patient.

Scoville and Milner (1957) also reported that their patients clustered into three groups based on the severity of their memory deficit, i.e. severe, moderate and mild. They believed that the severity of amnesia roughly correlated with the extent of damage to the hippocampal region. With so few patients and the paucity in the prior literature, there were clear uncertainties about the identity of the critical neural system responsible for this memory deficit. Already Milner had some good reasons to doubt the relevance of the amygdala and temporal neocortex, but the relative importance of the hippocampus versus the adjacent parahippocampal gyrus was uncertain.

The intact working memory system in patients like HM has led to a very consistent body of work demonstrating the independences of working memory systems from the hippocampus. Many experiments with rats, monkeys and humans demonstrate that that the hippocampal memory system and the working memory systems can function and be manipulated independently of one another. Nevertheless, the precise nature of the memories that can and cannot be formed after removal of the hippocampal region has been a source of great debate and extensive theoretical speculation.

It has been established experimentally that circuitry involving the amygdala, thalamus, parts of the cortex and brainstem is able to create new fear memories/habits through Pavlovian conditioning procedures in the absence of the hippocampal region. Conditioning of discrete skeletal muscle responses depends upon the cerebellum and its connections and proceeds normally without a hippocampus. Instrumental conditioning of cue-based habits engages a striatal system in the absence of the hippocampal region. These spared learning and memory abilities have been grouped variously under the headings of habitual, procedural, implicit, nonrelational, and so on. They all have the character of being based on a relatively simple cue being unambiguously associated with a particular consequence or response.

In contrast, memories needing recall of more complex representations, especially involving some form of (spatial, temporal or conceptual) relationship among events that comprise the learning episode, appear to depend more critically on the hippocampal region. In work with humans there has been an emphasis on the idea that hippocampal-dependent memories involve awareness of the elements of the learning episode, to the point that they can be verbally reported and flexibly retrieved.

TEMPORAL GRADIENTS OF AMNESIA

Retrograde amnesia is one of the key features of the memory deficit described by Scoville and Milner. HM is described as having a "partial retrograde amnesia for the three years leading up to his operation". In humans, Milner and Penfield (1955) were the first to describe retrograde amnesia as a consequence of selective medial temporal lobectomy. A patient (PB) in 1951, and a second (FC) a year later, presented with anterograde amnesia after surgery, and in addition, both exhibited partial retrograde amnesia. The measurement of retrograde amnesia is now more refined. One can see a trend for increasing precision in the methods over the past five decades. Compare Scoville and Milner (1957) with Squire et al. (1975) and then with Gilboa et al. (2006).

The retrograde amnesia these patients displayed was described as severe for recent memories, but seemingly remote memories were relatively spared (Milner, 2005). This pattern of partial retrograde amnesia involving a disruption of recent memories and sparing of relatively more remote memories is termed a *temporal gradient* (the equivalent disruption of recent and remote memory is termed a *flat gradient*). In the following years many more amnesics who had sustained bilateral medial temporal lobe damage were described; all displayed similar temporally graded retrograde amnesia.

There is a very wide range of estimates for the length of the pre-injury interval from which memories are lost. There is a trend for the duration of retrograde amnesia to increase with the increasing rigour in procedures to measure it. HM's interval initially was suggested to affect only the preceding few years. In subsequent studies this interval lengthened to encompass two decades and in the most recent evaluation the investigators concluded that severe retrograde amnesia showed no temporal gradient (Steinvorth et al., 2005). We might consider a major update of Scoville and Milner with Corkin's (2002) simple statement: "H.M. was unable to supply an episodic memory of his mother or his father – he could not narrate even one event that occurred at a specific time and place." Experimental hippocampal damage in nonhuman animals mirrors this situation.

The full effect of the more recent experimental results concerning the temporal gradient of memory is still playing out within the theoretical domain. There are two theories. The standard model of systems consolidation theory has it that episodic and semantic memories initially depend on the hippocampal region, but after a sufficient period of interaction between hippocampus and neocortex these memories become independent of the hippocampus. After selective hippocampal damage, retrograde amnesia should always be temporally graded depending in part upon how recently the memory-related event occurred and how long it takes to transfer it to a permanent storage location. On this view, flat gradients are hypothesized to be due to extrahippocampal damage.

An alternate theory, parallel storage theory (Sutherland et al., 2010), has it that any memories from a single episode are stored concurrently in a number of networks, only one of which involves the hippocampus. The rate of memory acquisition and time-course of persistence depend on the specific parameters of experience-dependent plasticity in each network. On this view, if a memory initially depended on the hippocampal region then it always will – the systems consolidation process, as a memory transfer, does not exist. Flat gradients for these hippocampal memories will be the norm.

A recent direct test favours the second theory. After extensive damage limited to the hippocampus in three memory tasks (OCampo et al., 2015), flat gradients were found in all three tasks, even though no extra hippocampal damage was evident.

A mixed theoretical view is that episodic memories (detailed with linked time and place context information) always depend on the hippocampus, but are subject to temporally graded retrograde amnesia if the damage is partial. Semantic memories exhibit temporally graded retrograde amnesia even after complete hippocampal damage (Nadel and Moscovitch, 1997).

ADDITIONAL SUPPORTING EVIDENCE

The study of memory in patients who receive elective surgery, in addition to disclosing a wealth of information concerning how the brain handles memories, stimulated many other lines of enquiry. These include *the question of neurosurgical risk*, alternate ways of inactivating the hippocampus in human patients, the role of the left and right hippocampus in memory, and memory-related interpretations of the activity of single cells in the hippocampus, and imaging of the brain's memory systems.

It is important to realize that Milner had received a doctorate in psychology and she found herself recommending a well-founded caution about risks around some neurosurgical procedures. It should not be too hard to imagine some of the push-back from neurosurgeons. At the time they had very wide latitude and some were pursuing experimental psychosurgery that today would be recognized as unethical.

Milner and colleagues won this debate, and the importance of a preoperative evaluation of memory in the two hemispheres eventually was established world-wide.

In 1955 Juhn Wada arrived at the Montreal Neurological Institute on a fellowship. He had previously developed the technique of unilateral intra-carotid infusion of amobarbital to reduce complications from electroconvulsive shock therapy. These infusions anesthetized mainly the hemisphere ipsilateral to the injection for a number of minutes. At the Montreal Neurological Institute his procedure was used preoperatively to establish the dominant hemisphere for language. Importantly, Milner realized that this method could also be adapted to preoperatively assess memory abilities of each hemisphere, and thereby estimate the risk of producing severe amnesia after unilateral temporal lobe surgery. As a result, decades later, despite dramatic improvements in neuroimaging modalities, this test is still applied prior to surgery to establish amnesia risk in clinics around the world.

Within the hippocampal region there is memory specialization. In the early 1950s Milner had already observed differences in the mild memory effects of damaging the left or right hippocampal regions, with left damage having a greater impact on verbal memory and right on nonverbal memory. There are now extensive continuing efforts in current functional imaging and selective lesion/inactivation projects to identify different contributions of components of the hippocampal region. There are solid results connecting parts of this cortical region to a form of path integration, scene/landmark recognition, and individual objects.

In experimental procedures in which single hippocampal neuron recordings are made in awake, behaving rats, monkeys and humans there is evidence that the hippocampus is important for encoding memory. The experiments show that there are bursts of spikes in relation to the subjects' location within an environment, to imagined trajectories through space, to looking at a certain location within a spatial array, and to trajectories within a virtual environment. Recordings from parts of the parahippocampal gyrus have shown cells that respond to head direction, specific objects, scenes, distance to boundaries, and to a mosaic (or grid) of locations across an environment. Discoveries of location-specific characteristics of firing of cells in the hippocampus and parahippocampal gyrus led to the 2014 Nobel Prize in Physiology or Medicine being awarded to John O'Keefe for place cells and to Edvard and May-Britt Moser for grid cells. Much of the significance and intense focus on their work originated from Scoville and Milner's findings.

CONCLUSION

Scoville and Milner (1957) reported on the first series of patients who received operations that included bilateral surgical excisions of the medial temporal lobe. They described a selective memory deficit involving an inability to create long-term memories from new experiences, spared intellectual abilities (including working memory), a likely critical lesion involving the hippocampal region (but not the amygdala or temporal neocortex), and a partial loss of memories created before the operation.

Milner's suggestion that the amygdala and temporal neocortex were not critical for the memory deficit has been confirmed. Likewise, it is now established that anterograde amnesia follows selective damage to the hippocampus or parahippocampal gyrus. When damage includes all of both of these regions the amnesia is especially severe. Furthermore, Milner's suggestion that the severity of the memory deficit is related to the extent of damage to the hippocampal region has been confirmed in multiple species in many laboratories. In fact, this is so strongly established now that if a report were to claim that a lesion or inactivation in the hippocampal region does not affect memory, a ready-to-hand response would be that the investigators had not damaged or inactivated a sufficient extent bilaterally or they were using an insensitive memory task.

Finally, the psychological evidence that was gained from the study of temporal lobe patients such as HM sounded an effective cautionary note to neurosurgeons about the risks to memory of bilateral hippocampal damage. Furthermore, the more than 100,000 follow-up papers have resolved some of the concerns raised about the fact that their patients were preoperatively pathological (psychotics, epileptics), while still supporting the idea that the hippocampus plays a role in memory. Nevertheless, puzzles still remain about the status of remote memories and the specific roles for hippocampal subregions in memory.

REFERENCES

Corkin, S. (2002) What's new with the amnesic patient HM? *Nature Reviews Neuroscience*, 3: 153–60.

Corkin, S. (2013) *Permanent Present Tense.* New York: Basic Books.

Gilboa, A., Winocur, G., Rosenbaum, R.S. et al. (2006) Hippocampal contributions to recollection in retrograde and anterograde amnesia. *Hippocampus*, 16: 966–80.

Milner, B. (1977) Memory mechanisms. *Canadian Medical Association Journal*, 18 June, 116: 1374–6.

Milner, B. (2005) The medial temporal-lobe amnesic syndrome. *Psychiatric Clinics of North America,* 28: 599–611.

Milner, B. and Penfield, W. (1955) The effect of hippocampal lesions on recent memory. *Transactions of the American Neurological Association,* 80: 42–8.

Nadel, L. and Moscovitch, M. (1997) Memory consolidation, retrograde amnesia and the hippocampal complex. *Current Opinion in Neurobiology*, 7 (2): 217–27.

OCampo, A.C., Hales, J.B., Saturday, S., Squire, L.R. and Clark, R.E. (2015) Targeting dorsal and ventral CA1 as a novel lesion approach to studying memory consolidation in rats. *Society for Neuroscience Abstract*, Chicago, October.

Scoville, W.B. and Milner, B. (1957) Loss of recent memory after bilateral hippocampal lesions. *Journal of Neurolology, Neurosurgery & Psychiatry,* 20: 11–21.

Squire, L.R., Slater, P.C. and Chace, P.M. (1975) Retrograde amnesia: temporal gradient in very long term memory following electroconvulsive therapy. *Science*, 187 (4171): 77–9.

Steinvorth, S., Levine, B. and Corkin, S. (2005) Medial temporal lobe structures are needed to re-experience remote autobiographical memories: evidence from H.M. and W.R. *Neuropsychologia,* 43: 479–96.

Sutherland, R.J., Sparks, F.T. and Lehmann, H. (2010) Hippocampus and retrograde amnesia in the rat model: a modest proposal for the situation of systems consolidation. *Neuropsychologia,* 48 (8): 2357–69.

Vanderwolf, C.H. and Cain, D.P. (1994) The behavioral neurobiology of learning and memory: a conceptual reorientation. *Brain Research Reviews,* 19: 264–97.

9 Revisiting MacLean: The limbic system and emotional behavior

Marie-H. Monfils

What brain structure gives rise to emotions? The answer to this question lies at the core of our understanding of brain function. The question is also important to translational research in which basic research on emotion is applied in a clinical setting. We have come a long way since investigators first pondered the question of where emotions might be stored within the brain.

An important essay entitled "The limbic system ('visceral brain') and emotional behavior," published by MacLean in 1955, provides a glimpse of the field of the brain and emotion at its inception. MacLean's article remains important because it opened a window onto how the concept of "emotion" was viewed around 1955 and still shapes our perspective today. It also offers a glimpse into the early days of the emergence of a field that has since gone on to be amongst the most exciting in neuroscience. Here, we approach MacLean's essay by first examining the scientific climate at the time he argued that the limbic system supported emotions. We then evaluate the anatomy and functions of the limbic system. Finally, we give an update on the neuroscience of emotions and revisit/question the limbic system's importance to the field.

THE SCIENTIFIC CLIMATE SUPPORTING "LIMBIC SYSTEM AS SEAT OF EMOTIONS"

In early anatomical research, Paul Broca (in the 1870s) noted that all mammals possessed a group of cortical areas on the medial surface of the cerebrum that were distinct from the surrounding cortex and the underlying brainstem. He named these areas the limbic lobe, after the Latin word *limbus*, which means "border" or "fringe." Later, Papez (1937), in what has been described as a *tour de force*, proposed that the limbic system, which had no proposed function, and emotion, which had no proposed neural substrate, were related. He proposed that these limbic structures were the emotional brain. Paul MacLean (1952) popularized the term "limbic system" for the limbic and cortical structures that govern the experience and expression of emotion.

Even though linking the limbic system to "emotion" was palpable in MacLean's writing, solid evidence was lacking and most of the evidence he presented was anecdotal. For example, in reference to reported synchronized activity recorded in the guinea pig hippocampus, MacLean writes "what these differences mean in terms of feeling (emotion) and intellect remains to be ascertained, but it is of no little significance to such states as pain and attention." In reference to the sensory experience reported to occur as a result of epileptiform activity, he notes "No experiment is so compelling as the one contrived by nature to support the thesis that this part of the brain [the hippocampus] allows a confluence of the bodily senses and imparts to them the quality of feeling."

Much of the work being reported at the time of MacLean's publication was anatomical and similarly behaviorally descriptive. In essence, the definitive anatomical descriptions of the limbic system were not accompanied by similarly definitive functions. Therefore the following sections will first describe the anatomy of the limbic system before turning to functions.

ANATOMY AND FUNCTIONS OF THE LIMBIC SYSTEM CIRCA 1955

ANATOMY OF THE LIMBIC SYSTEM

A schematic representation of the limbic system structures and connections is shown in Figure 9.1. The limbic system was thought to be comprised of the amygdala, the hippocampal formation and its associated structures, the hypothalamus and mammillary bodies, the septal nuclei and nucleus accumbens, the limbic cortex, and a number of fiber tracts (the medial forebrain bundle, the fornix, the stria terminalis and ventral amygdalofugal bundle). As research on memory and emotion has progressed, additional neocortical regions have often been included in the limbic system. One of these regions is the orbitofrontal cortex, mainly due to its role in emotion and reward (Kringelbach, 2005). This region of the prefrontal cortex has direct and indirect (via the thalamus) connections with the hippocampus and associated neocortex, the amygdala and hypothalamus. In short, these various anatomical structures were viewed as forming a circuit in which the structures of the hypothalamus that produced the physical manifestations of emotion (sweating, increased heart rate and so forth) projected through the limbic structures to the areas of the neocortex that represent the cognitive appreciation of emotion (the location, the person and so forth is dangerous). In turn, the structures of the neocortex that held cognitive representations could, through these same limbic structures, influence the hypothalamic areas to produce the physical manifestations of emotion.

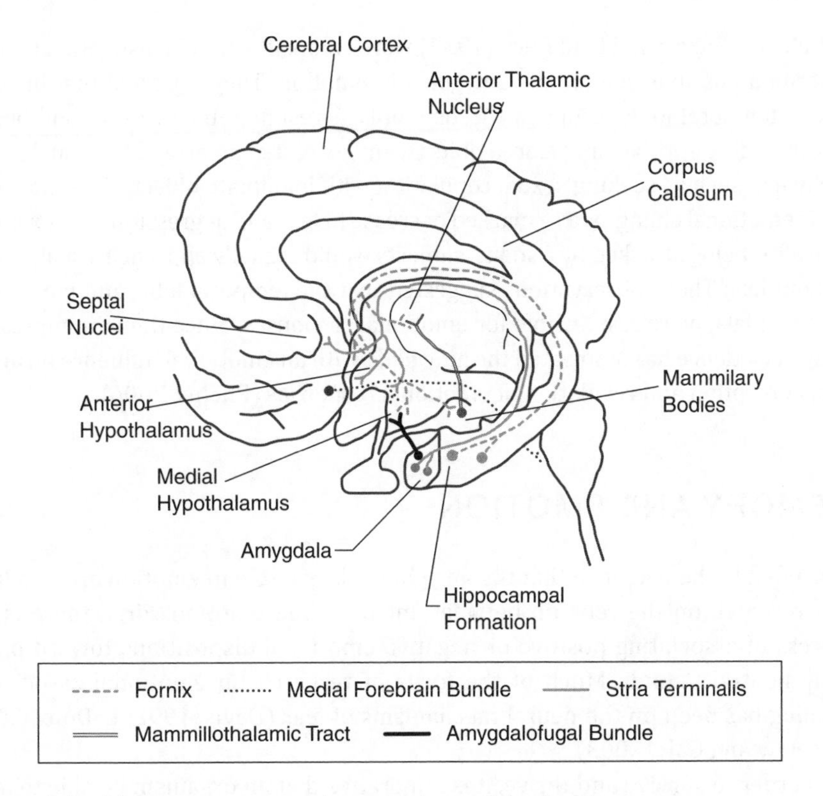

Figure 9.1 Major structures and connections of the limbic system

FUNCTIONS OF THE LIMBIC SYSTEM: A FOCUS ON EMOTION

The more the neural circuitry driving emotion, cognition and behavior has been examined, the more support has been found for the extensive role of limbic system structures in emotion. This is, in part, because *so many* structures have been included as being part of the limbic system and perhaps also because the definition of emotion was frequently vague! Simply looking at each structure, and performing a literature search on its function, suggests that the limbic system is involved in . . . everything! Approaching the task in this way in somewhat unhelpful, and takes us away from the original premise of MacLean's paper. For this reason it is more useful to consider whether any of these structures plays a formative role in "emotion" in its purest sense.

Emotion is a challenging concept to define, and there is considerable debate as to what it should include (see LeDoux, 2012, for a review). For the purposes of this chapter, emotion can be generally defined as a mental state that arises without conscious effort, and that often includes physiological and behavioral changes. Emotions can be thought to include/give rise to "feelings." In essence, feelings can be thought of as the psychological awareness of a physiological state. Emotions can also give rise to physiological responses (e.g. our heart pounding when we experience fear or feelings of love). Whether physiological responses should be ascribed as a definition to an emotion like fear has recently been questioned (see LeDoux, 2013, for a commentary on the topic).

Heinrich Klüver and Paul Bucy (1937) were among the first investigators to demonstrate a possible role for the amygdala in emotion. They reported that bilateral medial temporal lobe lesions in rhesus monkeys resulted in a range of odd behaviors, including approaching feared objects, and exhibited an array of unusual sexual behaviors, such as seeking sexual copulation with inanimate objects. The most dramatic emotional change was a marked decrease in fear and aggression – for example, even after being attacked by a snake, subjects would casually approach a snake again without fear. These observations suggested that the temporal lobe, and most likely the amygdala, were responsible for emotional responses. Since then an impressive array of evidence has associated the amygdala with an emotional influence on attention, perception and regulation of emotional responses (Phelps, 2006).

MEMORY AND EMOTION

Included in the evidence that the amygdala plays a role in emotion are two lines of research on the relation between memory and emotion. First, there is the process of associating positive or negative emotional dispositions toward previously neutral stimuli. Much of the focus of research on emotional association learning has been on the neural mechanisms of fear (Davis, 1992; LeDoux, 2000; Fanselow and Gale, 2003).

In order to survive and thrive, it is imperative that an organism be able to form affective associations between sensory information and context. This is particularly true for learned fear. Through socialization or painful experience, an organism must learn to avoid certain behaviors in order to escape/avoid harm. These fear associations are mediated by the amygdala independently of the events in which the disposition was acquired (Kim and Fanslow, 1992; Phillips and LeDoux, 1992). What is germane to this line of research is that normal emotional conditioning can occur even if declarative memory for the learning situation is impaired. For example, damage to the hippocampus may impair declarative memory while sparing emotional memory of a situation.

In a fearful situation, sensory information reaches the basolateral nucleus of the amygdala, and the signal is then relayed to the central nucleus. Efferents from the central nucleus project to the hypothalamus and other subcortical areas which control a broad range of fear-related responses, including activation of the sympathetic nervous system and motor behaviors (e.g. increases in heart rate, blood pressure, perspiration, the startle response, somatomotor immobility and ultrasonic vocalizations) (for reviews, see Kapp et al., 1992; McGaugh et al., 1992; Sarter and Markowitsch, 1985). After these associations are established, if the organism later finds itself in a situation resembling the one in which the fear association was created, it will generate fear-related responses, even in the absence of the fear-inducing stimulus itself.

The second line of research on emotion and learning is the widely observed phenomenon that memories for events that resulted in emotional arousal tend to be more accurate and longer lasting than memories of neutral events (Berlyne, 1969; Heuer and Reisberg, 1992; Kleinsmith and Kaplan, 1963). The amygdala also plays a role in modulating memories of emotionally arousing experiences. An extensive review of the processes involved is beyond the scope of this chapter, but details may be found in other articles (McGaugh, 2004; Phelps, 2006; Schultz, 2006; Shors, 2006). Briefly, when an emotionally arousing event occurs, epinephrine and glucocorticoids are released by the adrenal glands. This leads to an increase in norepinephrine and, ultimately, increased amygdala activity. Evidence from a large number of non-human animal studies suggests that the amygdala's connections with the striatum, hippocampus, neocortex and nucleus basalis magnocellularis allows it to modulate memory following an emotional event. In other words, what would otherwise be a simple declarative memory is enhanced in situations that evoke strong emotions.

STATUS UPDATE: WHAT DOES THE "LIMBIC SYSTEM" HAVE TO DO WITH EMOTIONS?

The title of MacLean's essay implies the most important function of the limbic system to be that of subserving emotions. This point is also emphasized throughout his essay; for example, he writes in concluding "from these and other experiments concerned with ablation and stimulation, inferences are derived that give additional support to the theory that this region of the brain is concerned with the experience and elaboration of emotion." Does the research performed since 1955 stand up to MacLean's conclusion? Yes and no. There are certainly elements from his article that continue to echo in today's thinking on the seat of emotions. The following quote is one that, to this day, remains extremely insightful: "In discussing the symbolic process, Kubie has emphasized that memory is dependent on the interplay of intrabodily and externally derived experience."

This, to me, is reminiscent of the way much of the field views and understands memory (including fear memories). There is acknowledgement of physiological arousal playing a role in the expression of fear memories, as well as an understanding that external cues previously paired with an unconditioned stimulus will subsequently trigger the retrieval of a memory.

What is perhaps most telling is the next phrase in MacLean's essay. In reference to Kubie's insights, he says "he refers to the former as the gut component of memory."

This is an important aspect of MacLean's essay and remains the aspect of the limbic system that is most widely known and acknowledged. Memories are often defined as being *either* explicit or implicit (or declarative vs non-declarative). Fear memories are thought to be a certain special kind of memory that encompasses both. The implicit component is essentially that "gut feeling" Kubie mentions, or the "visceral brain" that MacLean refers to in his title. Contextual elements, specificity of memory associations, are thought to be more explicit components.

Throughout his career LeDoux has discussed the implicit and explicit elements of fear memories, and even mapped their likely neural circuits in systems he referred to as the "low road" and the "high road" (see Figure 9.2). Both roads terminate in the lateral amygdala. The low road (a much faster circuit) goes directly from the thalamus to the lateral amygdala. The high road goes through the thalamus, then to the cortex, before reaching the amygdala.

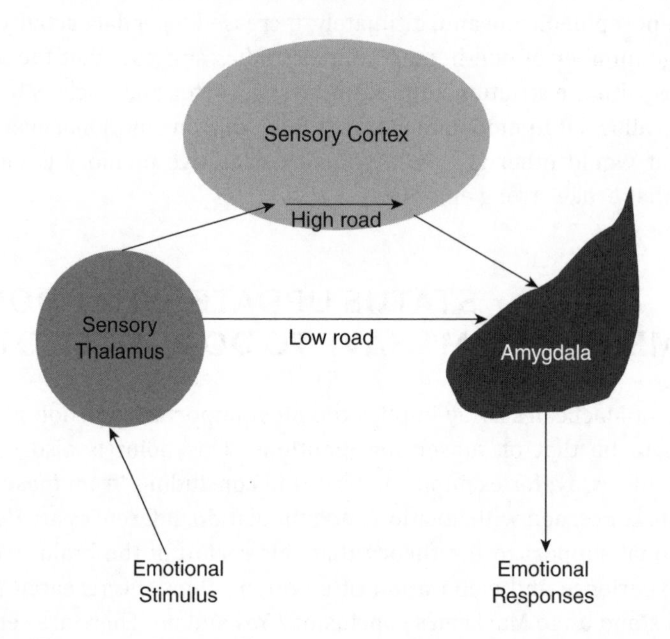

Figure 9.2 The direct (low road) and indirect (high road) pathways to the amygdala (*Source*: drawing adapted from LeDoux, 1996)

The low road is much faster, but lacks detailed information. The high road takes a few milliseconds longer, but provides more refined details. To imagine the impact of these two parallel and overlapping pathways, one can think of the following scenario. Imagine walking in a forest, and stumbling upon a recoiled item that, at first glance, might seem to be a snake in striking position. Most individuals' response to such a sight would be to get sweaty palms and an increased heart rate, the initiation of a fight or flight response. The quick response, if the item is in fact a snake, might be very adaptive and help the individual escape harm. If, upon further examination, the item in question happens to simply be a piece of rope that was mistaken for a snake, to have engaged the fight or flight response preemptively is generally of little cost, and well worth the trade-off. The cortical pathway will simply send information to the amygdala to quench the fear response, and the individual may resume their normal course of behavior.

Ultimately, there is general agreement for the existence of a neural network that supports emotions, but is the limbic system that network? MacLean himself eventually points out what many researchers have concluded over the years, i.e. that

"To date, experiments aimed at the ablation of the hippocampus alone, or severance of its projections by way of the fornix, have revealed disappointingly little by way of behavioral changes."

WHAT HAPPENED TO THE HIPPOCAMPUS IN EMOTION?

McLean and many early investigators of emotion stressed the role of the hippocampus in emotion. Why was there such a focus on this in MacLean's essay? Despite much of the evidence implicating the hippocampus in emotions being anecdotal at the time, there was a strong appeal in studying the structure. An important appeal of studies of the hippocampus has been its laminar structure, so inviting to electrophysiological recordings. The amygdala is much less so, being organized in a more archaic fashion, composed of multiple nuclei (and thus organized into a nuclear structure), and less conducive to allowing readily interpretable electrophysiological signals.

In essence, there was, of course, an obvious appeal to study the hippocampus on the basis of its anatomy and its physiological signature. The clear drawback (the extent of which could not have been predicted at the time) was and continues to be the difficulty in establishing the hippocampus as an engram for a particular behavior.

It would nevertheless be unfair to suggest that the hippocampus is not involved at all in the processing of emotions. To contextualize fear experience, there is evidence that the hippocampus is necessary. We might also point out that, in recent years, there has been a resurgence to include a broader network of structures in pursuing an understanding and an explanation of the fear system (the most widely studied emotion). Effectively, when attempting to account for the experience of emotion/feelings, one must also take into account the cortex. To relieve/recount/re-experience/explicitly recollect a fear-inducing moment, the cortex is necessary. Of course, both the cortex and the hippocampus have extensive connections with the amygdala. In addition, there is a large body of literature that implicates a role for the hippocampus in the stress-related modulation of memories. Stress may be conceived as being a form of emotion. It generally refers to a more prolonged/chronic state. Stress itself may have an impact on the formation and maintenance of explicit memories. It may also serve to activate stress hormones which, in turn, impact neural structures that underlie emotions (Rodrigues et al., 2009). Stress leads to the release of glucocorticoids from the adrenal cortex and catecholamines from the adrenal medulla and sympathetic nerves, and the hippocampus is rich in glucocorticoid receptors. Generally, findings seem to suggest that most negative effects of stress on memory affect hippocampus-dependent tasks (for a review, see Kim et al., 2015). Interestingly, tasks that do not depend on the hippocampus generally show improvement in the presence of a similar stress (Kim et al., 2001; Kim et al., 2015;

Schwabe and Wolf, 2012). Still, I would contend that the role of the hippocampus in emotions remains more of a modulatory rather than a necessary one.

CLINICAL APPLICATIONS OF UNDERSTANDING THE FEAR SYSTEM

One of the most important accounts to detail how fear may be acquired was reported by Pavlov (1927). According to his model of classical conditioning, fear of a stimulus can be acquired by pairing an initially neutral stimulus (called the conditioned stimulus; CS) to another, inherently aversive stimulus (called the unconditioned stimulus; US). When the CS is later presented on its own, it elicits a fear response. The necessary point of convergence of the CS and US is known to be the amygdala (LeDoux, 2000).

In addition to classical conditioning, research has also clearly demonstrated that fear can be acquired through alternate mechanisms, such as vicarious learning, or learning through observation (Askey and Field, 2008).

Appropriate fear responses are essential to survival, but this adaptive mechanism sometimes goes awry, and the associated memories are extremely persistent. As such, reducing fear responses is often a difficult task both in the laboratory and in clinical settings (e.g. for anxiety and fear disorders, including post-traumatic stress disorder and specific phobias). Many aspects of animal models of fear conditioning resemble pathological fear and anxiety conditions seen in humans (Rosen and Schulkin, 1998; Wolpe, 1981; for a review, see Delgado et al., 2006), so optimizing fear reduction in these models is critical to our efforts to understand and treat fear conditions in the clinic. One of the most established approaches to reducing fear is through extinction or exposure therapy. In extinction, repeatedly presenting the CS in the absence of the US leads to a progressive decrease in expression of fear to the stimulus (Bouton and Bolles, 1979; Pavlov, 1927; Rescorla and Heth, 1975). Extinction (and exposure therapy) is typically effective – in the short term – in reducing fear, though the possibility of relapse generally remains. New techniques have surfaced in recent years, and hold promise in optimizing avenues to reduce fear more permanently (Monfils et al., 2009; Nader et al., 2000; Schiller et al., 2010).

CONCLUSION

The neuroscience of emotion has come a long way since 1955. While it is clear that the early scientific view of emotion had the limbic system as being the center of emotions, ultimately, many of its subregions became better known for other functions. Narrowing down/simplifying/reducing pathways to the core elements necessary for fear processing opened the field to a more thorough understanding of how valence-laden experiences are encoded and remembered. The reductionistic approach to the field of fear conditioning was also enticing to a great number of researchers who have made great strides in understanding the contributions of the amygdala.

There was a time during which speaking of understanding emotions in non-human animals was an unimaginable feat. In pursuing fear as a universal emotion, and focusing on the amygdala as the most important structure for this emotion, the field became much more manageable. The field has also been one for which the direct translational relevance has been the fastest.

REFERENCES

Berlyne, D.E. (1969) Arousal, reward and learning. *Annals of the New York Academy of Sciences*, 159 (3): 1059–70.

Bouton, M.E. and Bolles, R.C. (1979) Role of conditioned contextual stimuli in reinstatement of extinguished fear. *Journal of Experimental Psychology: Animal Behavior Processes*, 5(4): 368–78.

Broca, P. (1878) Anatomie comparee des circonvolutions cerebrales: le grand lobe limbique et la scissure limbique dans la serie des mammifers. *Review of Anthropology* 1 (Ser. 2): 385–498.

Davis, M. (1992) The role of the amygdala in fear and anxiety. *Annual Review of Neuroscience*, 15 (1): 353–75.

Delgado, M.R., Olsson, A. and Phelps, E.A. (2006) Extending animal models of fear conditioning to humans. *Biological Psychology*, 73(1):39–48.

Fanselow, M.S. and Gale, G.D. (2003) The amygdala, fear, and memory. *Annals of the New York Academy of Sciences*, 985 (1): 125–34.

Heuer, F. and Reisberg, D. (1992) Emotion, arousal, and memory for detail. In S.Å. Christianson (ed.), *The Handbook of Emotion and Memory: Research and Theory*. Hillsdale, NJ: Erlbaum. pp. 151–180.

Kapp, B.S., Whalen, P.J., Supple, W.F. and Pascoe, J.P. (1992) Amygdaloid contributions to conditioned arousal and sensory information processing. In J.P. Aggleton (ed.), *The Amygdala*. New York: Wiley. pp. 229-253.

Kim, J.J. and Fanselow, M.S. (1992) Modality-specific retrograde amnesia of fear. *Science*, 256 (5057): 675–7.

Kim, E.J., Pellman, B. and Kim, J.J. (2015) Stress effects on the hippocampus: a critical review. *Learning and Memory*, 22 (9): 411–6.

Kleinsmith, L.J. and Kaplan, S. (1963) Paired-associate learning as a function of arousal and interpolated interval. *Journal of Experimental Psychology*, 65 (2): 190.

Klüver, H. and Bucy, P.C. (1937) "Psychic blindness" and other symptoms following bilateral temporal lobectomy in Rhesus monkeys. *American Journal of Physiology*, 119: 352–3.

Kringelbach, M.L. (2005) The human orbitofrontal cortex: linking reward to hedonic experience. *Nature Reviews Neuroscience*, 6 (9): 691–702.

LeDoux, J.E. (2000) Emotion circuits in the brain. *Annual Review of Neuroscience*, 23 (1): 155–84.

LeDoux, J.E. (2012) Coming to terms with fear. *Proceedings of the National Academy of Sciences*, 111 (8): 2871–8.

LeDoux, J.E. (2013) The slippery slope of fear. *Trends in Cognitive Neuroscience*, 17 (4): 155–6.

MacLean, P.D. (1952) Some psychiatric implications of physiological studies on frontotemporal portion of limbic system (visceral brain). *Electroencephalography & Clinical Neurophysiology*, 4 (4): 407–18.

MacLean, P.D. (1955) The limbic system ("visceral brain") and emotional behavior. *Archives of Neurology & Psychiatry*, 73 (2): 130–4. PMID:13227663

McGaugh, J.L., Introini-Collison, I.B., Cahill, L., Kim, M. and Liang, K.C. (1992) Involvement of the amygdala in neuromodulatory influences on memory storage. In J.P. Aggleton (ed.), *The Amygdala*. New York: Wiley. pp. 431–51.

McGaugh, J.L. (2004) The amygdala modulates the consolidation of memories of emotionally arousing experiences. *Annual Review of Neuroscience*, 27: 1–28.

Monfils, M.-H., Cowansage, K.K., Klann, E. and LeDoux, J.E. (2009) Extinction-reconsolidation boundaries: key to persistent attenuation of fear memories. *Science*, 324: 951–5.

Nader, K., Schafe, G.E. and Le Doux, J.E. (2000) Fear memories require protein synthesis in the amygdala for reconsolidation after retrieval. *Nature*, 406: 722–6.

Papez, J.W. (1937) A proposed mechanism of emotion. *Archives of Neurology and Psychiatry,* 38: 725–43.

Pavlov, I. (1927). *Conditioned reflexes: An Investigation of the Physiological Activity of the Cerebral Cortex*. Translated and edited by G.V. Anrep. London: Oxford University Press.

Phelps, E. (2006) Emotion and cognition: insights from studies of the human amygdala. *Annual Review of Psychology*, 57: 27–53.

Phillips, R.G. and LeDoux, J.E. (1992) Differential contribution of amygdala and hippocampus to cued and contextual fear conditioning. *Behavioral Neuroscience*, 106 (2): 274–85.

Rescorla, R.A. and Heth, C.D. (1975). Reinstatement of fear to an extinguished conditioned stimulus. *Journal of Experimental Psychology: Animal Behavior Processes*, 1: 88–96.

Rodrigues, R.M., LeDoux, J.E. and Sapolsky, R.M. (2009) The influence of stress hormones on fear circuitry. *Annual Review of Neuroscience*, 32: 289–313.

Rosen, J.B. and Schulkin, J. (1998) From normal fear to pathological anxiety. *Psychological Review*, 105(2): 325-50.

Sarter, M. and Markowitsch, H.J. (1985) Involvement of the amygdala in learning and memory: a critical review, with emphasis on anatomical relations. *Behavioral Neuroscience*, 99 (2): 342.

Schiller, D., Monfils, M.-H., Raio, C.M., Johnson, D.C., LeDoux, J.E. and Phelps, E.A. (2010) Preventing the return of fear in humans using reconsolidation update mechanisms. *Nature*, 463: 49–53.

Schwabe, L. and Wolf, O.T. (2012) Stress modulates the engagement of multiple memory systems in classification learning. *The Journal of Neuroscience*, 32(32): 11042–9.

Schultz, W. (2006) Behavioral theories and the neurophysiology of reward. *Annual Review of Psychology*, 57: 87–115.

Shors, T.J. (2006) Stressful experience and learning across the lifespan. *Annual Review of Psychology*, 57: 55–85.

Wolpe, J. (1981) The dichotomy between classical conditioned and cognitively learned anxiety. *Journal of Behavior Therapy and Experimental Psychiatry*, 12, 35–42.

10 Revisiting Phineas Gage: Lessons we learned from damaged brains

Antoine Bechara

A great deal of contemporary decision research in economics, business, psychology, and neuroscience now accepts the idea that emotions play a significant role in influencing decision making. The case of Phineas Gage in 1848 paved the way for the notion that the frontal lobes were linked to judgment, decision making, social conduct, and personality. A number of cases of frontal lobe damage with defects similar to those of Phineas Gage appeared in the literature but received little attention. A revival of interest in the decision-making and social aspects of "frontal lobe syndrome" was triggered in part by the description of a modern counterpart to Phineas Gage in 1985, patient EVR. Almost 25 years ago, insights from studies on brain lesion patients set the cornerstone for this stream of research and led to the formulation of the somatic marker theory. Despite some debate, this theory is still providing a unique neuroanatomical and cognitive framework that helps explain the role of emotion in decision making. In this chapter I review evidence that the ventromedial prefrontal cortex (vmPFC) plays a role in eliciting visceral responses that are related to the value of objects and events in the world. I review the results of lesion studies in humans showing that the vmPFC is necessary for eliciting visceral responses to certain forms of emotional stimuli, and for guiding decision making in the face of uncertain reward and punishment.

INTRODUCTION

Most of us are taught from early on that logical, rational calculation forms the basis of sound decisions. Emotion has no IQ, and it can only cloud the mind and interfere with good judgment. But what if we were wrong? What if sound, rational decision making in fact depended on prior accurate emotional processing? For instance, we often talk about a "gut feeling" when we meet a person for the first time. We're told to "trust our gut" and "follow our heart" when faced with difficult decisions about our future, that it's "gut check time" when faced with a situation that tests our nerve and determination. Do our emotions and gut feelings really have something to say when making good judgment? Studies of decision making in neurological patients who can no longer process emotional information normally suggest just that.

The case of Phineas Gage in 1848 (Harlow, 1848) paved the way for the notion that the frontal lobes were linked to judgment, decision making, social conduct, and personality. A number of cases of frontal lobe damage with defects similar to those of Phineas Gage appeared in the literature but received little attention (Ackerly and Benton, 1948; Brickner, 1932). A revival of interest in the decision-making and social aspects of "frontal lobe syndrome" was triggered in part by the description of a modern counterpart to Phineas Gage in 1985, patient EVR (Eslinger and Damasio, 1985b). Almost 25 years ago (i.e. since the 1990s, shortly after patient EVR), insights from studies on brain lesion patients set the cornerstone for a stream of research that led to the formulation of the somatic marker theory. The theory is that emotion-related signals (somatic markers), which are indexed changes in the visceral state, such as changes in heart rate, blood pressure, gut motility, and glandular secretion, assist cognitive processes in implementing decisions (Damasio, 1994). The theory was primarily inspired by the observed decision-making deficits in patients with frontal lobe lesions, especially the medial sectors of the frontal lobes.

Patients with damage in the ventromedial prefrontal cortex (vmPFC) make poor decisions in part because they are unable to elicit somatic (visceral) responses that "mark" the consequences of their actions as positive or negative. The medial prefrontal region functions as a trigger of visceral responses that reflect the anticipated value of the choices. Changes in the visceral state may be considered a form of anticipation of the bodily impact of objects and events in the world. Visceral responses to biologically relevant stimuli allow an organism to maximize the survival value of situations that may impact the state of the internal milieu. These include events that promote homeostasis, such as an opportunity to feed or engage in social interaction, as well as events that disrupt homeostasis, such as a physical threat or a signal of social rejection. Visceral responses are a sub-component of a broader emotional response system that also includes changes in the endocrine and skeletomotor systems, as well as changes within the brain that alter the perceptual processing of biologically relevant stimuli (Damasio, 1994).

Thus, the somatic marker theory attributes the inability of certain patients with frontal lobe damage to make advantageous decisions in real life. Their defect is the absence of an emotional (somatic) mechanism that rapidly signals the prospective consequences of an action, and accordingly assists in the selection of an advantageous response option (Damasio, 1994). Because the somatic marker signal is derived from prior life experiences with rewards and/or punishments, their "affective" or "emotional" past is no longer able to anticipate or forecast the future.

Despite some debate, the somatic marker theory is still providing a unique neuroanatomical and cognitive framework that helps explain the role of emotion in decision making. In this chapter I review evidence that the ventromedial prefrontal cortex (vmPFC) plays a role in eliciting visceral responses that are related to the incentive value of objects and events in the world, and which play a role in guiding decision making in the face of uncertain reward and punishment. Thus the evidence supports the view that the process of decision making is influenced in many important ways by neural substrates that regulate homeostasis, emotion, and feeling.

A HISTORICAL OVERVIEW: PHINEAS GAGE AND THE PASSAGE OF AN IRON ROD THROUGH HIS HEAD

One of the first and most famous cases of the "frontal lobe syndrome" is Phineas Gage, described by Harlow (1848, 1868). Phineas Gage was a railroad construction worker, and survived an explosion that blasted an iron tamping bar through the front of his head. Before the accident, Gage was a man of normal intelligence, energetic and persistent in executing his plans of operation. He was responsible, sociable, and popular among peers and friends. After the accident, his medical recovery was remarkable. He survived the accident with normal intelligence, memory, speech, sensation, and movement. His behavior changed, however. He became irresponsible, untrustworthy, and impatient of restraint or advice when it conflicted with his desires. Using modern neuroimaging techniques, Damasio and colleagues have reconstituted the accident by relying on measurements taken from Gage's skull (Damasio et al., 1994). The key finding of this neuroimaging study was that the most likely placement of Gage's lesion included the vmPFC region, bilaterally. However, the original report of Harlow, and CT scans of the actual skull of Phineas Gage, imply that the damage was only in the left hemisphere.

The case of Phineas Gage paved the way for the notion that the frontal lobes were linked to social conduct, judgment, decision making, and personality. A number of instances similar to the case of Phineas Gage have appeared in the literature (Ackerly and Benton, 1948; Brickner, 1932). Interestingly, all these cases received little attention for many years. The revival of interest in various aspects of the "frontal lobe syndrome" was triggered in part by the patient EVR, a modern counterpart to Phineas Gage (Eslinger and Damasio, 1985b).

Over the years, our research group has studied patients with this type of lesion. Those with damage to the vmPFC develop severe impairments in personal and social decision making, in spite of otherwise largely preserved intellectual abilities. These patients had normal intelligence and creativity before their brain damage. After the damage, they begin to have difficulties planning their work day and future, and difficulties in choosing friends, partners, and activities. The actions they elect to pursue often lead to losses of diverse types, e.g. financial, social standing, family and friends. The choices they make are no longer advantageous, and are remarkably different from the kinds of choices they were known to make in the pre-morbid period. They also often decide against their best interests and are unable to learn from previous mistakes, as reflected by repeated engagement in decisions that lead to negative consequences. In striking contrast to this real-life decision-making impairment, their problem-solving abilities in laboratory settings remain largely normal.

While these patients often show that they are normal on most conventional neuropsychological tests (Bechara et al., 1998; Tranel, 2000), clinical observations reveal that they show abnormalities in expressing appropriate emotions and experiencing certain feelings.

These observations were the origins of the somatic marker theory (Damasio, 1994). The theory posits that the neural basis of the decision-making impairment characteristic of patients with vmPFC damage is defective activation of the somatic states (emotional signals) that attach value to given options and scenarios. These emotional signals function as covert, or overt, biases for guiding decisions. Deprived of these signals, patients must rely on slow cost–benefit analyses of various conflicting options. These options may be too numerous and their analysis may be too lengthy to permit rapid, on-line decisions to take place appropriately. Patients may resort to deciding based on the immediate reward of an option, or may fail to decide altogether if many options have the same basic value.

A NOTE ON ANATOMICAL TERMS

The terms vmPFC and orbitofrontal cortex (OFC) are often used interchangeably, even though they do not refer to identical regions. For this reason it is necessary to clarify exactly what we mean when we use these terms. The OFC is the entire cortex occupying the ventral surface of the frontal lobe, dorsal to the orbital plate of the frontal bone. The term vmPFC designates a region that encompasses medial portions of the OFC along with ventral portions of the medial prefrontal cortex. The vmPFC is an anatomical designation that has arisen because lesions that occur in the basal portions of the anterior fossa, which include meningiomas of the cribiriform plate and falx cerebri, and aneurysms of the anterior communicating and anterior cerebral arteries, frequently lead to damage in this area (see Figure 10.1). Often this damage is bilateral. With respect to the cytoarchitectonic fields identified in the human orbitofrontal and medial prefrontal cortices by Price and colleagues (Ongur and Price, 2000), the vmPFC comprises Brodmann area (BA) 14 and medial portions of BA 11 and 13 on the orbital surface, and BA 25 and 32 and caudal portions of BA 10 on the mesial surface. The vmPFC excludes lateral portions of the OFC, namely BA 47/12, as well as more dorsal and posterior regions of BA 24 and 32 of the medial prefrontal cortex. The vmPFC is thus a relatively large and heterogeneous area.

Viewing the vmPFC as a single region may blur the distinction between functions subserved by the OFC on the one hand and the ventral portion of the medial prefrontal cortex on the other. Evidence from rodents (Chudasama and

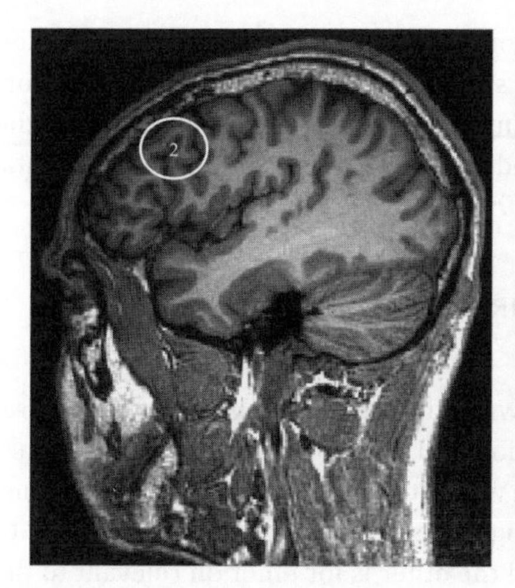

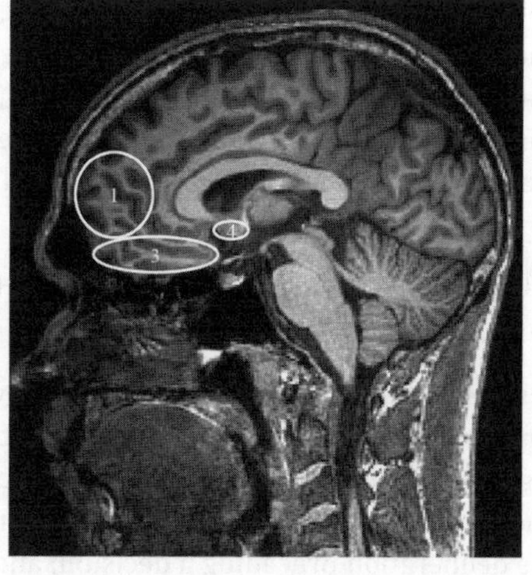

Figure 10.1 Approximate locations of relevant brain areas in the medial sector of the prefrontal cortex: (1) ventromedial prefrontal cortex, (2) dorsolateral prefrontal cortex, (3) medial orbitofrontal cortex, (4) amygdala

Robbins, 2003) and nonhuman primates (Pears et al., 2003) suggests that these regions subserve distinct motivational and learning functions. Thus lesions of the vmPFC in humans may disrupt more than one process. This is important to keep in mind when inconsistencies arise between animal studies and human studies. These differences are also important when comparing human lesion studies, which tend to examine the functions of relatively large regions, and functional imaging studies, which reveal more focused patterns of activity.

THE GENESIS OF SOMATIC MARKER THEORY

Previous research has studied patients with lesions of the frontal lobes. Although many of these patients retained a normal intellect, they demonstrated impairments in decision making (Bechara et al., 1998; Damasio et al., 1990; Eslinger and Damasio, 1985a, b). The behaviors and decision-making deficits observed in patients with vmPFC lesions presented a puzzling defect. It is difficult to know whether their defect is in knowledge pertinent to the situation or to general intellectual problems (Anderson et al., 1999; Saver and Damasio, 1991). Nevertheless, their deficits cannot be explained in terms of defects in language comprehension or expression, or in working memory or attention (Anderson et al., 1999; Bechara et al., 1998; Saver and Damasio, 1991). For many years, the condition of these patients has posed a double challenge. First, although the decision-making impairment is obvious in the real life of these patients, there had been no laboratory probe to detect and measure the impairment. Second, there had been no satisfactory account of the neural and cognitive mechanisms underlying the impairment. While these vmPFC patients were intact on nearly all neuropsychological tests, they did have compromised ability to express emotion and to experience feelings in appropriate situations. In other words, despite normal intellect there were abnormalities in emotion and feeling, along with the abnormalities in decision making. Based on these observations, the somatic marker theory was proposed (Damasio, 1994; Damasio et al., 1991).

OUTLINE OF THE NEURAL CIRCUIT THROUGH WHICH THE VMPFC INFLUENCES DECISIONS

Several neural structures have been shown to be key components of the neural circuitry underlying somatic state activation and decision making. Decision making is a complex process that relies on the integrity of at least two neural systems: one for memory which is required to bring online knowledge and information to the deliberation preceding a decision; and the other is for emotion relevant to the situation. Emotion requires effector structures such as the hypothalamus and autonomic brainstem nuclei that produce changes in internal milieu and visceral structures, along with other effector structures such as the ventral striatum, periaqueductal gray and other brainstem nuclei, which produce changes in facial expression and specific approach or withdrawal behaviors. Emotion also requires cortical structures that receive afferent input from the viscera and internal milieu, such as the insular cortex (see Figure 10.2).

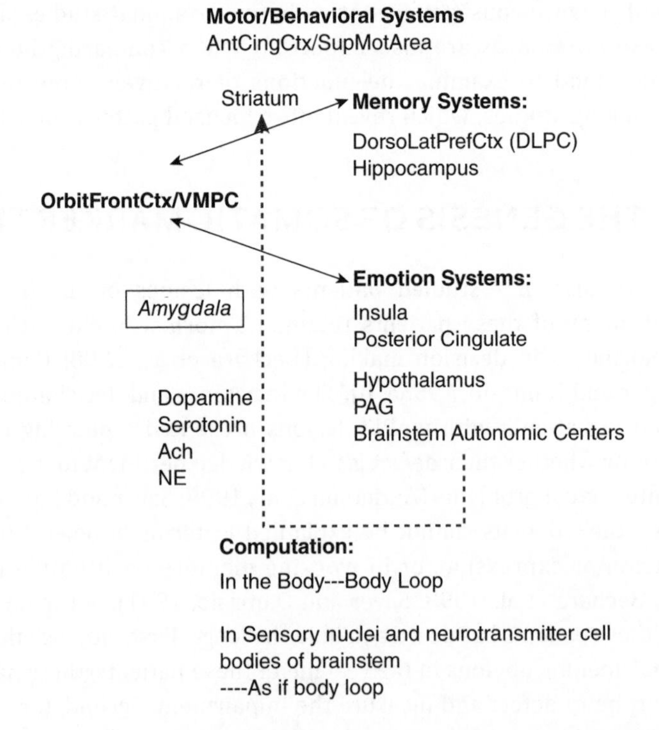

Figure 10.2 A schematic of all the brain regions involved in decision making according to the somatic marker hypothesis

During the process of pondering decisions, the immediate prospects of an option may be driven by more subcortical mechanisms (e.g. via the amygdala) that do not require the prefrontal cortex. However, weighing the future consequences requires the prefrontal cortex to trigger somatic responses about possible future consequences. Specifically, when pondering the decision, the immediate and future prospects of an option may trigger numerous somatic responses that conflict with each other (i.e. positive and negative somatic responses). The end result, though, is that an overall positive or negative signal emerges (a "go" or "stop" signal, as it were). There is a debate as to where this overall somatic state may be computed. We have argued that this computation occurs in the body proper (via the so-called body loop), but it can also occur in the brain, in areas that represent "body" states such as the dorsal tegmentum of the midbrain, or areas such as the insula and posterior cingulate (via the so-called as-if-body loop) (see Figure 10.2). The controversy has largely been in relation to the body loop, with certain investigators arguing that decision making is not necessarily dependent on "somatic markers" expressed in the body (e.g. see Maia and McClelland, 2004; but also see Bechara et al., 2005 and Persaud et al., 2007 for counter arguments). Irrespective of whether this computation occurs in the body itself, or within the brain, an emerging positive or negative somatic state will consequently bias the decision one way or the other (Bechara and Damasio, 2005).

In order for somatic signals to influence cognition and behavior, they must act on the appropriate neural systems. One target for somatic state action is the striatum. A large number of channels convey body information (that is, somatic signals) to the central nervous system (e.g. spinal cord, vagus nerve, and humoral signals). Evidence suggests that the vagal route is especially critical for relaying somatic signals (Martin et al., 2004). The next link in this body–brain channel involves neurotransmitter systems (Bechara and Damasio, 2005). Indeed, the cell bodies of the neurotransmitters dopamine, serotonin, noreadrenaline, and acetylcholine are located in the brainstem; the axon terminals of these neurotransmitter neurons synapse on cells throughout the cortex and striatum. When somatic state signals are transmitted to the cell bodies of dopamine or serotonin neurons, for example, the signaling influences the pattern of dopamine or serotonin release on neurons subserving behavior and cognition within the cortex. This chain of neural action provides a way for somatic states to exert a biasing effect on decisions. The functional neuroimaging work on the role of dopamine in reward processing and error prediction provides support for the proposed neural framework (Montague and Berns, 2002; Schultz et al., 1997).

One of the predictions of the somatic marker hypothesis is that working memory (and other executive processes of working memory, such as response inhibition and reversal learning) is a key process in decision making. Consequently, damage to neural structures that impair working memory, such as the dorsolateral prefrontal cortex, also leads to impaired decision making. Nonetheless, one criticism of the theory is that deficits in decision making as measured by the Iowa Gambling Task may not be specific to the ventromedial prefrontal cortex (Manes et al., 2002) or may be explained by deficits in other processes, such as reversal learning (Fellows and Farah, 2003). Nevertheless, research has demonstrated that the relationship between decision making on the one hand and working memory or reversal learning on the other are asymmetrical (e.g. see Bechara, 2004; Bechara and Damasio, 2005, for reviews). In other words, working memory and/or reversal learning are not dependent on intact decision making (i.e. subjects can have normal working memory and normal reversal learning in the presence or absence of deficits in decision making). Some patients with ventromedial prefrontal cortex lesions who are severely impaired in decision making (i.e. are abnormal in the Iowa Gambling Task) have superior working memory, and perform normally on simple reversal learning tasks. In contrast, decision making seems to be influenced by the intactness or impairment of working memory and/or reversal learning, i.e. decision making is worse in the presence of abnormal working memory and/or poor reversal learning. Patients with right dorsolateral prefrontal cortex lesions and severe working memory impairments showed low normal results in the Iowa Gambling Task (Bechara et al., 1998). Patients with damage to the more posterior sector of the ventromedial prefrontal cortex (which includes the basal forebrain), included in the study by Fellows and Farah (2003), show impairments on reversal learning tasks, but similar patients with similar lesions also showed poor performance on the Iowa Gambling Task (Bechara et al., 1998).

EMPIRICAL TESTS OF SOMATIC MARKER THEORY

At the core of somatic marker theory is the insight that decision makers encode the consequences of alternative choices affectively. For many years, these vmPFC patients presented a puzzling defect. Although the decision-making impairment was obvious in their real-world behavior, there was no effective laboratory probe to detect and measure the impairment. Bechara's development of what became known as the "Iowa Gambling Task" enabled the detection of these patients' elusive impairment in the laboratory, and facilitated investigating its possible causes (Bechara et al., 1994). Work using the Iowa Gambling Task provides the key empirical support for the proposal that somatic markers significantly influence decision making.

THE IOWA GAMBLING TASK (IGT)

The IGT mimics real-life decisions closely. The task is carried out in real time and it resembles real-world contingencies. It factors reward and punishment (i.e. winning and losing money) in such a way that it creates a conflict between an immediate, luring reward and a delayed, probable punishment. As in real-life choices, the task offers choices that may be risky, and each of these is full of uncertainty because a precise calculation or prediction of the outcome of a given choice is not possible. The way that a person can do well on this task is to follow their hunches and gut feelings. The IGT was the first neuropsychological task that introduced money as a form of reward and punishment in cognitive testing. This work has drawn attention to the potential value in studying the neural basis of decision making, and in bringing this question to the laboratory through the use of structured decision-making tasks involving choices that mimic real-life situations by factoring in uncertainty, reward and punishment.

The IGT uses four decks of cards. The goal in the task is to maximize profit on a loan of play money. Subjects can select one card at a time from any deck they choose. The subject's decision to select from one deck versus another is largely influenced by various schedules of immediate reward and future punishment. These schedules are pre-programmed and known to the examiner, but not to the subject, and they entail the following principles: two of the decks (decks A and B) yield a relatively high immediate reward in comparison to the two other decks (C or D), i.e. the subject can get $100 versus $50. In each of the four decks, subjects encounter unpredictable punishments in the form of money loss. The punishment is higher in the high-paying decks A and B, and lower in the low-paying decks C and D. Hence, decks A and B are disadvantageous because they cost more in the long run, i.e. one loses $250 every 10 cards. Decks C and D are advantageous because they result in an overall gain in the long run, i.e. one wins $250 every 10 cards.

We have investigated the performance of control subjects and patients with vmPFC lesions on this task. Control subjects avoided the bad decks A and B and preferred the good decks C and D. In sharp contrast, the vmPFC patients preferred decks A and B.

From these results, we suggested that the patients' performance profile is comparable to their real-life inability to decide advantageously. This is especially true in personal and social matters, a domain for which in life, as in the task, an exact calculation of the future outcomes is not possible, and choices must be based on hunches and gut feelings.

EVIDENCE THAT EMOTIONAL SIGNALS GUIDE DECISIONS

In light of the finding that the IGT detects the decision-making impairment of vmPFC patients in the laboratory, we went on to address the question of whether the impairment is linked to a failure in somatic signaling (Bechara et al., 1996).

To address this question, we added a physiological measure to the IGT. The goal was to assess somatic state activation while subjects were making decisions during performance of the task. We studied two groups: control subjects and vmPFC patients. They performed the IGT while we recorded their electrodermal activity (skin conductance responses; SCRs). As the body begins to change after a thought, and as a given somatic state begins to be enacted, the autonomic nervous system begins to increase the activity in the skin's sweat glands. Although sweating is not observable to the naked eye, it can be amplified and recorded by a polygraph as a wave.

Both control subjects and vmPFC patients generated SCRs when they picked a card and were told that they had won or lost money. The most important difference, however, was that control subjects, as they became experienced with the task, began to generate SCRs *prior* to the selection of a card, i.e. during the time when they were pondering from which deck to choose. These anticipatory SCRs were more pronounced before picking a card from the disadvantageous decks A and B, when compared to the advantageous decks C and D. In other words, these anticipatory SCRs were like gut feelings that warned the subject against picking from the bad decks. Patients with vmPFC damage failed to generate such SCRs before picking a card. This failure to generate anticipatory SCRs *before* picking cards from the bad decks correlates with their failure to avoid these bad decks and choose advantageously in this task. These results provide strong support for the notion that decision making is guided by emotional signals (gut feelings) that are generated in anticipation of future events.

An important question regarding the information content of visceral responses is whether these are elicited by the vmPFC. If somatic markers are to be useful in guiding decision-making processes involving uncertain reward and punishment, then they should provide information about both the valence of an anticipated outcome (e.g. whether a choice will result in winning or losing money) as well as information about the magnitude of the anticipated outcome (e.g. how much money will be won or lost). Our results using the IGT show that the vmPFC triggers anticipatory visceral responses to both advantageous and disadvantageous decks. The responses are larger for disadvantageous decks than for advantageous decks, though they are still deployed for the advantageous decks. Further experiments from our laboratory (Bechara et al., 2002) and others (Tomb et al., 2002) show that when the reward–punishment contingencies are reversed, with the disadvantageous decks paying out a lower quantity of reward rather than doling out a higher punishment, the SCRs are greater to the advantageous decks than to the disadvantageous decks. This suggests that SCR is not merely an index of the potential "badness" of choices. Rather, SCR can index the magnitude of both the anticipated negative outcome of a choice as well as the magnitude of the anticipated positive outcome of a choice.

The finding that the SCR does not differentiate the anticipated valence of the outcomes is consistent with other work (Lang et al., 1993): it does not differentiate the hedonic valence of emotional stimuli but does index the magnitude of the arousal that they elicit. This would mean that some other signal is required in order to assess the valence of the anticipated outcome. Although our laboratory (Rainville et al., 2006) has provided preliminary evidence that cardiovascular responses, such as changes in heart rate, can provide information that distinguishes between positive and negative emotional states, the fact remains that the preponderance of evidence speaks to the lack of such a distinction at the peripheral visceral level (Cacioppo et al., 2000). Although it is possible that such signals can combine with those reflected in the SCR to provide information about both the perceived valence of the future outcome of a choice as well as its perceived magnitude, there is a strong likelihood that this discrimination is not achieved until the signals reach the brain. Indeed, the somatovisceral afference model of emotion (SAME) does provide an explanation for how undifferentiated visceral responses might produce distinguishable emotions (Cacioppo et al., 1993). Perhaps somatic markers operate in a fashion that is consistent with that model.

EMOTIONAL SIGNALS NEED NOT BE CONSCIOUS

According to the somatic marker theory, the vmPFC mediates an implicit representation of the anticipated value of choices that is distinct from an explicit awareness of the correct strategy. To test this idea we performed a study (Bechara et al., 1997) in which we examined the development of SCRs over time in relation to subjects' knowledge of the advantageous strategy in the IGT. In the study the IGT was administered as before, but the task was interrupted at regular intervals and the subjects were asked to describe their knowledge about "what was going on" in the task and about their "feelings" about the task. Control subjects began to choose preferentially from the advantageous decks before they were able to report why these decks were preferred over the disadvantageous decks. They then began to form "hunches" about the correct strategy, which corresponded to their choosing more from the advantageous decks than from the disadvantageous decks. Finally, some subjects reached a "conceptual" stage, where they possessed explicit knowledge about the correct strategy (i.e. to choose from decks C and D because they pay less, but also result in less punishment). As before, control subjects developed SCRs preceding their choices that were larger for the disadvantageous decks than for the advantageous decks. The SCR discrimination between advantageous and disadvantageous decks also preceded the development of conceptual knowledge of the correct strategy. In fact, the SCR discrimination between advantageous and disadvantageous decks even preceded the development of hunches about the correct strategy. In contrast to the control subjects, those with damage in the vmPFC failed to switch from the disadvantageous decks to the advantageous decks. In addition, they failed to develop anticipatory responses that discriminated between the disadvantageous and advantageous decks. They never developed "hunches" about the correct strategy. Taken together, the results suggest that anticipatory visceral responses governed by the vmPFC precede emergence of advantageous choice behavior, which itself precedes explicit knowledge of the advantageous strategy. In short, signals generated by the vmPFC, reflected in visceral states, may function as a non-conscious bias toward the advantageous strategy (see Figure 10.3).

An intriguing observation was that not all the normal control subjects were able to figure out the task, in the sense that they did not reach a conceptual understanding of it. Although a number of controls did not reach the conceptual stage, they still performed advantageously. In contrast, almost half the sample of the vmPFC lesion patients were able to reach the conceptual understanding and state which decks were good and which ones were bad and why. Nevertheless these same vmPFC lesion patients still performed disadvantageously (see Figure 10.3). This suggests that the anticipatory SCRs represent unconscious biases derived from prior experiences with reward and punishment. These biases

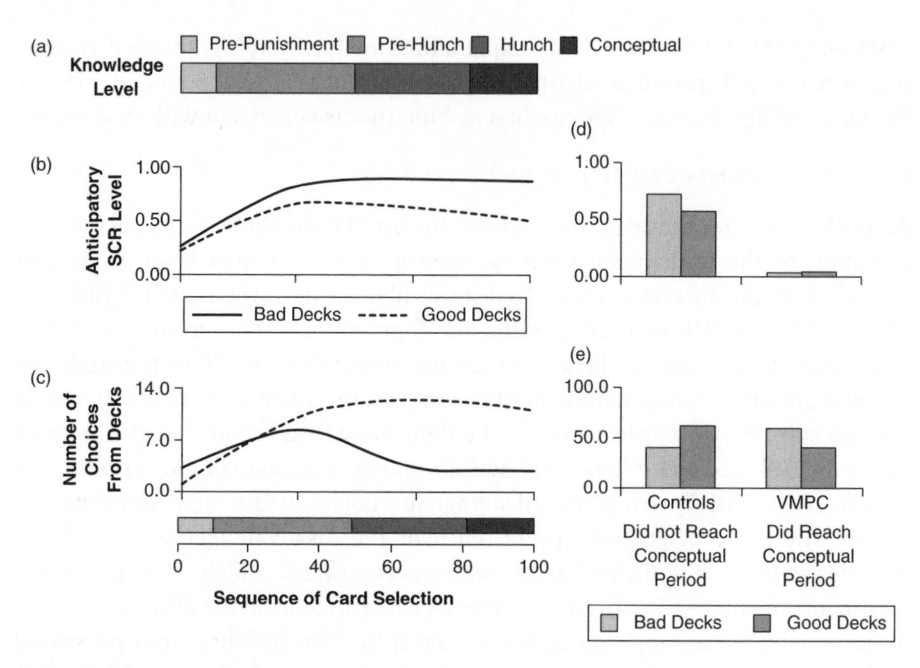

Figure 10.3 Anticipatory visceral responses that discriminate between advantageous and disadvantageous choices precede knowledge of the correct strategy. The pre-punishment period covered the start of the game when subjects sampled the decks and before they encountered the first loss. The pre-hunch period consisted of the next series of cards when subjects continued to choose cards from various decks but professed no notion of what was happening in the game. The hunch period (never reached in patients) corresponded to the period when subjects reported "liking" or "disliking" certain decks, and "guessed" which decks were risky or safe, but were not sure of their answers. The conceptual period corresponded to the period when subjects were able to articulate accurately the nature of the task and tell for certain which were the good and bad decks, and why they were good or bad.

help deter the control subject from pursuing a course of action that is disadvantageous in the future. This occurs even before the subject becomes aware of the goodness or badness of the choice about to be made. Without these biases, the knowledge of what is right and what is wrong may still become available. By itself, this knowledge is not sufficient to ensure an advantageous behavior. Therefore, although the vmPFC lesion patients may manifest as declarative of what is right and what is wrong, they fail to act accordingly. They may "say" the right thing, but they "do" the wrong thing.

Critique of the Somatic Marker Theory

Despite the fact that the original studies on somatic markers and their impact on the decision-making field have inspired many subsequent studies, and have driven the emergence of the new fields of neuroeconomics, decision neuroscience and consumer neuroscience, some have questioned the theory's utility as an explanatory framework for decision making (Maia and McClelland, 2004). Whereas the somatic marker theory incorporates the roles of many different brain structures relevant to decision making (Dunn, 2006), most of the critiques have been leveled on only one specific component of the somatic marker neural circuitry, i.e. the role of peripheral body signals in decision making. Although the somatic marker theory postulates an "as-if-body loop" which bypasses the peripheral route altogether, most critiques are about the role of these body signals in decision making. The fact remains, however, that there are no currently available neurological theories that provide an alternative to the somatic marker theory. Most often, the somatic marker theory is critiqued in models that are not neurological in perspective. Competing theories, mostly psychological, cognitive or behavioral, derive from the viewpoints of their own domains and schools of thought, while somatic marker theory was established on neurological evidence based on the comparison of decision-making strategies of patients with specific and focal brain lesions to healthy subjects. Because of the theoretical origin of the different approaches, i.e. neurology versus nonneurological, a conceptual comparison simply leads to a comparison of "apples and oranges" and therefore does not yield meaningful insights.

The somatic marker theory is a neurological theory that specifically details the different neural steps that take place inside the brain before the execution of a decision. Certainly, there are currently numerous neuroscientific studies on decision making that address the role of a whole variety of events such as expectation, conflict monitoring, gains, losses, and error detection in decision making. None of these theories are comprehensive neurological theories, and they usually focus on only one specific process of the more complex phenomenon of decision making. For example, the dopaminergic system has been implicated in decision making (Montague et al., 1996; Schultz et al., 1997) and is clearly considered a neuroscientific approach to decision making. Yet the dopamine story is very specific and constrained, and does not explain decision making and its influence by emotions in a comprehensive manner as somatic marker theory does. Moreover, the somatic marker theory is inclusive of the role of dopamine in decision making (Bechara and Damasio, 2005). Therefore, the dopamine link to reward, error prediction, and decision making is not an alternate view to the somatic marker theory, but rather one specific link in a broader neural model of decision making described under the somatic marker theory.

CONCLUSION

Emotion is a major factor in the interaction between environmental conditions and human decision processes. Emotion (underlying somatic state activation) provides valuable implicit or explicit knowledge for making fast and advantageous decisions. Historically, most theories of choice have been cognitive in perspective, and have assumed that decisions derive from an assessment of the future outcomes of various options and alternatives through some type of cost–benefit analyses (see Loewenstein et al., 2001 for a review). The somatic marker theory provides neurobiological support for the notion that people make judgments not only by evaluating the consequences and their probability of occurring, but also (and even sometimes primarily) at a "gut" level (Damasio, 1994; Loewenstein et al., 2001; Schwartz and Clore, 1983; Zajonc, 1984). It is important to note that emotion is not always beneficial to decision making and sometimes it can be disruptive (Bechara and Damasio, 2005). Thus, the process of decision making depends in many ways on neural substrates that regulate homeostasis, emotion and feeling. In other words, the process of deciding is not just logical and computational but also emotional.

Finally, details of Gage's behavioral impairment that appeared in many books and articles may have occasionally been exaggerated, or perhaps a bit embellished to suit the theory of the storyteller. For instance, Gage's behavior has been described as psychopathic, aggressive and violent. He was also described as someone who made inappropriate sexual comments and as promiscuous, irresponsible and untrustworthy; someone who lacked forethought, concern for the future, the capacity for embarrassment, and the ability to make ethical decisions; and someone who lost all respect for social conventions. Although most of these behaviors are observed in patients with extensive bilateral damage of the prefrontal cortex, or patients with frontal dementia (degeneration of the frontal lobes), there is a question as to whether Gage really exhibited all these behaviors. There are reports on his physical and mental condition shortly before his death implying that his most serious mental changes were temporary, so that in later life he was far more functional, and socially far better adapted, than in the years immediately following his accident. Perhaps there is a neuropsychological explanation for Gage's recovery of function. Based on the few case studies we conducted in patients with unilateral (as opposed to bilateral) frontal lobe lesions, we found that women are more sensitive to left prefrontal damage whereas men are more sensitive to right prefrontal damage (Tranel et al., 2005). In other words, women who suffer left prefrontal damage exhibit impairments similar to patients with bilateral damage, which is characterized by many of the behavioral impairments listed above. Women with right side damage tend to recover and appear almost normal. The opposite is true for men. Men who suffer right prefrontal damage exhibit impairments similar to patients with bilateral damage. Those with only left side damage tend to recover well. It appears that Phineas Gage's brain damage was mostly related to the left prefrontal cortex (albeit Hanna Damasio and colleagues' reconstruction study suggested that the damage was bilateral). If Phineas Gage's brain damage was only on the left side, then this could explain the recovery of his behavioral impairments. Thus the stories of Phineas Gage may have been embellished, but they are not fundamentally wrong. Unquestionably, the case of Phineas Gage influenced neurology, psychology and philosophy discussions about the mind and brain, and particularly the debate on cerebral localization, and it was the first to suggest that damage to specific parts of the brain might induce specific personality changes.

REFERENCES

Ackerly, S.S. and Benton, A.L. (1948) Report of a case of bilateral frontal lobe defect. *Proceedings of the Association for Research in Nervous and Mental Disease* (Baltimore), 27: 479–504.

Anderson, S.W., Bechara, A., Damasio, H. et al. (1999) Impairment of social and moral behavior related to early damage in the human prefrontal cortex. *Nature Neuroscience,* 2 (11): 1032–7.

Bechara, A. (2004) The role of emotion in decision-making: evidence from neurological patients with orbitofrontal damage. *Brain and Cognition,* 55: 30–40.

Bechara, A. and Damasio, A.R. (2005) The somatic marker hypothesis: a neural theory of economic decision. *Games and Economic Behavior,* 52 (2): 336–72.

Bechara, A., Damasio, A.R., Damasio, H. and Anderson, S.W. (1994) Insensitivity to future consequences following damage to human prefrontal cortex. *Cognition,* 50: 7–15.

Bechara, A., Damasio, H., Tranel, D. and Anderson, S.W. (1998) Dissociation of working memory from decision making within the human prefrontal cortex. *Journal of Neuroscience,* 18: 428–37. Retrieved from <Go to ISI>://000071265600040

Bechara, A., Damasio, H., Tranel, D. and Damasio, A.R. (1997) Deciding advantageously before knowing the advantageous strategy. *Science,* 275 (5304): 1293–5. Retrieved from <Go to ISI>://A1997WK64400038

Bechara, A., Damasio, H., Tranel, D. and Damasio, A.R. (2005) The Iowa gambling task (IGT) and the somatic marker hypothesis (SMH): some questions and answers. *Trends in Cognitive Sciences,* 9 (4): 159–62.

Bechara, A., Dolan, S. and Hindes, A. (2002) Decision-making and addiction (Part II): myopia for the future or hypersensitivity to reward? *Neuropsychologia,* 40 (10): 1690–1705.

Bechara, A., Tranel, D., Damasio, H. and Damasio, A.R. (1996) Failure to respond autonomically to anticipated future outcomes following damage to prefrontal cortex. *Cerebral Cortex,* 6: 215–25.

Brickner, R.M. (1932) An interpretation of frontal lobe function based upon the study of a case of partial bilateral frontal lobectomy: localization of function in the cerebral cortex. *Proceedings of the Association for Research in Nervous and Mental Disease* (Baltimore), 13: 259–351.

Cacioppo, J., Berntson, G., Larsen, J. et al. (2000) The psychophysiology of emotion. In M. Lewis and J. Haviland-Jones (eds), *The Handbook of Emotion* (2nd edn). New York: Guiford. pp. 173–91.

Cacioppo, J., Klein, D.J., Berntson, G.G. and Hatfield, E. (1993) The psychophysiology of emotion. In M. Lewis and J.M. Haviland (eds), *The Handbook of Emotion* (1st edn). New York: Guilford. pp. 119–148.

Chudasama, Y. and Robbins, T.W. (2003) Dissociable contributions of the orbitofrontal and infralimbic cortex to Pavlovian autoshaping and discrimination reversal learning: further evidence for the functional heterogeneity of the rodent frontal cortex. *Journal of Neuroscience,* 23 (25): 8771–80.

Damasio, A.R. (1994) *Descartes' Error: Emotion, Reason, and the Human Brain*. New York: Grosset/Putnam.

Damasio, A.R., Tranel, D. and Damasio, H. (1990) Individuals with sociopathic behavior caused by frontal damage fail to respond autonomically to social stimuli. *Behavioural Brain Research,* 41 (2): 81–94.

Damasio, A.R., Tranel, D. and Damasio, H. (1991) Somatic markers and the guidance of behavior: theory and preliminary testing. In H.S. Levin, H.M. Eisenberg and A.L. Benton (eds), *Frontal Lobe Function and Dysfunction.* New York: Oxford University Press. pp. 217–29.

Damasio, H., Grabowski, T., Frank, R. et al. (1994) The return of Phineas Gage: clues about the brain from the skull of a famous patient. *Science,* 264: 1102–4.

Dunn, B.D. (2006) The somatic marker hypothesis: a critical evaluation. *Neuroscience and Biobehavioral Reviews,* 30 (2): 239–71. Retrieved from <Go to ISI>://WOS:000236130000008

Eslinger, P. and Damasio, A.R. (1985a) Behavioral disturbances associated with rupture of anterior communicating artery aneurysms. *Seminars in Neurology,* 4: 385–9.

Eslinger, P. and Damasio, A.R. (1985b) Severe disturbance of higher cognition after bilateral frontal lobe ablation: patient EVR. *Neurology,* 35 (12): 1731–41.

Fellows, L.K. and Farah, M.J. (2003) Ventromedial frontal cortex mediates affective shifting in humans: evidence from a reversal learning paradigm. *Brain,* 126: 1830–7.

Harlow, J.M. (1848) Passage of an iron bar through the head. *Boston Medical and Surgical Journal,* 39: 389–93.

Harlow, J.M. (1868) Recovery from the passage of an iron bar through the head. *Publications of the Massachusetts Medical Society,* 2: 327–47.

Lang, P.J., Bradley, M.M., Cuthbert, B.N. and Patrick, C.J. (1993) Emotion and psychopathology: a startle probe analysis. In L.J. Chapman, J.P. Chapman and D.C. Fowles (eds), *Experimental Personality and Psychopathology Research* (Vol. 16). New York: Spring Publishing Co. pp. 163–99.

Loewenstein, G.F., Weber, E.U., Hsee, C.K. and Welch, N. (2001) Risk as feelings. *Psychological Bulletin,* 127(2): 267–86.

Maia, T.V. and McClelland, J.L. (2004) A reexamination of the evidence for the somatic marker hypothesis: what participants really know in the Iowa Gambling Task. *Proceedings of the National Academy of Sciences,* 101: 16075–80.

Manes, F., Sahakian, B., Clark, L. et al. (2002) Decision-making processes following damage to the prefrontal cortex. *Brain,* 125: 624–39.

Martin, C., Denburg, N., Tranel, D. et al. (2004) The effects of vagal nerve stimulation on decision-making. *Cortex,* 40: 1–8.

Montague, P.R. and Berns, G.S. (2002) Neural economics and the biological substrates of valuation. *Neuron,* 36 (2): 265–84.

Montague, P.R., Dayan, P. and Sejnowski, T.J. (1996) A framework for mesencephalic dopamine systems based on predictive hebbian learning. *Journal of Neuroscience,* 16 (5): 1936–47.

Ongur, D. and Price, J.L. (2000) The organization of networks within the orbital and medial prefrontal cortex of rats, monkeys and humans. *Cerebral Cortex,* 10 (3): 206–19.

Pears, A., Parkinson, J.A., Hopewell, L. et al. (2003) Lesions of the orbitofrontal but not medial prefrontal cortex disrupt conditioned reinforcement in primates. *Journal of Neuroscience,* 23 (35): 11189–201.

Persaud, N., McLeod, P. and Cowey, A. (2007) Post-decision wagering objectively measures awareness. *Nature Neuroscience,* 10 (2): 257–61.

Rainville, P., Bechara, A., Naqvi, N. and Damasio, A.R. (2006) Basic emotions are associated with distinct patterns of cardiorespiratory activity. *International Journal of Psychophysiology,* 61 (1): 5–18. doi:10.1016/j.ijpsycho.2005.10.024

Saver, J.L. and Damasio, A.R. (1991) Preserved access and processing of social knowledge in a patient with acquired sociopathy due to ventromedial frontal damage. *Neuropsychologia,* 29: 1241–9.

Schultz, W., Dayan, P. and Montague, P.R. (1997) A neural substrate of prediction and reward. *Science,* 275: 1593–9.

Schwartz, N. and Clore, G.L. (1983) Mood, misattribution, and judgements of well-being: information and directive functions of affective states. *Journal of Personality and Social Psychology,* 45: 513–23.

Tomb, I., Hauser, M., Deldin, P. and Caramazza, A. (2002) Do somatic markers mediate decisions on the gambling task? *Nature Neuroscience,* 5 (11): 1103–4.

Tranel, D. (2000) Frontal lobe disorders. In A.E. Kazdin (ed.), *Encyclopedia of Psychology.* Oxford: Oxford University Press.

Tranel, D., Damasio, H., Denburg, N.L. and Bechara, A. (2005) Does gender play a role in functional asymmetry of ventromedial prefrontal cortex? *Brain,* 128: 2872–81. doi:10.1093/brain/awh643

Zajonc, R.B. (1984) On the primacy of affect. *American Psychologist,* 39: 117–23.

11 Revisiting Tulving et al.: Priming of semantic autobiographical knowledge: A case study of retrograde amnesia

Melanie J. Sekeres, Gordon Winocur and Morris Moscovitch

INTRODUCTION

In what has become a seminal paper since it was published as a case study in 1988, observations by Tulving and colleagues of patient K.C. have profoundly influenced memory research, challenging then current theories and changing the way we think about the nature of memory formation and its long-term representation.

A closed head injury at the age of 30 left K.C. with extensive damage to the medial temporal lobes (MTL) and to other cortical regions. Although neuropsychological testing showed that intellectual function was largely preserved, K.C.'s anterograde and retrograde memory were severely impaired. On the basis of early CT scans, his memory deficits were originally attributed to extensive damage throughout the frontal and temporal lobes (Moscovitch, 1982; Milner et al., 1984; Schacter, 1987). High resolution imaging of K.C. and other patients with similar impairments later suggested that his particular pattern of *anterograde* and *retrograde amnesia* is the result of damage to the MTL, including the hippocampus and related prefrontal and parietal cortical areas that are commonly activated during episodic memory retrieval (Rugg and Vilberg, 2013).

At the time of Tulving et al.'s case study, the impact that K.C. would have in addressing central issues in the field of memory research was not fully appreciated. For example, at that time, the consensual view of retrograde amnesia was that all declarative memories are temporally graded, with older memories surviving hippocampal damage. Tulving et al.'s observations of K.C. challenged this idea, but it was many years later that the implications of these observations for understanding key memory processes were recognized.

As the full range of research inspired by Tulving et al.'s paper is beyond the scope of this chapter (see the review by Rosenbaum et al., 2005, 2014), we will focus on the impact of this ground-breaking paper on research into multiple memory systems, memory consolidation theory, and the role of the MTL in spatial memory, autobiographical memory, recognition, and future imagining.

MULTIPLE MEMORY SYSTEMS: EPISODIC AND SEMANTIC MEMORY

In 1972, Tulving made the insightful observation that conscious *declarative memory* incorporates two distinctly different types of memory: the recollection and re-experiencing of contextually-defined personal events (*episodic memory*, i.e. recalling your visit to the Statue of Liberty), and memory for knowledge about the world and general facts (*semantic memory*, i.e. knowing that the Statue of Liberty is located in New York). Along with advancing our understanding of non-declarative (e.g. procedural) memory, Tulving's observation reinforced the principle of multiple memory systems that was beginning to receive wide acceptance. Against this background, Tulving et al.'s objective, in their initial investigation of K.C., was to explore the possibility of dissociating different types of memory in an individual whose damaged brain prevented the expression of normal memory processes. Examination of K.C.'s retrograde memory was particularly informative. Initially, it appeared that K.C. exhibited a *temporally-graded retrograde amnesia* (TGRA) with recently learned memories virtually abolished but those acquired years before the onset of injury relatively spared. Careful probing, however, revealed that this was too simplistic an interpretation. In exploring K.C.'s recall of premorbidly experienced autobiographical events, it became clear that all of his preserved memories were entirely semantic in nature, characterized by a general knowledge about the world and his personal life that was devoid of contextual detail and emotional reactivity. His selective incapacity for episodic memory was also reflected in his inability to form new memories (see also Sanders and Warrington, 1971; Kinsbourne and Wood, 1975).

K.C.'s pattern of anterograde memory loss was similar to that of the bitemporal patient H.M. (Scoville and Milner, 1957; Penfield and Milner, 1958). Following surgical removal of his hippocampus, H.M. developed a profound anterograde amnesia, yet intelligence, language and motor abilities were preserved. He was even able to acquire new skills, despite having no recollection of performing the tasks (Corkin, 1968; Milner et al., 1968). These dissociations were important, as they pointed to separate systems mediating memory, language and motor abilities. Also like K.C., H.M. displayed severe TGRA in that he recognized famous faces from many years prior to his MTL resection, but very few from the years leading up to, and following, his surgery (Marslen-Wilson and Teuber, 1975). This finding was taken as support for the idea that, if sufficient time passed, declarative memories no longer require the hippocampus; H.M.'s pattern of impaired recent memory but spared remote memory was attributed to the *age* of the memory rather than the *type* of memory.

Prompted by studies of other MTL amnesics (Nadel and Moscovitch, 1997), Suzanne Corkin systematically investigated H.M.'s retrograde memory and identified the same pattern of lost episodic memory and preserved semantic memory as seen in K.C. (Corkin, 2002; Steinvorth et al., 2005). Consistent with Tulving's (1972) assertion that semantic and episodic memories are mediated by different systems, H.M.'s recognition of people who had been famous for a long time could be supported by semantic knowledge, unlike more recently famous people for whom memories had not yet become semanticized. Like K.C., H.M. was unable to recall a specific event or episode from a given epoch.

A significant observation in Tulving et al.'s original study of K.C. was his ability to benefit from *semantic priming*, in which various events and semantic information from his past gradually became accessible to him. Initially, K.C. was completely unaware of his previous occupation, but after several weeks of interviews during which he was told about his former work at a manufacturing plant, K.C. successfully recognized semantic information related to his job. He could identify the company name and the name of his closest co-worker, his former role as a delivery truck driver and even describe the function of specialized manufacturing equipment. As well, despite having no recollection of experiencing them, K.C. eventually was able to acquire new memories that were semantically-related to his previous knowledge.[1] No such facilitation occurred for episodic events, offering further evidence that the two types of memory are mediated by different systems. Through the subsequent development of *schema assimilation* theories (Tse et al., 2007; van Kesteren et al., 2012; for a review see Ghosh and Gilboa, 2014), we now understand that K.C.'s ability to learn new concepts was likely due to the integration of new information into a schematic knowledge base supported by non-hippocampal regions (Sharon et al., 2011; see also discussion in Corkin, 2013 of similar abilities in H.M.).

The dissociation between episodic and semantic memory in amnesia provided neuropsychological evidence consistent with Tulving's distinction of two separable systems. This sparked new research into the neural mechanisms underlying this dissociation, and led to considerable theoretical debate, particularly with respect to the process of consolidating memories over the long-term (see also the discussion in Corkin, 2013).

MEMORY CONSOLIDATION AND TEMPORALLY-GRADED RETROGRADE AMNESIA

The concept of memory consolidation is based on the principle that memories need time to become 'fixed' in neuronal networks of the brain (Müller and Pilzecker, 1900; Burnham, 1904). Consolidation is thought to be a dual but continuous process, involving a reorganization of neuronal networks supporting a memory. The first stage, occurring in the seconds and minutes following encoding, involves a rapid *synaptic consolidation* (also called cellular consolidation), which requires protein synthesis-dependent biochemical and morphological changes in individual neurons, leading to long-lasting changes in the strength of synaptic connections of

[1]Evidence of intact semantic priming was observed also in patient H.M. (Keane et al., 1987, 1995; Corkin, 2002).

the neuronal circuitry (Davis and Squire, 1984; Kandel, 2001; Dudai, 2004). The second stage (*systems consolidation*), in which the memory becomes gradually reorganized and distributed across multiple brain regions, is a slower process that can take weeks, months, and possibly years following new learning (see Dudai, 2012, for a review). A critical question inspired by work with K.C. and other patients with MTL damage relates to the permanence of consolidated memories. Specifically, is a consolidated memory identical to the original version, or does the consolidation process entail a change in the nature of the memory? More broadly, people began to question the nature of memory consolidation, as certain observations of K.C.'s preserved and lost memories ran counter to the commonly held view of memory consolidation and challenged its core principles.

STANDARD CONSOLIDATION THEORY

The dominant view of systems-level memory consolidation at the time of Tulving et al.'s report, the *Standard Consolidation Theory* (SCT), proposed that the hippocampus is critical for the initial consolidation of all declarative memories (*equivalence* of event/episodic and fact/semantic information), although its direct role in storage and retrieval is time-limited (Scoville and Milner, 1957; Squire, 1992; Squire and Alvarez, 1995; Squire and Bayley, 2007). SCT holds that following initial encoding, the identical memory eventually becomes represented in neocortical regions (*duplication*), is immutable, and insensitive to hippocampal disruption (*resilience*). The TGRA observed in MTL patients was taken as strong support for this position (Scoville and Milner, 1957; Marslen-Wilson and Teuber, 1975; Rempel-Clower et al., 1996; Squire and Bayley, 2007).

Tulving et al.'s critical observation that K.C. was unable to recall very remote episodic memories posed a serious challenge to SCT. This finding, confirmed in subsequent studies of other MTL amnesics without accompanying damage to prefrontal and other neocortical regions (Damasio et al., 1985; Kopelman et al., 1989; Barr et al., 1990; Steinvorth et al., 2005; see Table 11.1), gave rise to new theoretical models that maintain that the pattern of retrograde amnesia in such patients is related not only to the age of the memory, but also to the type of memory being tested (Nadel and Moscovitch, 1997; Winocur and Moscovitch, 2011).

MULTIPLE TRACE THEORY

Multiple Trace Theory (MTT), as proposed by Nadel and Moscovitch (1997), challenged the premise that remote episodic and semantic memories are simply transferred to neocortical regions and protected from hippocampal damage. MTT introduced the idea that the frequency of retrieval or re-activation influences the degree to which memories are retained following subsequent MTL damage. A basic tenet of MTT is that after initial acquisition, interactions between the hippocampus and neocortical regions allow for the formation of new traces that add new information and strengthen the original memory. As the memory is repeatedly reactivated, the common featural elements, through statistical regularity, are incorporated into schematic/semantic memory networks in the neocortex, rendering those elements resistant to loss following damage to the network. The temporal and contextual details of the memory form distributed traces involving the hippocampus, and continue to depend on this structure for their storage and retrieval. According to this position, the extent of the gradient in episodic and semantic memory depends on the number of reactivations and the richness of the traces in the hippocampus and neocortex. Newly acquired memories, and infrequently retrieved old memories, will have formed few traces and are vulnerable to even modest hippocampal damage; memories with multiple traces are vulnerable only if damage to the hippocampus is extensive (for a review see Winocur and Moscovitch, 2011; Lane et al., 2014).

TRACE TRANSFORMATION THEORY

Building on MTT, Winocur and Moscovitch developed *Trace Transformation Theory* (TTT) (Winocur, Moscovitch and Bontempi, 2010; Winocur and Moscovitch, 2011), which posits that events are initially encoded as episodic memories in rich contextual detail. The hippocampus is essential to this process and necessary for the recall of such memories, *regardless of how long ago they were formed*. With time and experience, and by the process outlined in MTT, episodic memories are transformed into less detailed schematic (semantic) memories that capture the essential features or gist of the original, and are represented in neocortical regions. A central element of TTT is that hippocampus-dependent, episodic memories that are retained over time co-exist with schematic memories, and engage in a dynamic interplay in which the type of memory that is retrieved will depend on situational demands of the task. It follows that episodic memories will always be susceptible to the effects of hippocampal damage, hence the non-graded retrograde amnesia (RA) for this type of memory that is reliably seen in MTL patients. Schematic/semantic memories, represented as they are outside the MTL, are unaffected by lesions to MTL/hippocampus (but see Duff and Brown-Schmidt, 2012; Race et al., 2015).

TTT, like MTT, fits with Tulving's ideas of multiple memory systems in the brain, and offers a parsimonious explanation for the expression of both temporally-graded and non-graded RA following hippocampal damage (see Table 11.1). TTT also offers an alternative account of the fascinating *reconsolidation* phenomenon in which a consolidated memory can be brought back to a hippocampus-dependent state through a process of reactivation (Debiec et al., 2002; for a review see Tronson and Taylor, 2007; Dudai, 2012). The ability to reactivate a hippocampus-dependent memory suggests that the hippocampus, though not always required in memory retrieval, can be re-engaged. This is contrary to SCT, which asserts that once consolidated in the neocortex, a memory is no longer susceptible to hippocampal manipulation (Squire, 1992; Squire and Alvarez, 1995; Squire and Bayley, 2007). The reconsolidation phenomenon challenges the idea that consolidation is a stable, linear process involving complete hippocampal disengagement and an immutable cortical memory. Rather, it reveals consolidation to be a dynamic process, in which memories not only undergo reorganization, but can even be returned to an unstable state where they are again susceptible to interference or disruption (Nader et al., 2000; Nader and Hardt, 2009; Dudai, 2012).

The development of novel genetic neurobiological tools has been useful in identifying underlying neural mechanisms mediating consolidation and reconsolidation processes, but our ability to study these processes in the living human brain is still limited (Agren, 2014). As well, the existing literature has not adequately identified those mechanisms mediating the qualitative changes to a memory as it ages and reorganizes across neuronal networks in the hippocampus and the cortex. As this line of research continues to develop, it will be important to identify the molecular cascades and neuronal modifications mediating the development of schematic/semantic memory networks (for review, see Sekeres et al., in press).

Table 11.1 Cases of graded and non-graded retrograde amnesia following medial-temporal lobe damage. * denotes studies in which both graded and non-graded retrograde amnesia are reported.

Temporally-graded retrograde amnesia	Non-graded retrograde amnesia
Scoville and Milner (1957)	Sanders and Warrington (1971)
Marslen-Wilson and Teuber (1975)	Cermack and O'Conner (1983)*
Cermack and O'Conner (1983)*	Damasio et al.(1985)
Corkin (1984)	Tulving et al. (1988)
Damasio et al.(1985)	Warrington and McCarthy (1988)*
Warrington and McCarthy (1988)*	Barr et al. (1990)
O'Conner et al. (1992)*	Victor (1990)
Kartsounis et al. (1995)*	O'Conner et al. (1992)*
Rempel-Clower et al. (1996)	Kartsounis et al. (1995)*
Reed and Squire (1998)*	Hirano and Noguchi (1998)
Kapur and Brooks (1999)	Kopelman et al. (1999)*
Kopelman et al. (1999)*	Viskontas et al. (2000)*
Viskontas et al. (2000)*	Cipolotti et al. (2001)
Bayley et al. (2003)	Steinvorth et al. (2005)
Rosenbaum et al. (2005)*	Rosenbaum et al. (2005)*
Bayley et al. (2006)	Bright et al. (2006)*
Wais et al. (2006)	Gilboa et al. (2006)*
Bright et al. (2006)*	Maguire et al. (2006)*
Gilboa et al. (2006)*	Chan et al. (2007)
Maguire et al. (2006)*	Noulhiane et al. (2007)
Hepner et al. (2007)*	Hepner et al. (2007)*
Hassabis et al. (2007)*	Hassabis et al. (2007)*

(Continued)

Table 11.1 (Continued)

Temporally-graded retrograde amnesia	Non-graded retrograde amnesia
Rosenbaum et al. (2008)*	Rosenbaum et al. (2008)*
Kirwan et al. (2008)	Waidergoren et al. (2012)*
Squire et al. (2010)	
Andelman et al. (2010)	
Waidergoren et al. (2012)*	
Smith (2014)	
Witt et al. (2015)	

REMOTE SPATIAL MEMORY

Tulving and colleagues' work with K.C. provided early evidence that under certain conditions, *allocentric spatial memory* is preserved following MTL damage. This observation, confirmed in studies of other MTL amnesics (Beatty et al., 1987; Teng and Squire, 1999; Corkin, 2002; Maguire et al., 2006), conflicted with the widely held view that the hippocampus plays an essential role in forming, retaining and using cognitive maps that mediate memories based on allocentric spatial relationships (O'Keefe and Nadel, 1978; Morris, 1984; Maguire et al., 1996).

Tulving et al. showed that K.C. could produce an adequate floorplan of his childhood house, though he could not identify his bedroom and other landmarks within the house. K.C. was also able to navigate successfully in his old, familiar neighborhood.[2] Rosenbaum and colleagues (2000) probed K.C.'s knowledge of distances, direction, landmarks and routes in that neighborhood, as well as his knowledge of well-known geographical locations and landmarks, to determine whether he was relying on a detailed, internal representation of the environment or on a more schematized topographical representation. Consistent with the latter interpretation, K.C. performed well on all spatial tasks, including those that ostensibly rely on cognitive maps, and recognized major, salient landmarks, but was unable to recognize or locate detailed elements of the environment (i.e. individual buildings). As with his preserved autobiographical memory, it appeared that his spared spatial abilities relied on a more general topographical, allocentric representation of the neighborhood (see Figure 11.1).

Functional neuroimaging studies of healthy controls have identified cortical areas, including the posterior parahippocampal cortex, posterior cingulate cortex, retrosplenial cortex, and the precuneus, that mediate retrieval of remote spatial memories (Hirshhorn et al., 2012). K.C. had preserved tissue in some of these key

[2]Similar findings of preserved spatial memory following MTL damage were observed in patients H.M. (Corkin, 2002), E.P. (Teng and Squire, 1999), and T.T. (Maguire et al., 2006).

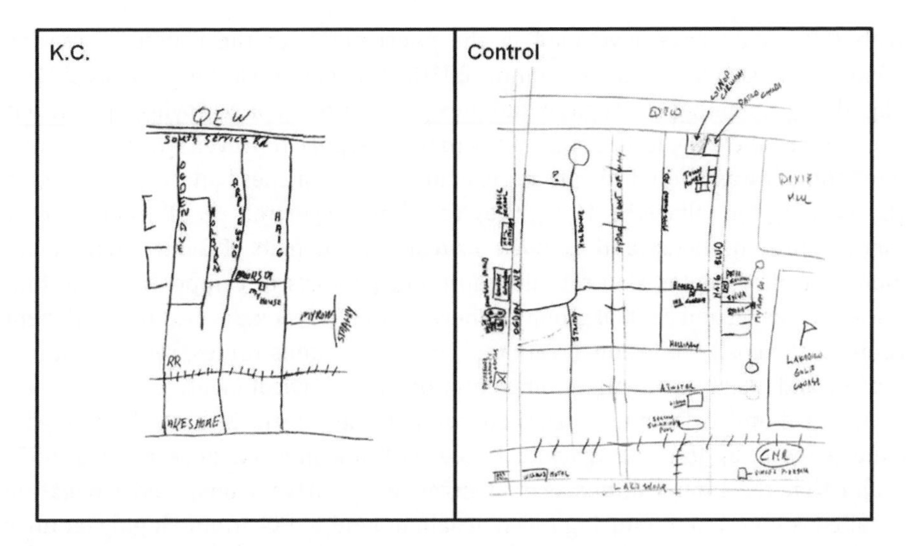

Figure 11.1 A map of his neighborhood drawn by K.C. (left) and by a healthy control (right) who moved away around the time of K.C.'s brain injury (adapted from Rosenbaum et al., 2005)

regions, including the parhippocampal and posterior cingulate cortex and the right parietal lobe, which may have supported his preserved topographical memory[3] (Rosenbaum, Winocur, et al., 2004).

Studies involving rodents have identified increased activity in the prefrontal cortex, notably the anterior cingulate cortex, and decreased activity in the hippocampus during retrieval of remote spatial memories. This finding suggests a time-dependent reorganization of the neural circuitry supporting the memory, which itself is likely to have been transformed in accordance with the representational capacities of the new tissue (Bontempi et al., 1999; Maviel et al., 2004). It is important to note that these studies did not directly assess the *quality* of remote spatial memory. Thus the question remains as to whether, like K.C., rodents rely more on cortically-based schematic representations in remote spatial memory tasks than on discrete local contextual cues that depend on the hippocampus.

To address this question, Winocur and colleagues modelled the conditions of K.C.'s familiar neighborhood by rearing rats for several months in a complex 'village' environment (Winocur et al., 2005; Winocur, Moscovitch et al., 2010). While living in the village, the rats experienced the various locations and spatial relationships from many perspectives, permitting them to develop a cognitive map of the environment. Lesioning the hippocampus *after* village-rearing did not impair their ability to find specific locations during subsequent memory testing. When the quality of the memory was carefully probed, it became clear that hippocampal lesioned rats were relying on schematic representations which, consistent

[3]E.P. had extensive damage to the parahippocampal cortex, but retained some remote spatial memory (Teng and Squire, 1999).

with TTT, would have developed in the neocortex over the course of rearing (Winocur, Moscovitch and Bontempi, 2010; Winocur, Moscovitch et al., 2010). Slightly manipulating local contextual cues (i.e. relocating or removing a particular cue) did not affect the lesion rats' ability to find the goal; however, when the spatial environment was altered such that the configuration of cues had to be re-mapped (i.e. rotating the village by 180°), they could no longer rely on allocentric schematic representations, and showed navigational deficits. Placing 'roadblocks' along specified routes also revealed different patterns of compensation in hippocampal lesion and control groups. Whereas controls took the next most efficient route to the goal, the lesioned rats took more circuitous routes, revealing a less flexible and integrated spatial representation of the environment.

Maguire and colleagues observed comparable effects in a study of highly experienced London taxi drivers. Using PET imaging, Maguire et al. (1997) found that the hippocampus was engaged when drivers imagined navigating detailed routes, but not during recall of salient landmarks around highly familiar environments. A subsequent study involving a taxi driver with bilateral hippocampal lesions determined that a functional hippocampus was not required for navigation between highly familiar locations if using 'main artery' routes, suggesting semanticized versions of spatial memories that were supported by extra-hippocampal cortical regions. When forced to navigate between these same familiar locations using alternate, side street routes, the patient was severely impaired (Maguire et al., 2006). In a related fMRI study, Hirshhorn and colleagues investigated changes in neural representations as spatial memories become increasingly familiar. Participants who had recently moved to Toronto highly engaged the hippocampus during a spatial navigation task in the city, but showed little hippocampal activity on the same task one year later when they were much more familiar with the spatial layout of the city (Hirshhorn et al., 2012). As the neural representations supporting the spatial memories changed over time, it is likely that they also underwent a qualitative change, becoming more schematic in nature.

Together, these findings in rodents and humans provide evidence that different memory systems support qualitatively different types of spatial memory that develop with time and experience. While non-hippocampal schematic spatial representations are sufficient to support successful navigation in many situations, the memories lack the relational quality required for efficient navigation (O'Keefe and Nadel, 1978) and the fine detail required for in-depth recollection (Corkin, 2002; Rosenbaum, Winocur et al., 2004; Rosenbaum, Ziegler et al., 2004; Spiers and Maguire, 2006).

ASSESSING AUTOBIOGRAPHICAL MEMORY

Tulving et al.'s observation that K.C.'s memory for detailed autobiographical experiences was virtually destroyed while his ability to recall semantic information was relatively intact, highlighted the need for better instruments to assess autobiographical memory. Kopelman and colleagues (1989) developed a structured interview task (Autobiographical Memory Interview; AMI), which was designed to assess the qualitative content of autobiographical memory from different time points (childhood, early adult life, recent events). In amnesic patients, they found impoverished episodic *and* semantic detail relative to controls, with a slight temporal gradient for both types of information. The AMI may not distinguish adequately between highly-detailed episodic memories that require the hippocampus, and less-detailed memories that depend more on extra-hippocampal structures. Moreover, in most people's narratives of personal events, episodic and semantic elements are intermingled, and do not fall neatly into the two categories specified by the AMI.

Levine and colleagues addressed these issues with their Autobiographical Interview (AI), which scores each narrative according to the details reported in the memory (Levine et al., 2002). The AI categorizes memory details as either internal/unique to the memory episode (time, place, event, perceptual, emotion thought information) or external (information related to the episode but not unique to it, or general comments about the episode). K.C.'s scores on both tests reflected his severely impaired recollection of recently acquired memories and confirmed his anterograde deficit. When his remote memories were probed, he exhibited temporally graded semantic memory, but was unable to recall a single autobiographical event that was specific to time or place from any point in his life (non-graded) (Rosenbaum, Winocur et al., 2004; Rosenbaum et al., 2005; Gilboa et al., 2006). Interestingly, K.C. did benefit from cuing/probing on the AI test. Similar to the semantic priming effects described above, however, the cues facilitated the retrieval of highly rehearsed memories, and suggest that any facilitation of K.C.'s episodic memory likely reflects the retrieval of a semanticized version of his autobiographical memory (Rosenbaum et al., 2005).

Addis, Moscovitch and McAndrews (2007) used the AI to probe the quality of autobiographical memory in MTL epilepsy patients and observed poor retrieval of internal (episodic) details relative to healthy controls, but no deficit in the number of external (semantic) details provided. This was accompanied by a reduction in fMRI hippocampal activity, and increased activation in cortical areas of the *recollection network*, which was interpreted as compensatory activity supporting the retrieval of semantic memory. Others have used the AMI and AI to test remote memory in patients with MTL damage, and similarly reported impaired episodic memory but spared semantic memory, reinforcing the importance of the hippocampus for detailed autobiographical memories (Viskontas et al., 2000; Steinvorth et al., 2005; Gilboa et al., 2006; St-Laurent et al., 2009; but see also Squire and Bayley, 2007, for an alternative interpretation which asserts that the extent of cortical damage extending *beyond* the hippocampus and adjacent MTL determines the degree of temporally-graded autobiographical memory loss).

Rosenbaum and colleagues found that the amount of MTL damage sustained was related to the degree of impairment in recall of internal/episodic detail for both recent and remote autobiographical memories, with extensive bilateral hippocampal damage producing the most severe deficits, and smaller hippocampal lesions producing less severe retrograde amnesia regardless of the extent of extra-hippocampal damage. The degree of hippocampal damage was unrelated to the amount of external/semantic detail recalled at any time point (Rosenbaum et al., 2008), again supporting Tulving's idea that different neural systems mediated episodic and semantic memory. It should be noted that semantic memory impairments are often observed in patients with damage extending to the lateral and anterior temporal lobe (TL), including those with frontotemporal dementia, suggesting a dissociation between the TL and MTL in representing semantic memory (Snowden et al., 1989; Hodges et al., 1992; Mummery et al., 2000; McKinnon et al., 2006).

RECOGNITION MEMORY: RECOLLECTION AND FAMILIARITY

Work on *recognition memory* also points to the conclusion that episodic and semantic memories are controlled by different systems. In early work, Tulving (1985) showed that recognition memory based on *recollection* (judging studied word items as 'remembered') declines dramatically over the course of a week, whereas recognition based on *familiarity* (judging an item as 'know') shows little decline. In accounting for these and similar findings (Wixted, 2007; Wixted and Mickes, 2010), Yonelinas (2002) proposed a dual process model in which recognition memory can reflect two distinct processes: recollection, which involves a 'sense of reliving', or a reinstantiation of contextual and perceptual details related to the cued event, and relies on the hippocampus and parahippocampal cortex (Aggleton et al., 2005); and familiarity, which involves judging that a cued event has previously occurred based on a feeling of 'knowing', and is thought to rely on the perirhinal cortex (Brown and Aggleton, 2001; Yonelinas, 2002; Norman and O'Reilly, 2003; Montaldi et al., 2006; Staresina et al., 2012). A 'remember/know' judgment is typically used to distinguish between recollection and familiarity (Tulving, 1985). While a 'remembered' item may be *both* recollected and familiar, a 'know' item lacks contextual details, and is assumed to be based solely on familiarity (Mickes et al., 2013; Rugg and Vilberg, 2013).

Several studies have suggested that different MTL regions play unique roles during recollection, with the parahippocampal cortex representing contextual information, the perirhinal cortex representing object information, and the hippocampus binding together context and object information during successful recollection (Diana et al., 2007; Montaldi and Mayes, 2010). Patients with damage limited to the hippocampus can successfully perform recognition tests on the basis of familiarity, but are impaired when recollection is required (Holdstock et al., 2002; Mayes et al., 2002; Barbeau et al., 2005). Given the critical role of the hippocampus in recollection, it follows that such patients should be impaired on recognition tests of recent and remote memory when recollection is required. While Tulving et al.'s original study did not compare recollection and familiarity-based recognition memory, it is now clear that a deficit in recollection was at the root of K.C.'s profoundly impaired remote autobiographical episodic memory. In a direct test of this idea, Gilboa et al. used the AI to compare recollective and recognition memory of retrograde events in K.C. (who had extensive damage throughout the MTL) and patient A.D. (who had damage restricted to the bilateral fornix). While K.C. exhibited impairment in both recollection and recognition memory, patient A.D. exhibited impaired recollection but normal familiarity-based recognition (Gilboa et al., 2006). This dissociation between recollection and familiarity following fornix lesions suggests that the extended hippocampal system is required for re-experiencing autobiographical memories, but not for recent or remote memories that are remembered on the basis of familiarity alone which is mediated by other MTL structures.

While these studies provide strong evidence for the role of the hippocampus in recollection, it remains unclear whether familiarity is also partially dependent on the hippocampus, or whether the two processes are mediated by separate temporal lobe structures. Work with patient N.B., who underwent a surgical resection of the left-anterior temporal lobe which removed the perirhinal cortex (sparing the hippocampus), provides strong support for the latter position. N.B. exhibited a selective impairment of familiarity-based memory, but preserved recollection (Bowles et al., 2007; Martin et al., 2011). This distinction has been compared to the distinction between episodic and semantic memory (Gardiner, 2001; Tulving et al., 2001), and in line with this, N.B.'s impaired familiarity processing is consistent with reports of semantic dementia in patients with anterior temporal lobe damage (Hodges et al., 1992; Mummery et al., 2000). Together, the findings of investigations on recognition memory support Tulving's idea of multiple memory systems, where the hippocampally-mediated recollective/episodic memory is sensitive to hippocampal damage, whereas the familiarity/semantic memory, which relies on extra-hippocampal cortical regions, is resistant to hippocampal disruption.

Following on from the principle of multiple systems, Sadeh and colleagues (2013) proposed the idea that recollected and familiarity-based memories are forgotten via different mechanisms. They suggest that, in the intact brain, forgetting can occur as a result of *decay* or *interference*, depending on the neural architecture and mnemonic processes supporting recognition memory (see review by Hardt et al., 2013). The hippocampus has unique pattern separation properties, in which sparse non-overlapping encoding processes enable it to distinguish between similar memories (Leutgeb et al., 2007; Bakker et al., 2008). Therefore, recollected memories relying on the hippocampus are less likely to interfere with one another, but may decay quickly. The perirhinal cortex lacks this pattern separation ability, and therefore familiarity-based memories relying on this structure are more likely to interfere with each other. According to this view, memory in patients with hippocampal lesions should be especially vulnerable to the effects of interference, as they are unable to rely on recollective processes to distinguish between similar memories (see Figure 11.2; for a comprehensive review of the literature on the dual process theory and its potential implications for differential mechanisms of forgetting, see Sadeh et al., 2013). In line with this idea, when interference was

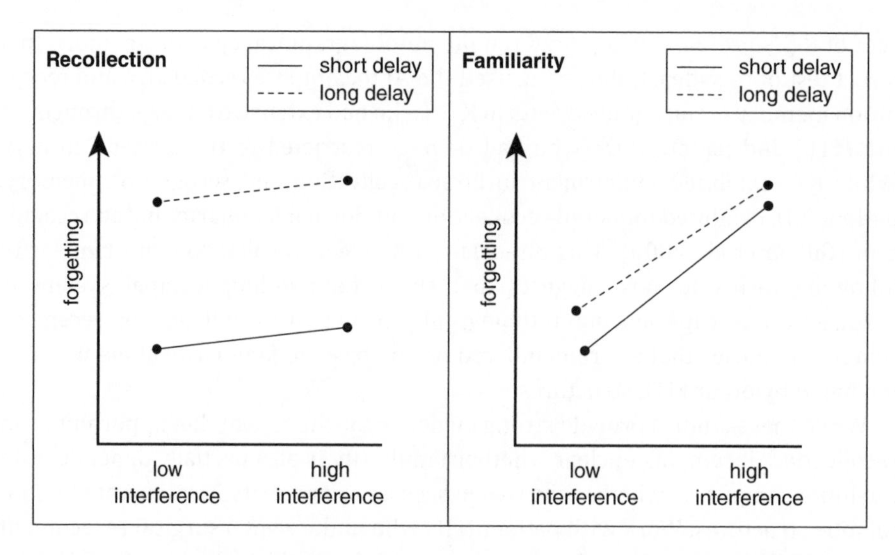

Figure 11.2 Hypothesized effects of interference and decay on recollection and familiarity – based memories (adapted from Sadeh et al., 2013). Recollection is vulnerable to decay over time, and is less sensitive to interference. The opposite pattern applies to familiarity.

minimal, K.C. was able to encode and retain new semantic knowledge (albeit over prolonged and repeated training sessions). When interference was introduced between cuing and subsequent retrieval, however, the disruptive effects of interference on K.C.'s non-hippocampally-mediated recognition memory led to predictable impairment (Hayman et al., 1993). It is possible that the repetitive training allowed a familiarity-based semantic memory representation to gradually develop. According to Sadeh et al. (2013), in the absence of interference, such a representation would have been accessible to K.C. during the cued recognition test, and could have supported his successful recognition memory.

PROSPECTIVE THINKING

The ability to recollect specific past experiences has been shown to be related to the process of imagining the future. One of K.C.'s deficits, briefly noted in the 1988 paper, was his complete inability to project himself into the future: as Tulving et al. stated, '"*He seems to live in a "permanent present"*"' (1988: 14). A similar inability to think prospectively was documented in patient H.M., and is discussed in the book *Permanent Present Tense* (Corkin, 2013). Subsequent study of K.C.'s inability to imagine himself in the future revealed an interesting distinction. He was able to generate plausible and relatively detailed future scenarios, but when instructed to imagine *himself* in one of these future scenarios, he was unable to do so[4]

[4]Amnesic patient D.B. exhibited a similar ability to imagine future public events, but an inability to prospectively imagine personal events (Klein et al., 2002).

(Rosenbaum et al., 2005, 2009), emphasizing a dissociation between episodic and semantic elements in both recollection and prospective thinking. Tulving has long argued that the episodic memory system supports both retrospective and prospective 'mental time travel' (Tulving, 1985; Wheeler et al., 1997). As in autobiographical memory retrieval, MTL patients' attempts at future simulation typically lack internal detail, suggesting that both processes are mediated by common underlying mechanisms (Kopelman et al., 1989; Gilboa et al., 2006; Rosenbaum et al., 2008).

Functional neuroimaging studies have revealed similarities in the core networks supporting the recollection of past events and the imagining of future events. These networks include the medial prefrontal cortex, lateral and medial temporal regions (hippocampus and parahippocampal cortex), and lateral and medial parietal regions (precuneus and retrosplenial cortex) (Okuda et al., 2003; Addis, Wong and Schacter, 2007; Schacter et al., 2007; Hassabis and Maguire, 2009; Spreng and Grady, 2010; Addis and Schacter, 2011). Many of these regions overlap with critical nodes of the *default mode network* (DMN) that is engaged during passive, internally-focused cognition, or rest (see Figure 11.3)

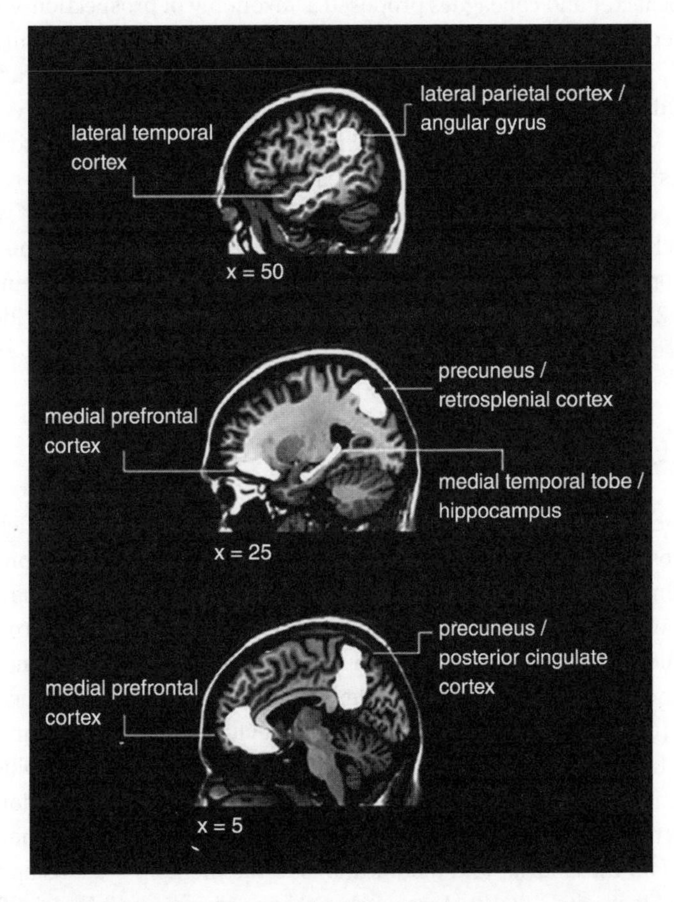

Figure 11.3 Network of common brain regions involved in memory recollection, future imagining, and the default-mode network

(Raichle et al., 2001; Buckner and Carroll, 2007; Buckner et al., 2008; Spreng et al., 2009; Andrews-Hanna et al., 2010). The overlap between the recollection network and the DMN has led to much speculation and research into what the brain is actually doing at rest. One possibility is that, during rest or mind wandering, the brain is engaged in recollection as well as prospective thinking (including planning and imagining), with both processes involving midline DMN regions (Spreng and Grady, 2010). Recent data from Irish et al. (2012) suggest that different portions of the DMN/recollection network may be differentially activated during each task, with the anterior temporal lobe necessary for future thought but not for remote memories. Further investigation is needed to untangle precise networks mediating recollection and future imagining, as well as the underlying synaptic plasticity required to support prospective thinking in the healthy brain.

Schacter and colleagues propose that this common network of brain regions adaptively functions to integrate and recombine associations from past experiences stored in episodic memory in order to predict possible future situations (Schacter and Addis, 2007a, b; see also Schacter, 2012, for further discussion of the 'constructive episodic simulation hypothesis'). Recognizing the complexity of the concept of prospective cognition, Schacter and colleagues proposed a 'taxonomy of prospection' which incorporates affective forecasting, prospective memory, temporal discounting, episodic simulation and autobiographical planning, and outlines modes of future thinking which include simulation, prediction, intention and planning (Szpunar et al., 2014). These ideas stress the importance of drawing upon stored episodic and semantic representations in order to predict or simulate future situations. Given this relationship, it is not surprising that disruption of key nodes of the network, several of which were damaged in K.C., produces severe deficits in both retrospective and prospective thinking, while sometimes sparing those aspects that depend on semantic memory (Kwan et al., 2012, 2013). As well, this work points to a neural basis for Tulving's ideas linking retrospective and prospective mental time travel to the episodic memory system.

CONCLUSION

In the 27 years since the publication of Tulving's case study of K.C., the field of cognitive neuroscience has expanded dramatically. The seminal observation that K.C.'s profound amnesia for context-specific events (episodic memory) did not extend to general knowledge of a public or personal nature (semantic memory) provided compelling evidence for Tulving's concept of multiple memory systems, each one defined by different characteristics and its own underlying neural network and process-specific assemblies (Cabeza and Moscovitch, 2013; Moscovitch, Cabeza, Winocur and Nadel, 2016). The timely development of modern neuroimaging techniques, which enables identification of the coordinated neural networks that comprise these various systems, provided further support for this principle and changed the course of memory research (for updates on these ideas, and modifications to the memory systems approach, see Cabeza and Moscovitch, 2013; Moscovitch, Cabeza, Winocur and Nadel, 2016). More recently, the emergence of sophisticated genetic and neurobiological tools has uncovered cellular and molecular mechanisms that point to memory as a dynamic process that impacts and interacts with other cognitive operations. K.C. died in 2014, but his legacy, fostered initially by Tulving's observations and enlarged by the work of subsequent investigators, will continue to influence the study of memory as the field evolves.

ACKNOWLEDGMENTS

This chapter was written with grant support to GW and MM from the Canadian Institutes for Health Research (CIHR) and the Natural Sciences and Engineering Research Council of Canada. MJS was supported by a post-doctoral fellowship from CIHR.

REFERENCES

Addis, D.R., Moscovitch, M. and McAndrews, M.P. (2007) Consequences of hippocampal damage across the autobiographical memory network in left temporal lobe epilepsy. *Brain,* 130 (Pt 9): 2327–42.

Addis, D.R. and Schacter, D.L. (2011) The hippocampus and imagining the future: where do we stand? *Frontiers in Human Neuroscience*, 5.

Addis, D.R., Wong, A.T. and Schacter, D.L. (2007) Remembering the past and imagining the future: common and distinct neural substrates during event construction and elaboration. *Neuropsychologia,* 45 (7): 1363–77.

Aggleton, J. P., Vann, S. D., Denby, C. et al. (2005) Sparing of the familiarity component of recognition memory in a patient with hippocampal pathology. *Neuropsychologia,* 43 (12): 1810–23.

Agren, T. (2014) Human reconsolidation: a reactivation and update. *Brain Research Bulletin,* 105: 70–82.

Andelman, F., Hoofien, D., Goldberg, I. et al. (2010) Bilateral hippocampal lesion and a selective impairment of the ability for mental time travel. *Neurocase,* 16 (5): 426–35.

Andrews-Hanna, J.R., Reidler, J.S., Sepulcre, J. et al. (2010) Functional-anatomic fractionation of the brain's default network. *Neuron,* 65 (4): 550–62.

Bakker, A., Kirwan, C.B., Miller, M. and Stark, C.E. (2008) Pattern separation in the human hippocampal CA3 and dentate gyrus. *Science,* 319 (5870): 1640–2.

Barbeau, E.J., Felician, O., Joubert, S. et al. (2005) Preserved visual recognition memory in an amnesic patient with hippocampal lesions. *Hippocampus,* 15 (5): 587–96.

Barr, W., Goldberg, E., Wasserstein, J. and Novelly, R. (1990) Retrograde amnesia following unilateral temporal lobectomy. *Neuropsychologia,* 28 (3): 243–55.

Bayley, P.J., Hopkins, R.O. and Squire, L.R. (2003) Successful recollection of remote autobiographical memories by amnesic patients with medial temporal lobe lesions. *Neuron,* 38 (1): 135–44.

Bayley, P.J., Hopkins, R.O. and Squire, L.R. (2006) The fate of old memories after medial temporal lobe damage. *Journal of Neuroscience,* 26 (51): 13311–17.

Beatty, W.W., Salmon, D.P., Bernstein, N. and Butters, N. (1987) Remote memory in a patient with amnesia due to hypoxia. *Psychological Medicine,* 17 (3): 657–65.

Bontempi, B., Laurent-Demir, C., Destrade, C. and Jaffard, R. (1999) Time-dependent reorganization of brain circuitry underlying long-term memory storage. *Nature,* 400 (6745): 671–5.

Bowles, B., Crupi, C., Mirsattari, S.M. et al. (2007) Impaired familiarity with preserved recollection after anterior temporal-lobe resection that spares the hippocampus. *Proceedings of the National Academy of Sciences,* 104 (41): 16382–7.

Bright, P., Buckman, J., Fradera, A. et al. (2006) Retrograde amnesia in patients with hippocampal, medial temporal, temporal lobe, or frontal pathology. *Learning & Memory,* 13 (5): 545–57.

Brown, M.W. and Aggleton, J.P. (2001) Recognition memory: what are the roles of the perirhinal cortex and hippocampus? *Nature Reviews Neuroscience,* 2 (1): 51–61.

Buckner, R.L. and Carroll, D.C. (2007) Self-projection and the brain. *Trends in Cognitive Sciences,* 11 (2): 49–57.

Buckner, R.L., Andrews-Hanna, J.R. and Schacter, D.L. (2008) The brain's default network. *Annals of the New York Academy of Sciences,* 1124 (1): 1–38.

Burnham, W.H. (1904) Retroactive amnesia: illustrative cases and a tentative explanation. *American Jounral of Psychology,* 14: 382–96.

Cabeza, R. and Moscovitch, M. (2013) Memory systems, processing modes, and components functional neuroimaging evidence. *Perspectives on Psychological Science,* 8 (1), 49–55.

Cermak, L.S. and O'Connor, M. (1983) The anterograde and retrograde retrieval ability of a patient with amnesia due to encephalitis. *Neuropsychologia,* 21 (3): 213–34.

Chan, D., Henley, S.M., Rossor, M.N. and Warrington, E.K. (2007) Extensive and temporally ungraded retrograde amnesia in encephalitis associated with antibodies to voltage-gated potassium channels. *Archives of Neurology,* 64 (3): 404–10.

Cipolotti, L., Shallice, T., Chan, D. et al. (2001) Long-term retrograde amnesia . . . the crucial role of the hippocampus. *Neuropsychologia,* 39 (2): 151–72.

Clark, R.E., Broadbent, N.J. and Squire, L.R. (2005) Hippocampus and remote spatial memory in rats. *Hippocampus,* 15 (2): 260–72.

Corkin, S. (1968) Acquisition of motor skill after bilateral medial temporal-lobe excision. *Neuropsychologia,* 6 (3): 255–65.

Corkin, S. (1984) *Lasting consequences of bilateral medial temporal lobectomy: clinical course and experimental findings in HM.* Paper presented at the Seminars in Neurology.

Corkin, S. (2002) What's new with the amnesic patient H.M.? *Nature Reviews Neuroscience,* 3 (2).

Corkin, S. (2013) *Permanent Present Tense: The Unforgettable Life of the Amnesic Patient* (Vol. 1000). New York: Basic Books.

Damasio, A.R., Eslinger, P.J., Damasio, H. et al. (1985) Multimodal amnesic syndrome following bilateral temporal and basal forebrain damage. *Archives of Neurology,* 42 (3): 252–9.

Davis, H.P. and Squire, L.R. (1984) Protein synthesis and memory: a review. *Psychological Bulletin,* 96 (3): 518–59.

Debiec, J., LeDoux, J.E. and Nader, K. (2002) Cellular and systems reconsolidation in the hippocampus. *Neuron,* 36 (3): 527–38.

Diana, R.A., Yonelinas, A.P. and Ranganath, C. (2007) Imaging recollection and familiarity in the medial temporal lobe: a three-component model. *Trends in Cognitive Sciences,* 11 (9): 379–86.

Dudai, Y. (2012) The restless engram: consolidations never end. *Annual Review of Neuroscience,* 35: 227–47.

Duff, M.C. and Brown-Schmidt, S. (2012) The hippocampus and the flexible use and processing of language. *Frontiers in Human Neuroscience,* 6.

Gardiner, J.M. (2001) Episodic memory and autonoetic consciousness: a first–person approach. *Philosophical Transactions of the Royal Society B: Biological Sciences,* 356 (1413): 1351–61.

Ghosh, V.E. and Gilboa, A. (2014) What is a memory schema? A historical perspective on current neuroscience literature. *Neuropsychologia*, 53: 104–14.

Gilboa, A., Winocur, G., Rosenbaum, R.S. et al. (2006) Hippocampal contributions to recollection in retrograde and anterograde amnesia. *Hippocampus,* 16 (11): 966–80.

Hardt, O., Nader, K. and Nadel, L. (2013) Decay happens: the role of active forgetting in memory. *Trends in Cognitive Sciences*, 17(3): 111–20.

Hassabis, D. and Maguire, E.A. (2009) The construction system of the brain. *Philosophical Transactions of the Royal Society B: Biological Sciences,* 364 (1521): 1263–71.

Hayman, C.G., Macdonald, C.A. and Tulving, E. (1993) The role of repetition and associative interference in new semantic learning in amnesia: a case experiment. *Journal of Cognitive Neuroscience,* 5 (4): 375–89.

Hepner, I.J., Mohamed, A., Fulham, M.J. and Miller, L.A. (2007) Topographical, autobiographical and semantic memory in a patient with bilateral mesial temporal and retrosplenial infarction. *Neurocase,* 13 (2): 97–114.

Hirshhorn, M., Grady, C., Rosenbaum, R.S. et al. (2012) Brain regions involved in the retrieval of spatial and episodic details associated with a familiar environment: an fMRI study. *Neuropsychologia,* 50 (13): 3094–106.

Hodges, J.R., Patterson, K., Oxbury, S. and Funnell, E. (1992) Semantic dementia. *Brain,* 115 (6): 1783–1806.

Holdstock, J., Mayes, A., Roberts, N. et al. (2002) Under what conditions is recognition spared relative to recall after selective hippocampal damage in humans? *Hippocampus,* 12 (3): 341–51.

Irish, M., Addis, D. R., Hodges, J. R. and Piguet, O. (2012) Considering the role of semantic memory in episodic future thinking: evidence from semantic dementia. *Brain,* 135 (7): 2178–91.

Kandel, E.R. (2001) The molecular biology of memory storage: a dialogue between genes and synapses. *Science,* 294 (5544): 1030–8.

Kapur, N. and Brooks, D.J. (1999) Temporally-specific retrograde amnesia in two cases of discrete bilateral hippocampal pathology. *Hippocampus,* 9 (3): 247–54.

Keane, M., Gabrieli, J. and Corkin, S. (1987) *Multiple relations between fact-learning and priming in global amnesia.* Paper presented at the Society for Neuroscience Abstracts.

Keane, M., Gabrieli, J.D., Mapstone, H.C. et al. (1995) Double dissociation of memory capacities after bilateral occipital-lobe or medial temporal-lobe lesions. *Brain,* 118 (5): 1129–48.

Kinsbourne, M. and Wood, F. (1975) Short-term memory processes and the amnesic syndrome. In D. Deutsch and J.A. Deutsch (eds), *Short-term Memory.* New York: Academic. pp. 258–91.

Kirwan, C.B., Wixted, J.T. and Squire, L.R. (2008) Activity in the medial temporal lobe predicts memory strength, whereas activity in the prefrontal cortex predicts recollection. *Journal of Neuroscience,* 28 (42): 10541–8.

Klein, S.B., Loftus, J. and Kihlstrom, J.F. (2002) Memory and temporal experience: the effects of episodic memory loss on an amnesic patient's ability to remember the past and imagine the future. *Social Cognition,* 20 (5): 353–79.

Kopelman, M., Wilson, B. and Baddeley, A. (1989) The autobiographical memory interview: a new assessment of autobiographical and personal semantic memory in amnesic patients. *Journal of Clinical and Experimental Neuropsychology,* 11 (5): 724–44.

Kopelman, M.D., Stanhope, N. and Kingsley, D. (1999) Retrograde amnesia in patients with diencephalic, temporal lobe or frontal lesions. *Neuropsychologia,* 37 (8): 939–58.

Kwan, D., Craver, C.F., Green, L. et al. (2012) Future decision-making without episodic mental time travel. *Hippocampus,* 22 (6): 1215–19.

Kwan, D., Craver, C.F., Green, L. et al. (2013) Dissociations in future thinking following hippocampal damage: evidence from discounting and time perspective in episodic amnesia. *Journal of Experimental Psychology: General,* 142 (4): 1355.

Lane, R.D., Ryan, L., Nadel, L. and Greenberg, L. (2014) Memory reconsolidation, emotional arousal and the process of change in psychotherapy: new insights from brain science. *Behavioral and Brain Sciences,* 38: 1–80. doi: 10.1017/S0140525X14000041

Leutgeb, J.K., Leutgeb, S., Moser, M.-B. and Moser, E.I. (2007) Pattern separation in the dentate gyrus and CA3 of the hippocampus. *Science,* 315 (5814): 961–6.

Levine, B., Svoboda, E., Hay, J.F. et al. (2002) Aging and autobiographical memory: dissociating episodic from semantic retrieval. *Psychology and Aging,* 17 (4): 677–89.

Maguire, E., Burke, T., Phillips, J. and Staunton, H. (1996) Topographical disorientation following unilateral temporal lobe lesions in humans. *Neuropsychologia,* 34 (10): 993–1001.

Maguire, E., Frackowiak, R. and Frith, C. (1996) Learning to find your way: a role for the human hippocampal formation. *Proceedings of the Royal Society of London B: Biological Sciences,* 263 (1377): 1745–50.

Maguire, E., Frackowiak, R. and Frith, C. (1997) Recalling routes around London: activation of the right hippocampus in taxi drivers. *Journal of Neuroscience,* 17 (18): 7103–10.

Maguire, E.A., Nannery, R. and Spiers, H.J. (2006) Navigation around London by a taxi driver with bilateral hippocampal lesions. *Brain,* 129 (11): 2894–907.

Marslen-Wilson, W.D. and Teuber, H.-L. (1975) Memory for remote events in anterograde amnesia: recognition of public figures from news photographs. *Neuropsychologia,* 13 (3): 353–64.

Martin, C.B., Bowles, B., Mirsattari, S.M. and Köhler, S. (2011) Selective familiarity deficits after left anterior temporal-lobe removal with hippocampal sparing are material specific. *Neuropsychologia,* 49 (7): 1870–8.

Maviel, T., Durkin, T.P., Menzaghi, F. and Bontempi, B. (2004) Sites of neocortical reorganization critical for remote spatial memory. *Science,* 305 (5680): 96–9.

Mayes, A., Holdstock, J., Isaac, C. et al. (2002) Relative sparing of item recognition memory in a patient with adult-onset damage limited to the hippocampus. *Hippocampus,* 12 (3): 325–40.

McKinnon, M.C., Black, S.E., Miller, B. et al. (2006) Autobiographical memory in semantic dementia: implications for theories of limbic-neocortical interaction in remote memory. *Neuropsychologia,* 44 (12): 2421–9.

Mickes, L., Seale-Carlisle, T.M. and Wixted, J.T. (2013) Rethinking familiarity: Remember/Know judgments in free recall. *Journal of Memory and Language,* 68 (4): 333–49.

Milner, B., Corkin, S. and Teuber, H.-L. (1968) Further analysis of the hippocampal amnesic syndrome: 14-year follow-up study of HM. *Neuropsychologia,* 6 (3): 215–34.

Milner, B., Petrides, M. and Smith, M. (1984) Frontal lobes and the temporal organization of memory. *Human Neurobiology,* 4 (3): 137–42.

Montaldi, D. and Mayes, A.R. (2010) The role of recollection and familiarity in the functional differentiation of the medial temporal lobes. *Hippocampus,* 20 (11): 1291–1314.

Montaldi, D., Spencer, T.J., Roberts, N. and Mayes, A.R. (2006) The neural system that mediates familiarity memory. *Hippocampus,* 16 (5): 504–20.

Morris, R. (1984) Developments of a water-maze procedure for studying spatial learning in the rat. *Journal of Neuroscience Methods,* 11 (1): 47–60.

Moscovitch, M. (1982) Multiple dissociations of function in amnesia. In L.S. Cermak (ed.), *Human Memory and Amnesia.* Hillsdale, NJ: Erlbaum. pp. 337–70.

Moscovitch, M., Cabeza, R., Winocur, G. and Nadel, L. (2016) Episodic memory and beyond: the hippocampus and neocortex in transformation. *Annual review of psychology*, 67, 105–34.

Müller, G.E. and Pilzecker, A. (1900) *Experimentelle beiträge zur lehre vom gedächtniss* (Vol. 1). Published by J.A. Barth.

Mummery, C.J., Patterson, K., Price, C. et al. (2000) A voxel-based morphometry study of semantic dementia: relationship between temporal lobe atrophy and semantic memory. *Annals of Neurology,* 47 (1): 36–45.

Nadel, L. and Moscovitch, M. (1997) Memory consolidation, retrograde amnesia and the hippocampal complex. *Current Opinions in Neurobiology,* 7 (2): 217–27.

Nader, K. and Hardt, O. (2009) A single standard for memory: the case for reconsolidation. *Nature Reviews Neuroscience,* 10 (3): 224–34.

Nader, K., Schafe, G.E. and Le Doux, J.E. (2000) Fear memories require protein synthesis in the amygdala for reconsolidation after retrieval. *Nature,* 406 (6797): 722–6.

Norman, K.A. and O'Reilly, R.C. (2003) Modeling hippocampal and neocortical contributions to recognition memory: a complementary-learning-systems approach. *Psychological Review,* 110 (4): 611.

Noulhiane, M., Piolino, P., Hasboun, D. et al. (2007) Autobiographical memory after temporal lobe resection: neuropsychological and MRI volumetric findings. *Brain,* 130 (12): 3184–99.

O'Keefe, J. and Nadel, L. (1978) *The Hippocampus as a Cognitive Map* (Vol. 3). Oxford: Clarendon.

Okuda, J., Fujii, T., Ohtake, H., et al. (2003) Thinking of the future and past: the roles of the frontal pole and the medial temporal lobes. *NeuroImage,* 19 (4): 1369–80.

Penfield, W. and Milner, B. (1958) Memory deficit produced by bilateral lesions in the hippocampal zone. *AMA Archives of Neurology and Psychiatry,* 79 (5): 475–97.

Race, E., Keane, M.M. and Verfaellie, M. (2015) Sharing mental simulations and stories: hippocampal contributions to discourse integration. *Cortex,* 63: 271–81.

Raichle, M.E., MacLeod, A.M., Snyder, A.Z. et al. (2001) A default mode of brain function. *Proceedings of the National Academy of Sciences,* 98 (2): 676–82.

Reed, J.M. and Squire, L.R. (1998) Retrograde amnesia for facts and events: findings from four new cases. *Journal of Neuroscience,* 18 (10): 3943–54.

Rempel-Clower, N.L., Zola, S.M., Squire, L.R. and Amaral, D.G. (1996) Three cases of enduring memory impairment after bilateral damage limited to the hippocampal formation. *Journal of Neuroscience,* 16 (16): 5233–55.

Rosenbaum, R.S., Gilboa, A., Levine, B. et al. (2009) Amnesia as an impairment of detail generation and binding: evidence from personal, fictional, and semantic narratives in KC. *Neuropsychologia*, 47 (11): 2181–7.

Rosenbaum, R.S., Gilboa, A. and Moscovitch, M. (2014) Case studies continue to illuminate the cognitive neuroscience of memory. *Annals of the New York Academy of Sciences,* 1316: 105–33.

Rosenbaum, R.S., Kohler, S., Schacter, D.L. et al. (2005) The case of K.C.: contributions of a memory-impaired person to memory theory. *Neuropsychologia,* 43 (7): 989–1021.

Rosenbaum, R.S., Moscovitch, M., Foster, J.K. et al. (2008) Patterns of autobiographical memory loss in medial-temporal lobe amnesic patients. *Journal of Cognitive Neuroscience,* 20 (8): 1490–506.

Rosenbaum, R.S., Priselac, S., Kohler, S. et al. (2000) Remote spatial memory in an amnesic person with extensive bilateral hippocampal lesions. *Nature Neuroscience,* 3 (10): 1044–8.

Rosenbaum, R.S., Winocur, G., Ziegler, M. et al. (2004) fMRI studies of remote spatial memory in an amnesic person. *Brain and Cognition,* 54 (2): 170–2.

Rosenbaum, R.S., Ziegler, M., Winocur, G. et al. (2004) "I have often walked down this street before": fMRI studies on the hippocampus and other structures during mental navigation of an old environment. *Hippocampus,* 14 (7): 826–35.

Rugg, M.D. and Vilberg, K.L. (2013) Brain networks underlying episodic memory retrieval. *Current Opinion in Neurobiology,* 23 (2): 255–60.

Sadeh, T., Ozubko, J.D., Winocur, G. and Moscovitch, M. (2013) How we forget may depend on how we remember. *Trends in Cognitive Sciences,* 18 (1): 26–36.

Sanders, H.I. and Warrington, E.K. (1971) Memory for remote events in amnesic patients. *Brain,* 94 (4): 661–8.

Schacter, D.L. (1987) Implicit expressions of memory in organic amnesia: learning of new facts and associations. *Human Neurobiology,* 6 (2): 107–18.

Schacter, D.L. (2012) Adaptive constructive processes and the future of memory. *American Psychologist,* 67 (8): 603.

Schacter, D.L. and Addis, D.R. (2007a) The cognitive neuroscience of constructive memory: remembering the past and imagining the future. *Philosophical Transactions of the Royal Society of London: Series B, Biological Sciences,* 362 (1481): 773–86.

Schacter, D.L. and Addis, D.R. (2007b) Constructive memory: the ghosts of past and future. *Nature,* 445 (7123): 27.

Schacter, D.L., Addis, D.R. and Buckner, R.L. (2007) Remembering the past to imagine the future: the prospective brain. *Nature Reviews Neuroscience,* 8 (9): 657–61.

Schacter, D.L., Addis, D.R., Hassabis, D. et al. (2012) The future of memory: remembering, imagining, and the brain. *Neuron,* 76 (4): 677–94.

Scoville, W.B. and Milner, B. (1957) Loss of recent memory after bilateral hippocampal lesions. *Journal of Neurology, Neurosurgery, and Psychiatry,* 20 (1): 11–21.

Sekeres, M., Moscovitch, M. and Winocur, G. (in press) Mechanisms of memory consolidation and transformation. In B. Rasch and N. Axmacher (eds.), *Cognitive Neuroscience of Memory Consolidation*. New York: Springer Publishing. p. 134.

Sharon, T., Moscovitch, M. and Gilboa, A. (2011) Rapid neocortical acquisition of long-term arbitrary associations independent of the hippocampus. *Proceedings of the National Academy of Sciences*, 108 (3): 1146–51.

Smith, C.N. (2014) Retrograde memory for public events in mild cognitive impairment and its relationship to anterograde memory and neuroanatomy. *Neuropsychology,* 28 (6): 959.

Snowden, J.S., Goulding, P. and Neary, D. (1989) Semantic dementia: a form of circumscribed cerebral atrophy. *Behavioural Neurology*, 2: 167–82.

Spiers, H. J. and Maguire, E.A. (2006) Thoughts, behaviour, and brain dynamics during navigation in the real world. *NeuroImage,* 31 (4): 1826–40.

Spreng, R.N. and Grady, C.L. (2010) Patterns of brain activity supporting autobiographical memory, prospection, and theory of mind, and their relationship to the default mode network. *Journal of Cognitive Neuroscience,* 22 (6): 1112–23.

Spreng, R.N., Mar, R.A. and Kim, A.S. (2009) The common neural basis of autobiographical memory, prospection, navigation, theory of mind, and the default mode: a quantitative meta-analysis. *Journal of Cognitive Neuroscience,* 21 (3): 489–510.

Squire, L.R. (1992) Memory and the hippocampus: a synthesis from findings with rats, monkeys, and humans. *Psychological Review,* 99 (2): 195–231.

Squire, L.R. and Alvarez, P. (1995) Retrograde amnesia and memory consolidation: a neurobiological perspective. *Current Opinions in Neurobiology,* 5 (2): 169–77.

Squire, L.R. and Bayley, P.J. (2007) The neuroscience of remote memory. *Current Opinions in Neurobiology,* 17 (2): 185–96.

Squire, L.R., van der Horst, A.S., McDuff, S.G. et al. (2010) Role of the hippocampus in remembering the past and imagining the future. *Proceedings of the National Academy of Sciences,* 107 (44): 19044–8.

Squire, L.R. and Wixted, J.T. (2011) The cognitive neuroscience of human memory since HM. *Annual Review of Neuroscience,* 34: 259.

St-Laurent, M., Moscovitch, M., Levine, B. and McAndrews, M.P. (2009) Determinants of autobiographical memory in patients with unilateral temporal lobe epilepsy or excisions. *Neuropsychologia,* 47 (11): 2211–21.

Staresina, B.P., Fell, J., Do Lam, A.T. et al. (2012) Memory signals are temporally dissociated in and across human hippocampus and perirhinal cortex. *Nature Neuroscience,* 15 (8): 1167–73.

Steinvorth, S., Levine, B. and Corkin, S. (2005) Medial temporal lobe structures are needed to re-experience remote autobiographical memories: evidence from H.M. and W.R. *Neuropsychologia,* 43 (4): 479–96.

Szpunar, K.K., Spreng, R.N. and Schacter, D.L. (2014) A taxonomy of prospection: introducing an organizational framework for future-oriented cognition. *Proceedings of the National Academy of Sciences, USA,* 111 (52): 18414–21.

Teng, E. and Squire, L.R. (1999) Memory for places learned long ago is intact after hippocampal damage. *Nature,* 400 (6745): 675–7.

Tronson, N.C. and Taylor, J.R. (2007) Molecular mechanisms of memory reconsolidation. *Nature Reviews Neuroscience,* 8 (4): 262–75.

Tse, D., Langston, R.F., Kakeyama, M. et al. (2007) Schemas and memory consolidation. *Science,* 316 (5821): 76–82.

Tulving, E. (1972) Episodic and semantic memory. In E. Tulving and W. Donaldson (eds), *Organization of Memory*. New York: Academic Press. pp. 381–403.

Tulving, E. (1985) Memory and consciousness. *Canadian Psychology/Psychologie Canadienne,* 26 (1): 1.

Tulving, E. (2001) Episodic memory and common sense: How far apart? *Philosophical Transactions of the Royal Society B: Biological Sciences,* 356 (1413): 1505–15.

Tulving, E., Schacter, D.L., McLachlan, D.R. and Moscovitch, M. (1988) Priming of semantic autobiographical knowledge: a case study of retrograde amnesia. *Brain and Cognition,* 8 (1): 3–20.

Tulving, E., Schacter, D.L. and Stark, H.A. (1982) Priming effects in word-fragment completion are independent of recognition memory. *Journal of Experimental Psychology: Learning, Memory, and Cognition,* 8 (4): 336.

van Kesteren, M.T., Ruiter, D.J., Fernández, G. and Henson, R.N. (2012) How schema and novelty augment memory formation. *Trends in Neurosciences*, 35 (4): 211–19.

Victor, M. and Agamanolis, D. (1990) Amnesia due to lesions confined to the hippocampus: a clinical-pathologic study. *Journal of Cognitive Neuroscience,* 2 (3): 246–57.

Viskontas, I.V., McAndrews, M.P. and Moscovitch, M. (2000) Remote episodic memory deficits in patients with unilateral temporal lobe epilepsy and excisions. *Journal of Neuroscience,* 20 (15): 5853–7.

Waidergoren, S., Segalowicz, J. and Gilboa, A. (2012) Semantic memory recognition is supported by intrinsic recollection-like processes: "The butcher on the bus" revisited. *Neuropsychologia,* 50 (14): 3573–87.

Wais, P.E., Wixted, J.T., Hopkins, R.O. and Squire, L.R. (2006) The hippocampus supports both the recollection and the familiarity components of recognition memory. *Neuron*, 49 (3): 459–466.

Warrington, E.K. and McCarthy, R.A. (1988) The fractionation of retrograde amnesia. *Brain and Cognition,* 7 (2): 184–200.

Wheeler, M.A., Stuss, D.T. and Tulving, E. (1997) Toward a theory of episodic memory: the frontal lobes and autonoetic consciousness. *Psychological Bulletin,* 121 (3): 331–54.

Winocur, G. and Moscovitch, M. (2011) Memory transformation and systems consolidation. *Journal of the International Neuropsychological Society*, 17 (5): 766–80.

Winocur, G., Moscovitch, M. and Bontempi, B. (2010) Memory formation and long-term retention in humans and animals: convergence towards a transformation account of hippocampal-neocortical interactions. *Neuropsychologia,* 48 (8): 2339–56.

Winocur, G., Moscovitch, M., Fogel, S. et al. (2005) Preserved spatial memory after hippocampal lesions: effects of extensive experience in a complex environment. *Nature Neuroscience,* 8 (3): 273–5.

Winocur, G., Moscovitch, M., Rosenbaum, R.S. and Sekeres, M. (2010) An investigation of the effects of hippocampal lesions in rats on pre- and postoperatively acquired spatial memory in a complex environment. *Hippocampus,* 20 (12): 1350–65.

Witt, J., Vogt, V.L., Widman, G. et al. (2015) Loss of autonoetic consciousness of recent autobiographical episodes and accelerated long-term forgetting in a patient with previously unrecognized glutamic acid decarboxylase antibody related limbic encephalitis. *Frontiers in Neurology,* 6: 130.

Wixted, J.T. (2007) Dual-process theory and signal-detection theory of recognition memory. *Psychological Review,* 114 (1): 152.

Wixted, J.T. and Mickes, L. (2010) A continuous dual-process model of remember/know judgments. *Psychological Review,* 117 (4): 1025–54.

Yonelinas, A.P. (2002) The nature of recollection and familiarity: a review of 30 years of research. *Journal of Memory and Language,* 46 (3): 441–517.

ABBREVIATIONS

AI: Autobiographical Interview; **AMI**: Autobiographical Memory Interview; **DMN**: Default Mode Network; **fMRI**: functional magnetic resonance imaging; **MTL**: medial temporal lobe; **MTT**: Multiple Trace Theory; **PET**: positron emission tomography; **RA**: retrograde amnesia; **SCT**: Standard Consolidation Theory; **TGRA**: temporally-graded retrograde amnesia; **TL**: temporal lobe; **TTT**: Trace Transformation Theory

GLOSSARY OF TERMS

Allocentric spatial memory: memory for a spatial environment based on a map-like spatial representation that depends on the layout of elements in the environment in relation to each other (object-to-object representation). This is often contrasted with a first-person, egocentric spatial memory (self-to-object representation).

Anterograde amnesia: inability to form new memories following the onset of injury or brain damage.

Autobiographical memory: memory for personal experiences and facts about one's self.

Decay: a proposed process of forgetting based on the weakening or fading of the neural substrate of a memory over time. The precise mechanisms regulating temporal decay are not well understood, but may be the result of a failure to fully consolidate the memory, disuse of the non-relevant memory, silencing of weakly potentiated neuronal synapses, or AMPA-receptor internalization.

Declarative memory: a category of explicit long-term memory comprised of semantic and episodic memory, including memory for facts, general knowledge, and personally experienced events.

Default Mode Network: a network of brain regions engaged during passive, internally-focused cognition (mind-wandering) or rest. These regions include the medial prefrontal cortex, parahippocampus, retrosplenial/posterior parietal cortex, precuneus, lateral parietal cortex/angular gyrus.

Episodic memory: a component of declarative memory involving conscious recollection of an event occurring at a specific time and place. Explicit memory is characterized by a sense of mentally re-experiencing the contextual and perceptual details of a personal event or episode.

Familiarity: judging that a cued event or item has previously occurred based on a feeling of 'knowing', but lacking specific contextual details. Familiarity is thought to be mediated by the perirhinal cortex.

Interference: a proposed mechanism of forgetting in which similar, competing information disrupts the memory. Interference can be proactive whereby past memories interfere with subsequent new memories, or it can be retroactive whereby new information interferes with previous memory traces. The hippocampus is thought to minimize interference by encoding similar memories in non-overlapping traces (pattern separation).

Multiple memory systems: a theory proposing that episodic and semantic components of declarative memory are mediated by functionally separate neural systems.

Multiple Trace Theory (MTT): a theory of memory consolidation which proposes that the hippocampus is required for the encoding and initial consolidation of episodic memories. With repeated retrieval over time, new memory traces form both in the hippocampus and neocortex, with each new trace adding new information and strengthening the original memory. The common featural elements are incorporated into schematic/semantic memory networks in the neocortex, while the temporal and contextual details of the memory form distributed traces in the hippocampus, and continue to depend on this structure for their storage and retrieval.

Recognition memory: recognition may be supported by either recollection, or familiarity, or both processes.

Recollection: involves a 'sense of reliving', or a reinstantiation of contextual and perceptual details related to the cued event. Recollection is thought to be mediated by the hippocampus and parahippocampal cortex.

Recollection network: a network of brain regions which are commonly activated during memory recollection. These regions include the hippocampus, parahippocampus, retrosplenial/posterior parietal cortex, lateral parietal cortex, and the medial prefrontal cortex.

Reconsolidation: a previously consolidated memory can be brought back to a labile state following reactivation, during which time it is susceptible to modification (updating, strengthening) or disruption by interfering with the protein-synthesis dependent re-stabilization period.

Retrograde amnesia: loss of memories acquired prior to the onset of injury or brain damage.

Schema: an associative network of information which is adaptable and developed though the abstraction of common information over the course of multiple episodes (i.e. knowing that a typical birthday party includes balloons, presents, cake, candles). Schemas are thought to be represented in the medial prefrontal cortex.

Schema assimilation: new information that is consistent with a pre-existing schematic knowledge base will be rapidly acquired and integrated into the memory network.

Semantic memory: a component of declarative memory including memory for knowledge about the world and general facts. Semantic memory can be explicitly recalled, but lacks contextual details related to first acquiring this knowledge.

Semantic priming: typically, semantic priming refers to the observation that the presentation of a priming cue (i.e. pencil) facilitates the response to a semantically related subsequent target cue (i.e. pen). K.C. exhibited semantic priming for his autobiographical memory, in which the repeated presentation of facts about his life (i.e. working at a manufacturing plant) facilitated his subsequent ability to recognize semantically related autobiographical information (i.e. driving a truck for the manufacturing plant).

Standard Consolidation Theory (SCT): a theory of memory which holds that the hippocampus is critical for the initial consolidation of all declarative memories (both episodic and semantic information). As the memory network reorganizes over time, the identical memory becomes represented in neocortical regions and no longer requires the hippocampus for its storage or retrieval.

Synaptic consolidation (cellular consolidation): rapid protein synthesis-dependent biochemical and morphological changes in individual neurons, leading to long-lasting changes in the strength of synaptic connections of the neuronal circuitry supporting long-term memory.

Systems consolidation: the gradual reorganization of networks supporting long-term memory. Soon after encoding, memories rely heavily on the hippocampus and the MTL, but over time (weeks, months, years), memory traces form in the neocortex (likely in the prefrontal cortex). Competing theories argue about the continuing involvement of the hippocampus as a declarative memory ages and reorganizes in the cortex.

Temporally-graded retrograde amnesia: a common finding following MTL damage in which memories acquired long before (many years in humans) the onset of brain injury are spared, but memories acquired more recently (months and weeks prior to damage) are lost. Different patterns of TGRA are often reported for semantic and episodic memories, with semantic memories showing a temporal gradient, but episodic memories typically showing a flat gradient (both recent and very remote episodic memories are typically lost following MTL damage).

Trace Transformation Theory (TTT): a theory of memory stemming from MTT, which proposes that episodic memories are consolidated in rich contextual detail within the hippocampus. The hippocampus is always required for the storage and retrieval of this contextually detailed memory. Over time, episodic memories are transformed into less detailed schematic (semantic) memories that capture the essential features or gist of the original, and are represented in neocortical regions. The detailed hippocampus-dependent version, and the transformed schematic version of the memory, can co-exist in the brain. The situational demands at the time of retrieval will mediate which version of the memory is expressed.

12 Revisiting O'Keefe: Place units in the hippocampus of the freely moving rat

Matthew Shapiro

O'Keefe described hippocampal place cells and their relationship to cognitive maps in rats 19 years after Milner and Scoville reported the global amnesia for recent events following a hippocampal resection in the case of H.M. Though these two discoveries in the modern neuroscience of memory have yet to be fully reconciled, 50 years of research have brought the resolution to within reach. Converging evidence from studies of nonhuman animals and humans demonstrated the behavioral homology of the hippocampus across species, allowing this chapter to describe a pathway from place cells to memory. Neuronal populations recorded in behaving animals, including people, revealed that place cells represent past and future as well as current locations, distinguish identical places by motivational states and task rules, and are likely time as well as place cells. Activity recorded from different subregions of the hippocampus and closely connected structures verified a broad range of spatial and nonspatial codes. Place units discovered in behaving rats were the first page of an atlas of cognitive maps.

INTRODUCTION: MEMORY AND BRAIN

Brenda Milner and William Scoville were likely thrilled and horrified when they realized that Henry Molaison was stuck in time. Though removing medial temporal lobe structures on both sides of the brain succeeded in reducing the severity and frequency of his epileptic seizures as intended, the surgery left H.M. unable to remember recent events (Scoville and Milner, 1957; Milner, 2005). H.M.'s other mental abilities were unimpaired after the surgery: his I.Q. test score increased; his perceptual, motor, and reasoning skills remained intact. Moreover, H.M. learned and retained new skills normally, such as in a mirror drawing task. For the rest of his life, however, H.M. described every moment as if he were waking from a dream. Though he could remember childhood events through his high school years, after surgery his brain no longer updated the internal autobiography described as episodic memory (Tulving and Markowitsch, 1998). The degree of memory deficits in other patients with temporal lobe damage was correlated consistently with hippocampal loss, and more recent cases have confirmed that damage to one layer of cells in the hippocampus, the CA1 layer, was sufficient to impair memory (Zola-Morgan et al., 1986; Rempel-Clower et al., 1996).

Scoville and Milner's discovery launched the modern era of memory neuroscience, with thousands of experiments aimed at understanding how the hippocampus helps the brain remember in people and other animals. Back then, learning and memory were considered to be one thing, a single biological function: learning was learning and memory was its repository. The defining features of human amnesia were not obvious in rats or monkeys with damage to the hippocampus who learned and remembered some, but not all, tasks perfectly well. Until the mid-1970s, a real possibility was that the human hippocampus had evolved memory functions that didn't exist in "lower" animals, and a prominent theory held that the hippocampus helped animals inhibit behavior, i.e. withhold learned responses (Douglas, 1967). Convincing animal models of human amnesia were only developed after the biological complexity of learning, memory, and animal cognition became more widely recognized. The discovery of place cells, their role in forming cognitive maps and the evidence that rats needed the hippocampus to remember places, played a key part in this history of neuroscience.

PLACE CELLS, SPATIAL MAPS, AND MENTAL REPRESENTATIONS

We now walk around with "smart phones" in our pockets, taking for granted that machines compute – represent and process information – and that at least some physical mechanisms for "thinking" are not limited to the human mind. In the 1940s, however, the notion that animals might think in any way was largely dismissed, and the predominant view was that animals learned by reinforcement that strengthened associations between stimuli and responses (Hull, 1932).

Because some animal behavior could not be explained simply by stimulus-response reinforcement learning, Tolman argued that rats (and people) interrelated stimuli, rewards and responses in an internal model of the environment (Tolman, 1948). This internal model Tolman called a "cognitive map" and this helped guide flexible, goal-directed behavior. For example, a rat trained to find food in a T-maze could learn either to make a body turn (a stimulus-response association) or learn where the food was located (a place). A probe test determined what the rat actually learned by putting it for the first time in the T-maze rotated in place 180 degrees: if the rat learned a body turn, it would enter the goal arm on the opposite side of the room; if the rat learned to approach a place, it would make the opposite body turn. In well-lit, open rooms, most rats approached the rewarded place rather than repeat the reinforced body turn.

Tolman argued that learning included more than "stimulus response connections," but also "in the building up in the nervous system of sets which function like cognitive maps." The claim that animals had cognitive maps was contentious for many decades, even after John O'Keefe and John Dostrovsky (1971) discovered "place cells," i.e. neurons in the hippocampus that responded "solely or maximally when the rat was situated in a particular [place] facing in a particular direction." The authors argued that place cells were the neural elements of a cognitive or spatial map, and that hippocampal circuits provided a neuronal mechanism for computing locations in an environment (O'Keefe and Dostrovsky, 1971).

CODING SPATIAL RELATIONSHIPS

Place cells respond to spatial relationships among environmental cues, and these relationships define the "sets" that Tolman proposed to form cognitive maps. O'Keefe investigated the coding properties of hippocampal neurons by recording their activity as rats explored a maze in an open room surrounded by experimenter-controlled cues (e.g. posters on curtains, etc.). In standard conditions, *place* cells were typically silent as a rat explored an environment until it entered one or two small patches in the environment where the cells fired at high rates, i.e. the cell's *place fields* (O'Keefe, 1976).

By moving or rearranging parts of the maze or changing room cues, O'Keefe discovered that place cells responded to the constellation of cues that surrounded the recording arena. For example, when all of the controlled cues were rotated around the maze by 90° as a constellation, the place fields rotated with the cues, the cells firing in the same location relative to the cues and one another, even as the rat occupied a different physical maze arm (O'Keefe and Conway, 1978). Changing the spatial relationships among the cues, e.g. by interchanging them, caused dramatic changes in neural activity, with some place cells becoming silent and other previously silent cells becoming active. Paradoxically, the place fields were typically unchanged if any one of the controlled cues were removed. The cells responded to sets of spatial relationships among the cues surrounding the maze, rather than individual or local cues, to signal places. Place cells did not respond to stimuli per se, but to sets of spatially consistent signals.

PLACE CELLS AND NAVIGATION

The Hippocampus as a Cognitive Map, written by John O'Keefe and Lynn Nadel (1978), integrated and extended these findings in a neuropsychological theory proposing that the hippocampus computes "an objective spatial framework within which the items and events of an organism's experience are located and interrelated." The framework is an absolute, two-dimensional Euclidean space that encloses but does not depend on objects, just as a room containing a particular chair is not defined by the chair. Place cells integrate external and internal inputs, and distances to and angles among distal stimuli heading direction and body movement, to signal unique locations.

In other words, active cell groups represent a place together with the objects perceived there and the movements used to get to the place. Hippocampal circuits were proposed to link place cells to one another so that neurons active in one location would trigger firing by cells coding "adjacent or subsequent spatial orientations," thereby representing stimuli expected to appear in anticipation of a given movement. Together, these hippocampal mechanisms were proposed to be a neural system for spatial navigation.

Distance and direction coding

O'Keefe and Nadel (1978) emphasized that their neuronal theory of hippocampal mapping was limited by available data and required postulating the existence of hypothetical neuronal mechanisms. Hippocampal cells were supposed to compute places by integrating the rat's movement speed, direction, distance, and other spatial signals that, at the time, had no neural basis. Since then, however, cells in closely connected medial temporal lobe (MTL) structures have been discovered that help elaborate the theory. Head-direction cells recorded in the presubiculum fire when a rat's head is pointed in a particular direction, independent of its location in a given environment (Ranck et al., 1987; Taube et al., 1990). Grid cells recorded in the medial entorhinal cortex fire in evenly spaced patches throughout a given environment in equidistant, hexagonal patterns (Fyhn et al., 2004). Boundary cells recorded in the subiculum fire along a wall or a barrier of environments (Lever et al., 2009). These cells provide direction, distance and movement signals, as well as the limits of movement in an environment, a neural substrate for spatial computation. Though the mechanisms that integrate these signals with one another or to spatial behavior remain unclear (Winter et al., 2015; c.f. Gibson et al., 2013), disrupting function in any one of these regions impairs performance in spatial memory tasks.

Yet MTL dysfunction impairs memory for recent events across sensory modalities, including faces, objects, pictures, sequences and scenes. These disconnected facts can be reconciled if parallel MTL circuits perform different, nonspatial computations. Different types of information are segregated to some degree through the MTL. Grid cells are found in the medial entorhinal cortex, not the lateral entorhinal cortex where cells respond to objects (Knierim et al., 2014). Neurons in the dorsal (septal) hippocampus have relatively small place fields that cover ~10–30% of the area of an environment, while those in the ventral (temporal) hippocampus have fields that take up the entire environment (Komorowski et al., 2013). Alternatively, more general "spatial" computations can apply to many types of information.

Coding abstract spaces

We typically navigate through the two-dimensional space shown on maps as distances and directions related to latitude and longitude. We use similar 2-D representations to describe relationships among arbitrary pairs of independent variables, the X and Y Cartesian coordinates related by math functions that map domain to range. The "central property" of cognitive maps, according to O'Keefe and

Nadel, is their ability "to order representations in a structured context." Recent findings suggest that the hippocampus may indeed perform general spatial computations that represent items and events as points in a space determined by task demands and ethology as well as physical place and time (Shapiro, 2015). The human hippocampus seems to represent social relationships, for example in a two-dimensional space defined by power and affiliation (Tavares et al., 2015). This generalized kind of spatial computation may explain how place cells support episodic memory.

FROM SPATIAL MAPS TO EPISODIC MEMORIES

SPATIAL MEMORY AND PLACE CELLS

Place fields predict spatial memory. Links between place cells, cognitive mapping and HM's memory deficit were missing until O'Keefe recorded hippocampal neurons in rats performing a spatial memory task (O'Keefe and Speakman, 1987).

The rats were trained to find food in a place indicated by a constellation of cues that surrounded the maze. Because the constellation was rotated into one of four 90° orientations from one trial to another, the rats had to ignore uncontrolled cues (e.g. sounds from the street outside the lab) and follow the controlled cues to find the food. Place cells recorded as the rats performed the task had fields that followed the cue constellation, i.e. the cells fired consistently with respect to the controlled cues, and rotated around the room in 90° increments with respect to the uncontrolled cues.

Next the rats were trained to *remember* the orientation of the cue constellation: a rat was locked in one arm in the presence of the cues and had to find the goal by memory after all of the controlled cues were removed. The rat and its place cells remembered the cue constellation. During correct memory trials, the rat entered the arm indicated by the recently removed cues, and the place fields were the same whether the rat followed available cues or selected the goal by memory.

During memory errors, though, the place fields were aligned with the rat's selected goal. That is, the same place cells were active, and the activated map was oriented in the rotation consistent with the rat's choice of goals. Recent memory errors could be attributed to the rat retrieving the wrong (rotated) map, and that "memory map" guided behavior (O'Keefe and Speakman, 1987). Though the mechanisms that maintain "memory maps" in recent memory are not fully clear (cf. Schmitt et al., 2003), synaptic plasticity mechanisms are needed to establish enduring memories and stable place field maps.

PLACE FIELD FORMATION AND STABILITY

We can remember momentary experiences for decades, suggesting that brain mechanisms encode information rapidly and store it stably. The hippocampus may be needed for such memories because its circuits support both readily induced and persistent synaptic plasticity. Indeed, the same neurochemical mechanisms are needed for enduring synaptic change, lasting new memories, and stable place fields.

When a rat first explores an unfamiliar environment, hippocampal cells develop place fields in minutes and the same fields are recorded the next day (Frank et al., 2004). Once formed, the place fields are remarkably stable and have been recorded in a familiar environment for weeks (Thompson and Best, 1990). Blocking synaptic change pharmacologically has no effect on place cells in familiar environments. If synaptic changes are blocked while the rat first explores an unfamiliar environment, however, place fields do not persist. Rather, the same cells recorded the next day fire in different patterns in the new room, even while they fire in their usual, stable fields in a familiar room (Kentros et al., 1998). The same pharmacological treatments have analogous effects on learning and memory. Blocking synaptic change impairs learning about unfamiliar environments, but does not prevent memory retrieval in familiar ones (Shapiro and O'Connor, 1992). The same neurochemical mechanisms that allow rapidly induced and enduring synaptic change stabilize new place fields and new memories.

Place field remapping

We distinguish memories for similar events, suggesting that we represent each memory with a distinct neural code. Analogously, independent subsets of CA1 cells fire in different environments, so that each environment is associated with its own "map." Moreover, sufficient changes to environmental features can cause "global remapping," so that completely independent sets of hippocampal neurons become active in the same room (Leutgeb et al., 2004). Interchanging room cues to disrupt the topology of the cue constellation (i.e. with no rotational symmetry), for example, can cause global remapping. In such cases, independent sets of active hippocampal cells represent different contexts. Distinguishing among contexts is important for memory guided behavior when small changes in the environment signal different outcomes, e.g. whether, for a thirsty animal, a freshwater spring includes the smell of predators. On the other hand, linking similar representations is important for generalizing across situations, and if the hippocampus only encoded independent maps, then how would these links be forged? Clearly, other brain structures could link independent representations in memory, but the hippocampus also encodes overlapping relationships among situations by "partial remapping." When new items are presented or new behaviors are required in a familiar room, some hippocampal cells maintain their original fields while others change (Muller and Kubie, 1987; Markus et al., 1995; Tanila et al., 1997; Shapiro et al., 1997). Partial remapping may provide a brain mechanism for encoding both the similarities and differences among overlapping situations (Eichenbaum et al., 1999).

Neuronal computation: population coding

Place cells were discovered before the digital age. Single units, the extracellular action potentials of one neuron, were recorded in behaving animals using electrodes made of fine wire and stored with a pen or tape recorder. One or two units were recorded at a time as an animal responded repeatedly (e.g. visited different maze arms) to identify reliable firing correlates.

The activity of a single neuron cannot describe population coding, i.e. how information is signaled by simultaneous or sequential firing patterns in neural groups. Thanks to advances in digital electronics, twenty-first-century laboratories record hundreds of single units simultaneously, and store the waveform of each action potential with microsecond resolution by computer. The massive increase in computer power during the past 30 years now analyzes these population codes in real time, using advanced statistics that were impractical until recently.

These "high density" methods have begun to reveal how neural groups contribute to memory guided behavior. Anatomically adjacent neurons have place fields in unpredictable locations, and cells throughout the hippocampus are active in any given place. Though the hippocampal map is not spatiotopic, i.e. organized as an analogue of a real map, the highly distributed population of active place cells is easily decoded into accurate predictions of an animal's whereabouts. High-density recordings show that the activity of about 60 place cells can predict the location of a rat's head to within one square inch within a square yard (Wilson and McNaughton, 1993). Fewer cells are needed to predict the rat's location accurately in mazes with more constrained trajectories.

NEURONAL COMPUTATION: DISTRIBUTED REPRESENTATIONS AND CONTENT ADDRESSABLE MEMORY

The properties of place cells and memories share similarities in being content-addressable memory systems, distributed representations that describe how groups of simple, neuron-like units represent and process information by associating collections of features (Hinton et al., 1986). To illustrate the idea, consider the word "dog." In a local and symbolic representation, like an ebook about animals, "dog" would be represented as a symbol, a letter string located in one particular place in computer storage where other symbols, words and perhaps pictures would describe the animal. Each item, an animal, is represented by one symbol and stored in one memory location by a separate processor, the CPU. Memory retrieval in this system requires finding the appropriate memory location, such as by scanning one dimensional alphabetized words in an index.

In distributed representations, though, "dog" might be represented by a collection of neuron-like "units" that respond to feature matches (e.g. furriness, size, color, shape, etc.). In a neural network system designed to recognize animals from video information, each unit might be activated as a function of how well the input matches the feature. A dog would be represented by the group of units – the collection of features – that are co-activated reliably by the dogs in the video input. Each item is represented by many units, and every unit takes part in representing many items.

Distributed representations store items using "activity-dependent" mechanisms in which connections among coactive units are strengthened, i.e. the collection of features that occur repeatedly in the same video frames would become associated. If furriness etc. were the features, then other furry animals, such as cats or hamsters, would activate many of the same units as dogs, and the activity pattern generated by wolves, dogs and coyotes would overlap. This kind of memory system is "content-addressable" because the same units that are activated during "perception" also "store" and "retrieve" items.

Perception, storage and retrieval of items occurs in parallel, by activating sets of feature units, not by scanning a list, and retrieval occurs by content-matching video input with modified connections among feature-detecting units. Because items are stored by many units, each item's representation is distributed across many locations rather than occupying a single memory location. The same self-modifying machinery represents and processes, perceives, stores and retrieves through patterns of activity distributed among the same elements.

Cognitive maps, as described O'Keefe and Nadel, can be considered as a special type of distributed, content-addressable memory system. Maps are flexible, high storage capacity representations that resist interference by coding events in a spatial context. Though the meaning, significance and appropriate response to a stimulus can be guided by place alone, these typically require further specification by time, motivation, and other features of a situation.

EPISODIC MEMORY IN CONTEXT

People remember events in episodes, framed by spatial, temporal, and personal context: we follow the reporter's dictum to remember who, what, where, when, why, and how, or would like to do so. Research on episodic-like memory has emphasized spatial and temporal context, on testing whether animals remember "what, where, and when," which some species do quite well. Scrub jays, for example, store food for later consumption. The birds remember not only where they hide food, but keep track of different kinds of food and how long ago these were hidden (Clayton et al., 2001). Rats have similar abilities to distinguish what, where, and when (Eacott et al., 2005), and these abilities require the hippocampus (Crystal et al., 2013).

Remembering an experience, we describe our view of an event, i.e. the who, what, where, when, why, and how of a past episode. The technical term for this everyday experience is "episodic memory," the ability to recall the past from our own perspective, framed by spatial, temporal, and personal context. Hippocampal dysfunction impairs memory for all kinds of episodes, whether these are defined by spatial context or not. People with hippocampal damage have poor memory for verbal and pictorial as well as visuospatial items (Zola-Morgan and Squire, 1986; Rempel-Clower et al., 1996).

The link between cells that code spatial information and episodic memory is not obvious. To account for how the hippocampus contributes to human episodic memory, the *The Hippocampus and the Cognitive Map* proposes generalizing spatial mapping to more abstract spaces (O'Keefe and Nadel, 1978). The "central property" of cognitive maps, according to O'Keefe and Nadel, is their ability "to order representations in a structured context." Experiences and memories are proposed to be contained, interrelated, and located in spatial context within hippocampal maps. Though the neuronal mechanisms needed for representing and processing structured contexts were unknown in 1978, O'Keefe and Nadel proposed that cognitive mapping by the human hippocampus included such dimensions as cause and time as well as Euclidean space. Hippocampal circuits may provide a more general cognitive map even in rats, one that includes other dimensions based on "ordered representations in a structured context" (O'Keefe and Nadel, 1978).

Spatiotemporal context: keeping memory on track

Beyond signaling places, hippocampal neurons encode the unfolding history of behavior, linking the "here and now" to "before and after," the start and end of goal-directed actions. Hippocampal neurons in rats fire differently when they follow identical spatial paths that either lead to different goals or begin at different starting points (Frank et al., 2000; Wood et al., 2000; Ferbinteanu and Shapiro, 2003). Ferbinteanu described these history-dependent firing patterns as *journey coding*.

In this study, rats were trained to find food in a plus-shaped maze by walking from one of two start arms through a choice point to one of two goal arms. Food was put in the same goal arm for each block of trials, which the rat had discovered by trial and error. After the rat chose the correct arm in eight out of ten trials, food was put into the other goal and the rat had to learn to choose the new goal arm. To perform well, the rats have to remember where and when, i.e. the correct arm during a given block of trials, and requires the hippocampus.

Note that in this task the rats use the same spatial path in each start arm on the way to the different goals, e.g. in the North start arm to either the West or East goal arms. Similarly, the rat uses identical paths to approach food in each goal arm after starting the trial in the North or South start arms. In both cases, CA1 cells coded journeys by firing at different rates depending on the past or future of behavior: *prospective* coding predicted which goal arm the rat was about to choose; *retrospective* coding retrodicted the arm that started the journey. Hippocampal cells signal past and future places in identical locations (Ferbinteanu and Shapiro, 2003; Smith and Mizumori, 2006), and more recent experiments discussed below show that indeed the cells encode sequences of spatial and non-spatial events in time.

Personal context 1: Why?

The biological purpose of memory is to inform adaptive behavior so that past experiences relevant to the present can help select actions to obtain goals. To be useful, in other words, memory must be guided by means, motive, and opportunity. Episodic memories are defined as events organized by a personal as well as a spatial and temporal context. Compared to "where" and "when," the mechanisms that link episodes to "why" and "how," i.e. motivation and skilled action, are less explored. Across species, memories distinguish past events that occur in identical places, events for which space is irrelevant yet which require hippocampal function. "Internal context" guides memory and is encoded by hippocampal neurons.

After the discovery of place cells, and before the publication of *The Hippocampus as a Cognitive Map*, Richard Hirsh published a theory of hippocampal function based on contextual memory retrieval (Hirsh, 1974). Just as stage settings in a play frame the story by indicating its place and time, the background of events establishes context, the larger and relatively stable perspective that helps us interpret foreground information. Spatial context, for example, can guide different responses to the same stimulus, e.g. answering a ringing telephone depending on whether the phone is in your house or not. Context is especially helpful for disambiguating items that by themselves are ambiguous or identical, such as the middle elements in "ply" and "212."

Hirsh's theory distinguishes between two types of learning, one rigid that binds each stimulus to a response, and the other flexible that allows multiple responses to a given stimulus. The flexible type stores links between stimuli and responses, associations that contain neither the content of the event nor obligatory responses. Like a card catalog in a library or the directory of a computer drive, each entry in this system points to an item in memory, but is not the item itself. Contextual associations add flexible links between a given stimulus and responses without necessarily eliciting any. Students of psychology may recognize the similarity between Hirsh's idea and others that distinguish memory systems: procedural versus declarative ("knowing how" vs "knowing that") (Cohen and Squire, 1980), horizontal and vertical associations (Wickelgren, 1979), and "taxon" and "locale" of memory (O'Keefe and Nadel, 1978).

Hirsh proposed that contextual associations stored by the hippocampus during learning guided flexible responses to remembered stimuli. One experiment tested rats who were both hungry and thirsty in a T-maze that contained food in one goal arm and water in the other. If a rat was later given water but not food it approached the food arm, and vice versa. The rats used internal context – deprivation state – to guide different responses in the same spatial environment. Damage to the hippocampal system impaired learning in this task, as predicted by contextual retrieval theory (Hirsh, 1974; Hirsh et al., 1979). Because other explanations were available as well, the results did not provide unequivocal support for the contextual theory. For example, hippocampal damage may have impaired spatial or internal state discrimination, both of which were necessary for good performance in the task.

More recent work provides further support for Hirsh's contextual retrieval theory. Pamela Kennedy demonstrates that contextual memory retrieval for nonspatial cues requires the hippocampus (Kennedy and Shapiro, 2004). In her studies, rats learned to approach one cue for water when they were thirsty, another cue for food when they are hungry, and to avoid a third cue that was never rewarded. The reward associated with each cue was consistent for each rat on all trials. Spatial discriminations were irrelevant and useless because the cues were moved among three locations on each trial; only the visual cues predicted the rewards.

Hippocampal damage reduced performance to near chance levels in this nonspatial task without altering cue or internal state discrimination. Rats with and without lesions discriminated the visual cues by approaching the rewarded cues rather than the non-rewarded one. When presented with food and water, all the rats approached and consumed the one of which they were deprived (Kennedy and Shapiro, 2004). In other words, the rats knew what they wanted and could tell the difference between the cues, but without the hippocampus they could not use deprivation state to guide nonspatial memory discriminations. Contextual memory retrieval requires hippocampal function.

To investigate how the hippocampus responded to changing internal contexts, Kennedy compared CA1 neuronal activity in hungry and thirsty rats while they performed the contextual retrieval task. Because the cues moved from trial to trial, and the rats approached the cues by walking along a narrow alley, spatial behavior was identical across deprivation conditions. CA1 activity correlated with deprivation state as rats followed identical trajectories, e.g. firing at higher rates when the rat was hungry and when it was thirsty (Kennedy and Shapiro, 2009). Memory retrieval based on internal context requires the hippocampus to distinguish between equally rewarded, nonspatial responses, and hippocampal codes distinguished identical places depending on whether the rats were hungry or thirsty. In retrospect, it would be surprising if goals did not inform memory or if the hippocampus was "blind" to internal context. Memory links the external and internal world so that we recognize opportunities to satisfy goals. We remember "what" in the context of "why" to choose actions: what and how well we remember is motivated.

PERSONAL CONTEXT 2: HOW?

The path to the desirable follows the way of the possible. Adaptive behaviors link goals with specific actions, the means of obtaining desired outcomes. Both sensory and motor processing in the cortex is described hierarchically. Joaquín Fuster (1995) describes memory as the construction of "perception action cycles" that link sensory and motor representations at equivalent hierarchical levels. Learning to distinguish small objects by touch, for example, would link neuronal activity patterns in primary somatosensory and motor cortices, whereas planning and remembering a vacation would engage medial temporal lobe (MTL) structures and prefrontal cortex (PFC) (Fuster, 1995).

The PFC computes plans and abstract rules that help us adapt rapidly to changing circumstances despite ingrained habits, such as switching from driving on the right side of the road in North America to the left side of the road in the UK (Miller and Cohen, 2001). Though "abstract rules" may seem vague, we can define these operationally as collections of multiple responses to the same contingencies. For example, "turn left" describes a generalized action that can be accomplished by many specific motor responses, e.g. while walking, swimming, riding a bicycle, or driving a car. Spatial contingencies, e.g. "go East", in a plus-shaped maze include multiple actions, e.g. turning right from the South but left from the North.

"Strategies" include even larger response categories, such as learning spatial or egocentric responses to "remember where" versus "remember turns," and remembering strategies requires the PFC in rats (Rich and Shapiro, 2009).

The PFC and the hippocampus are considered the highest order structures in the motor and sensory processing streams (Felleman and Van Essen, 1991). The perception-action cycle argues that the PFC and the hippocampus should interact strongly and implies that rules should be included in episodic memory and vice versa. If hippocampal representations include abstract rules, then hippocampal codes should distinguish actions guided by different rules during otherwise identical behaviors, just as they do for different spatial goals, and they do.

Janina Ferbinteanu trained rats to switch among different rules in the plus maze (Ferbinteanu et al., 2010). The spatial task was as described above: the rat learned by trial and error which of the two goal arms contained food during a block of trials, and the spatial goal was switched after it reached the performance criterion. The same rats also learned to approach a visible cue that was moved pseudorandomly between the goal arms. The spatial and cue approach tasks were tested each day in different blocks of trials. Hippocampal lesions impaired performance in the spatial but not the cue approach task. Note that the rats performed identical behaviors in each maze arm, two different responses guided by two different tasks. In each start arm, the rat approached the choice point guided either by spatial memory ("go East" or "go West") or by the visual cue ("turn left" or "turn right").

Hippocampal activity recorded in this experiment distinguished tasks and journeys. Single neurons fired differently depending on both the particular journey and the task rules that guided identical responses, and population codes discriminated among the four conditions throughout the maze. Firing rates in a start arm, in other words, predicted not only which goal arm the rat would enter, but also whether the goal was defined by spatial memory or the presence of the cue. Firing rates in a goal arm distinguished the starts of different journeys as well as the two tasks (Ferbinteanu et al., 2010). The results show that the hippocampus encodes abstract rules along with spatiotemporal information.

Taken together, the experiments suggest that hippocampal activity represents a wide range of internal contextual features, including motivation and abstract rules along with spatial and temporal context. As in people, the rat hippocampus supports memory for episodes by encoding events within a spatial, temporal, and internal context.

MEMORY TIME

Time, along with space, frames all experience, and hippocampal neurons code time across different scales. As animals undergo classical conditioning, hippocampal neurons gradually model the temporal structure of the milliseconds scale conditioned response (Berger et al., 1976; Solomon et al., 1986; Moyer, Jr. et al., 1990; McEchron and Disterhoft, 1997). As rodents explore a familiar spatial environment repeatedly and the same population of CA1 place cells is recorded over days, the active subpopulation drifts in rats (Mankin et al., 2012) and mice

(Ziv et al., 2013). Each cell's place field is consistent, but a given CA1 cell may or may not fire in any one recording session, and the subset of active CA1 cells becomes uncorrelated over time (Mankin et al., 2012). CA3 cells, in contrast, had stable place fields throughout the same recording sessions (Mankin et al., 2012). This difference in temporal stability provides a superb opportunity for investigating the mechanisms of stability and plasticity of spatial representations over time.

Memory tasks that require bridging 5–30-second delays reveal that CA1 cells fire reliably at specific intervals as a rat occupies one place. Such "time cells" are recorded in both spatial alternation (Pastalkova et al., 2008) and nonspatial associative memory tasks (MacDonald et al., 2011; Kraus et al., 2013). In each trial of an object–odor delayed association task in a modified T-maze, a rat was placed in a starting area, presented briefly with one of two objects, allowed to enter a waiting area for a 6–10-second delay, and then approach a scented, sand-filled flowerpot. Each object–odor pair was associated with a different response. In "go" trials, the rat could find its reward by digging in the flowerpot; in "no-go" trials the rat could find its reward in a different place by not digging. Thus, the rat had to remember which object had been presented before the delay.

CA1 cells responded to each of these memory components. Many had place fields, others distinguished between the objects, the odors, the go versus no-go responses, or had conjunctive properties, e.g. firing most when a specific odor was presented after a particular sample object. The most surprising finding was that ~50% of the CA1 cells that were active during the delay fired in specific periods as though the neurons coded the passage of time. The active population changed smoothly, with different subsets of neurons firing maximally at the start, middle, and end of the delay (MacDonald et al., 2011). The cells were signaling psychological and not physical time, because the activity was specific to the particular sample and matched stimuli that framed the delay, as though signaling the association between events in time. Just as some place cells remap when the distances among controlled cues change (Muller and Kubie, 1987), some time cells "re-timed" when the delay interval between sample and match changed, while others fired consistently at the start or the end of different delays (MacDonald et al., 2011).

Neurons in the medial temporal lobes signal both place and time. Thus, when rats walk on a treadmill at different speeds, CA1 ensembles distinguish both elapsed time and distance traveled (Kraus et al., 2013). Movement is not needed for CA1 time coding, however, and is prominent when rats are immobile while performing recent memory tasks using odor cues (MacDonald et al., 2013: 14607–16).

Neurons in the medial entorhinal cortex that fire in classical grid patterns as rats explore open fields (Fyhn et al., 2004) also fire during specific delay intervals between choices in a spatial alternation task (Kraus et al., 2015). Hippocampal time coding has been described in nonhuman primates and people. In monkeys trained to remember sequences of visual cues, most hippocampal neurons signaled the order of stimulus presentations rather than the stimuli per se (Naya and Suzuki, 2011).

Neurons in medial temporal lobe circuits that interconnect strongly with the hippocampus signaled item more than order. If hippocampal time signals contribute to coding behavioral episodes (Kraus et al., 2013), then lesions of the hippocampus should impair memory for event sequences. Indeed, lesions of CA1 or CA3 impair memory for recently presented odors after ten seconds (Farovik et al., 2010), suggesting that the time cell signal may be crucial for remembering recent nonspatial events. The hippocampus encodes recent events as sequences in time together with the temporal boundaries that frame episodes.

DYNAMIC REPRESENTATIONS "REPLAY"

The brain takes time to guide behavior, but not much. Sensation, perception, memory, active decision making, and initiating action all occur in less than 1 second. In reaction time tasks, for example, people identify and respond to stimuli in about 300 ms; rats placed at the start of a maze take less than a second to reach a choice point and approach a goal. Brain activity during these intervals implements the computations that link the opportunity, motivation, and means to behave adaptively in that particular situation.

Sometimes rats appear to stop and think as they approach the choice point of a maze, pausing and looking back and forth before selecting a goal arm, and Tolman proposed exactly that. Looking back and forth the rat was displaying *vicarious trial and error* considering the available options (Tolman, 1948). The evidence suggests that brain activity indeed represents alternative future trajectories. As rats paused at the choice points of T-mazes, hippocampal CA3 place cells fired in alternating, forward "sweeps" into the goal arms, as though the active population represented future journeys toward food (Johnson and Redish, 2007).

In other spatial navigation tasks, hippocampal populations distinguish spatial sequences from the recent past toward potential futures, as well as from current locations backward toward the origin of journeys (Foster and Wilson, 2006; Pfeiffer and Foster, 2013). The mechanisms that drive these "forward and reverse replays" are unknown, as is their behavioral significance. The physiological mechanisms that accompany replay may indeed support memory. Shantanu Jadhav used a closed-loop feedback system to detect and then disrupt replays during ongoing behavior by delivering current to the ventral hippocampal commissure, a fiber bundle that connects bilateral hippocampal circuits (Jadhav et al., 2012). Stimulation that disrupted replay events impaired spatial learning, but identical stimulation delayed until after replays did not, showing that replays are necessary for spatial memory (Jadhav et al., 2012). The results imply that replay is a crucial memory mechanism, perhaps driving persistent memory storage in cortical synapses to support consolidation.

Other possibilities remain. For example, disrupting replays could impair memory through other mechanisms, such as adding noise to hippocampal circuits when synaptic weights are changing independent of replay. Presumably replay events would specify those weight changes, so establishing more precise links between replay and memory will have to wait until experiments can distinguish between the effects of firing sequences during replay and other signaling mechanisms that accompany those events. Either way, active populations of hippocampal neurons signal more than the current location occupied by a rat. Firing correlates corresponding to vicarious trial and error in particular, and forward and backward replay in general, show that hippocampal neurons code sequences of events that extend in place and time.

CONCLUSION

By describing place cells as elements of a spatial map, O'Keefe suggested how neuronal activity recorded in behaving animals could reveal brain mechanisms of mental representation. He advanced the available technology for single unit recording, combined with Hebb's (1949) suggestions that neuronal groups could represent concepts and Tolman's cognitive maps. The approach led to the 2014 Nobel Prize for Physiology or Medicine being awarded to John O'Keefe, May Britt and Edvard Moser for discovering "a cellular basis for higher cognitive function."

Together with other MTL brain structures, hippocampal neurons perform computations that are critical for remembering recent events, episodic memories framed by personal and temporal as well as spatial context. The neural machinery for spatial computations is known in more detail than are the mechanisms for episodic memory, but recent discoveries have begun to reveal how the brain may construct a memory space. Place and grid cells encode time, and populations of hippocampal cells represent past and future as well as present locations, suggesting a mechanism for signaling the temporal context. Firing sequences that correspond to past, present, and future places occur several times a second and provide a mechanism that could allow memory to guide action. Motives and task rules retrieve different hippocampal representations during identical spatial behaviors that indicate a potential mechanism for signalling "personal" contexts. The same computations that calculate distances and angles among physical places may also compute distances and directions in other spaces, including social space.

A cognitive mapping theory that describes episodic memory coding in terms as rigorous and straightforward as place units describe location coding will require far better understanding of how the brain integrates means, motives, and opportunities. That is, we need to learn how relevant information from the past guides selective memory retrieval that filters the present and predicts the outcomes of potential actions. These computations are performed by widely distributed neural networks in brain regions including the PFC that are not well understood individually and that communicate using unknown mechanisms. Emerging technologies can record hundreds or thousands of neurons across several brain regions simultaneously. Comparing activity patterns within and between regions recorded during identical behaviors guided by different motives, rules, and strategies will reveal new and unexpected brain mechanisms of cognition.

REFERENCES

Berger T.W., Alger, B. and Thompson, R.F. (1976) Neuronal substrate of classical conditioning in the hippocampus. *Science,* 192: 483–5.

Clayton, N.S., Griffiths, D.P., Emery, N.J. and Dickinson, A. (2001) Elements of episodic-like memory in animals. *Philosophical Transactions of the Royal Society of London B: Biological Sciences*, 356: 1483–91.

Cohen, N.J. and Squire, L.R. (1980) Preserved learning and retention of pattern-analyzing skill in amnesia: dissociation of knowing how and knowing what. *Science,* 210: 207–9.

Crystal, J.D., Alford, W.T., Zhou, W. and Hohmann, A.G. (2013) Source memory in the rat. *Current Biology*, 23: 387–91.

Douglas, R.J. (1967) The hippocampus and behavior. *Psychological Bulletin*, 67: 416–42.

Eacott, M.J., Easton, A. and Zinkivskay, A. (2005) Recollection in an episodic-like memory task in the rat. *Learning & Memory*, 12: 221–3.

Eichenbaum, H., Dudchenko, P., Wood, E., Shapiro, M. and Tanila, H. (1999) The hippocampus, memory, and place cells: Is it spatial memory or a memory space? *Neuron,* 23: 209–26.

Farovik, A., Dupont, L.M. and Eichenbaum, H. (2010) Distinct roles for dorsal CA3 and CA1 in memory for sequential nonspatial events. *Learning & Memory*, 17: 12–17.

Felleman, D.J. and Van Essen, D.C. (1991) Distributed hierarchical processing in the primate cerebral cortex. *Cerebral Cortex*, 1: 1–47.

Ferbinteanu, J. and Shapiro, M.L. (2003) Prospective and retrospective memory coding in the hippocampus. *Neuron,* 40: 1227–39.

Ferbinteanu, J., Shirvalkar, P.R. and Shapiro, M.L. (2010) *Memory demands differentially alter prospective and retrospective coding.*

Foster, D.J., and Wilson, M.A. (2006) Reverse replay of behavioural sequences in hippocampal place cells during the awake state. *Nature*, 440: 680–3.

Frank, L.M., Brown, E.N. and Wilson, M. (2000) Trajectory encoding in the hippocampus and entorhinal cortex. *Neuron*, 27 (1): 169–78.

Frank, L.M., Stanley, G.B. and Brown, E.N. (2004) Hippocampal plasticity across multiple days of exposure to novel environments. *Journal of Neuroscience*, 24: 7681–9.

Fuster, J.M. (1995) *Memory in the Cerebral Cortex*. Cambridge, MA: MIT Press.

Fyhn, M., Molden, S., Witter, M.P., Moser, E.I. and Moser, M.B. (2004) Spatial representation in the entorhinal cortex. *Science,* 305: 1258–64.

Gibson, B., Butler, W.N. and Taube, J.S. (2013) The head-direction signal is critical for navigation requiring cognitive map but not for learning a spatial habit. *Current Biology*, 23 (16): 1536–40. doi: 10.1016/j.cub.2013.06.030

Hebb, D.O. (1949) *The Organization of Behavior*. New York: Wiley.

Hinton, G.E., McClelland, J.L. and Rumelhart, D.E. (1986) Distributed representations. In J.L. McClelland and D.E. Rumelhart (eds), *Parallel Distributed Processing: Explorations in the Microstructure of Cognition, Volume 1: Foundations.* Cambridge, MA: MIT Press. pp. 77–109.

Hirsh, R. (1974) The hippocampus and contextual retrieval of information from memory: a theory. *Behavioral Biology*, 12: 421–44.

Hirsh, R., Davis, R.E. and Holt, L. (1979) Fornix-thalamus fibers, motivational states, and contextual retrieval. *Experimental Neurology*, 65: 373–90.

Hull, C.L. (1932) The goal gradient hypothesis and maze learning. *Psychological Review*, 39: 25–43.

Jadhav, S.P., Kemere, C., German, P.W. and Frank, L.M. (2012) Awake hippocampal sharp-wave ripples support spatial memory. *Science*, 336: 1454–8.

Johnson, A. and Redish, A.D. (2007) Neural ensembles in CA3 transiently encode paths forward of the animal at a decision point. *Journal of Neuroscience* 27: 12176–89.

Kennedy, P.J. and Shapiro, M.L. (2004) Retrieving memories via internal context requires the hippocampus. *Journal of Neuroscience* 24: 6979–85.

Kennedy, P.J. and Shapiro, M.L. (2009) Motivational states activate distinct hippocampal representations to guide goal-directed behaviors. *Proceedings of the National Academy of Sciences, USA,* 106: 10805–10.

Kentros, C., Hargreaves, E.L., Hawkins, R.D., Kandel, E.R., Shapiro, M. and Muller, R.V. (1998) Abolition of long-term stability of new hippocampal place cell maps by NMDA receptor blockade. *Science,* 280: 2121–6.

Knierim, J.J., Neunuebel, J.P. and Deshmukh, S.S. (2014) Functional correlates of the lateral and medial entorhinal cortex: objects, path integration and local-global reference frames. *Philosophical Transactions of the Royal Society of London: B Biological Sciences*, 369: 20130369.

Komorowski, R.W., Garcia, C.G., Wilson, A., Hattori, S., Howard, M.W. and Eichenbaum, H. (2013) Ventral hippocampal neurons are shaped by experience to represent behaviorally relevant contexts. *Journal of Neuroscience*, 33: 8079–87.

Kraus, B.J., Brandon, M.P., Robinson, R.J., Connerney, M.A., Hasselmo, M.E. and Eichenbaum, H. (2015) During running in place, grid cells integrate elapsed time and distance run. *Neuron,* 88: 578–89.

Kraus, B.J., Robinson, R.J., White, J.A., Eichenbaum, H. and Hasselmo, M.E. (2013) Hippocampal "time cells": time versus path integration. *Neuron,* 78: 1090–1101.

Leutgeb, S., Leutgeb, J.K., Treves, A., Moser, M.B. and Moser, E.I. (2004) Distinct ensemble codes in hippocampal areas CA3 and CA1. *Science*, 305: 1295-8.

Lever, C., Burton, S., Jeewajee, A., O'Keefe, J. and Burgess, N. (2009) Boundary vector cells in the subiculum of the hippocampal formation. *Journal of Neuroscience*, 29: 9771–7.

MacDonald, C.J., Carrow, S., Place, R. and Eichenbaum, H. (2013) Distinct hippocampal time cell sequences represent odor memories in immobilized rats. *Journal of Neuroscience*, 33: 14607–16.

MacDonald, C.J., Lepage, K.Q., Eden, U.T. and Eichenbaum, H. (2011) Hippocampal "time cells" bridge the gap in memory for discontiguous events. *Neuron,* 71: 737–49.

Mankin, E.A., Sparks, F.T., Slayyeh, B., Sutherland, R.J., Leutgeb, S. and Leutgeb, J.K. (2012) Neuronal code for extended time in the hippocampus. *Proceedings of the National Academy of Sciences, USA,* 109: 19462–7.

Markus, E.J., Qin, Y.L., Leonard, B., Skaggs, W.E., McNaughton, B.L. and Barnes, C.A. (1995) Interactions between location and task affect the spatial and directional firing of hippocampal neurons. *Journal of Neuroscience*, 15: 7079–94.

McEchron, M.D. and Disterhoft, J.F. (1997) Sequence of single neuron changes in CA1 hippocampus of rabbits during acquisition of trace eyeblink conditioned responses. *Journal of Neurophysiology*, 78: 1030–44.

Miller, E.K. and Cohen, J.D. (2001) An integrative theory of prefrontal cortex function. *Annual Review of Neuroscience*, 24: 167–202.

Milner, B. (2005) The medial temporal-lobe amnesic syndrome. *Psychiatrix Clinics of North America*, 28: 599–611.

Moyer, J.R., Jr., Deyo, R.A. and Disterhoft, J.F. (1990) Hippocampectomy disrupts trace eye-blink conditioning in rabbits. *Behavioral Neuroscience*, 104: 243–52.

Muller, R.U. and Kubie, J.L. (1987) The effects of changes in the environment on the spatial firing of hippocampal complex-spike cells. *Journal of Neuroscience*, 7: 1951–68.

Naya, Y. and Suzuki, W.A. (2011) Integrating what and when across the primate medial temporal lobe. *Science*, 333 (6043):773-6. doi: 10.1126/science.1206773

O'Keefe, J. (1976) Place units in the hippocampus of the freely moving rat. *Experimental Neurology*, 51: 78–109.

O'Keefe, J. and Conway, D.H. (1978) Hippocampal place units in the freely moving rat: why they fire where they fire. *Experimental Brain Research*, 31: 573–90.

O'Keefe, J. and Dostrovsky, J. (1971) The hippocampus as a spatial map: preliminary evidence from unit activity in the freely-moving rat. *Brain Research*, 34: 171–5.

O'Keefe, J. and Nadel, L. (1978) *The Hippocampus as a Cognitive Map*. Oxford: Oxford University Press.

O'Keefe, J. and Speakman, A. (1987) Single unit activity in the rat hippocampus during a spatial memory task. *Experimental Brain Research*, 68: 1–27.

Pastalkova, E., Itskov, V., Amarasingham, A. and Buzsaki, G. (2008) Internally generated cell assembly sequences in the rat hippocampus. *Science,* 321:1322–27.

Pfeiffer, B.E. and Foster, D.J. (2013) Hippocampal place-cell sequences depict future paths to remembered goals. *Nature*, 497: 74–79.

Ranck, J.B., Muller, R.U. and Taube, J.S (1987) Head direction cells recorded from post subiculum in freely moving rats. *Neuroscience* 22 (suppl. 528p): 177.

Rempel-Clower, N.L., Zola, S.M., Squire, L.R. and Amaral, D.G. (1996) Three cases of enduring memory impairment after bilateral damage limited to the hippocampal formation. *Journal of Neuroscience*, 16: 5233–55.

Rich, E.L. and Shapiro, M. (2009) Rat prefrontal cortical neurons selectively code strategy switches. *Journal of Neuroscience*, 29: 7208–19.

Schmitt, W.B., Deacon, R.M., Seeburg, P.H., Rawlins, J.N. and Bannerman, D.M. (2003) A within-subjects, within-task demonstration of intact spatial reference memory and impaired spatial working memory in glutamate receptor-A-deficient mice. *Journal of Neuroscience*, 23: 3953–9.

Scoville, W.B. and Milner, B. (1957) Loss of recent memory after bilateral hippocampal lesions. *Journal of Neurology, Neurosurgery, and Psychiatry*, 20:11–21.

Shapiro, M. (2015) A limited positioning system for memory. *Hippocampus,* 25: 690–6.

Shapiro, M. and O'Connor, C. (1992) N-methyl-D-aspartate receptor antagonist MK-801 and spatial memory representation: working memory is impaired in an unfamiliar environment but not in a familiar environment. *Behavioral Neuroscience*, 106: 604–12.

Shapiro, M., Tanila. H. and Eichenbaum, H. (1997) Cues that hippocampal place cells encode: dynamic and hierarchical representation of local and distal stimuli. *Hippocampus,* 7: 624–42.

Smith, D.M. and Mizumori, S.J. (2006) Learning-related development of context-specific neuronal responses to places and events: the hippocampal role in context processing. *Journal of Neuroscience*, 26: 3154–63.

Solomon, P.R., Vander Schaaf, E.R., Thompson, R.F. and Weisz, D.J. (1986) Hippocampus and trace conditioning of the rabbit's classically conditioned nictitating membrane response. *Behavioral Neuroscience*, 100: 729–44.

Tanila, H., Shapiro, M.L. and Eichenbaum, H. (1997) Discordance of spatial representations in ensembles of hippocampal place cells. *Hippocampus,* 7: 613–23.

Taube, J.S., Muller, R.U. and Ranck, J.B. (1990) Head direction cells recorded from the postsubiculum in freely moving rats. I. Description and quantitative analysis. *Journal of Neuroscience* 10 (2): 420–35.

Tavares, R.M., Mendelsohn, A., Grossman, Y., Williams, C.H., Shapiro, M., Trope, Y. and Schiller, D. (2015) A map for social navigation in the human brain. *Neuron*, 87: 231–43.

Thompson, L.T. and Best, P.J. (1990) Long-term stability of the place-field activity of single units recorded from the dorsal hippocampus of freely behaving rats. *Brain Research* 509: 299–308.

Tolman, E.C. (1948) Cognitive maps in rats and men. *Psychological Review*, 56: 144–55.

Tulving, E. and Markowitsch, H.J. (1998) Episodic and declarative memory: role of the hippocampus. *Hippocampus,* 8: 198–204.

Wickelgren, W.A. (1979) Chunking and consolidation: a theoretical synthesis of semantic networks, configuring in conditioning, S-R versus congenitive learning, normal forgetting, the amnesic syndrome, and the hippocampal arousal system. *Psychological Review*, 86: 44–60.

Wilson, M.A. and McNaughton, B.L. (1993) Dynamics of the hippocampal ensemble code for space [see comments] [published erratum appears in *Science,* 264 (5155): 16]. *Science,* 261: 1055–8.

Winter, S.S., Clark, B.J. and Taube, J.S. (2015) Spatial navigation: disruption of the head direction cell network impairs the parahippocampal grid cell signal. *Science,* 347: 870–4.

Wood, E.R., Dudchenko, P.A., Robitsek, R.J. and Eichenbaum, H. (2000) Hippocampal neurons encode information about different types of memory episodes occurring in the same location [In Process Citation]. *Neuron,* 27: 623–33.

Ziv, Y., Burns, L.D., Cocker, E.D., Hamel, E.O., Ghosh, K.K., Kitch, L.J., El, G.A. and Schnitzer, M.J. (2013) Long-term dynamics of CA1 hippocampal place codes. *Nature Neuroscience*, 16: 264–6.

Zola-Morgan, S. and Squire, L.R. (1986) Memory impairment in monkeys following lesions limited to the hippocampus. *Behavioral Neuroscience*, 100 (2): 155–60.

Zola-Morgan, S., Squire, L.R. and Amaral, D.G. (1986) Human amnesia and the medial temporal region: enduring memory impairment following a bilateral lesion limited to field CA1 of the hippocampus. *Journal of Neuroscience*, 6 (1): 2950–67.

PART 3

Chemicals and Behaviour

13 Revisiting Phoenix, Goy, Gerall and Young: The organizational/activational theory of steroid-mediated sexual differentiation of brain and behavior

Sarah Raza and Robbin Gibb

THE ORIGIN OF THE ORGANIZATIONAL/ ACTIVATION THEORY

In North America, the first half of the twentieth century was a challenging time for scientists studying the physiology of sex and reproduction. While it was clear that by improving our understanding of reproductive physiology yields in the livestock industry could be enhanced, the implications for human reproduction and sex behaviors were largely ignored or suppressed. Indeed, experiments addressing such topics were described as "health research" to avoid censorship and funding sanctions (Phoenix, 2009). Although polite society was not prepared to accept scientific interference in the taboo domain of human sexuality, in the 1950s pregnant women with a history of miscarriage were prescribed androgenic hormonal treatment to prevent pregnancy loss. The masculinizing effect of this treatment on female infants was obvious and unexplained, and the first case reports of this phenomenon were published in 1955 (Gross and Meeker, 1955).

It was at this time, and in this context, that a group of researchers (Charles Phoenix, Robert Goy and Arnold Gerall) working in the William Young lab conducted a series of experiments in which they injected pregnant guinea pigs with testosterone, thereby exposing the developing fetuses to excess androgen in the prenatal period. They discovered that although the androgen exposure had no apparent effect on the male offspring, it produced genetic females that were phenotypically difficult to distinguish from males. They also showed male typical sexual behavior in adulthood (Phoenix et al., 1959). This seminal paper was important for several reasons.

In addition to attributing the action of gonadal hormones on the brain, this work demonstrated that fetal masculinization of the body and behavior were permanent. This work also provided evidence of a "critical period" for the masculinizing action of testosterone (Arnold, 2009). Perhaps most importantly, Phoenix and colleagues established a conceptual framework for the study of sex differences in brain and behavior. The authors dichotomized hormonal effects into two modes of action: organizational and activational. During the prenatal period, gonadal hormones (e.g. testosterone) act to "organize" or differentiate neural tissue in the direction of masculinization or feminization. In adulthood, gonadal hormones "activate" neural tissue already present and organized prenatally (Phoenix et al., 1959; Arnold and Breedlove, 1985; Arnold, 2009).

The central thesis of the original organizational/activational theory – that the fetal actions of hormones exert permanent effects on the neural substrate by which gonadal hormones act on in adulthood – provided the first heuristic framework for explaining gonadal and non-gonadal sexual differentiation of phenotype (Arnold, 2009). Phoenix et al. (1959) postulated that the dichotomy between organizational and activational effects is derived from three criteria: (1) the developmental age at steroid action, (2) the duration of the effect, and (3) the presence of critical periods. The authors proposed that early in development, during a critical period, hormones induce long-term, permanent alterations in the organization of neural pathways or in the steroid-responsiveness of neurons. Conversely, in adulthood (absence of critical periods), short-term, impermanent effects emerge in response to steroid hormones, with no implications for neural pathway reorganization (Arnold and Breedlove, 1985).

In this chapter, we discuss advances in our understanding of sexual differentiation of brain and behavior, including factors now known to play a role in the process. We also present challenges to the original theoretical framework and discuss controversies that arise from the organizational/activational theory.

CURRENT THINKING ABOUT SEX DETERMINATION AND DIFFERENTIATION

Phoenix and colleagues referred to the developmental effects of hormones as organizational, and this organization induced sexual differentiation in both body and brain. We now know that the process of sexual differentiation is secondary to sex determination, as the genes encoded on the sex chromosomes – X and Y – are the primary agents that ultimately give rise to the actions of gonadal hormones. In other words, sexual differentiation is the process by which the genetic information carried on the sex chromosomes determines the morphological pathway (testis or ovary) a bi-potential gonad will take. For example, the Sry gene contained on the Y chromosome induces testicular formation and then the hormone (testosterone) produced by the testes initiates the process of sex differentiation (McCarthy et al., 2009). (A detailed discussion on the role of genetics in sex differentiation is found below.)

It appears that although some feminizing effects of estrogen in the prenatal period have been demonstrated, the organizing effect of steroidal hormones is largely due to testosterone exposure. The presence of testosterone directs the body and brain to assume a male phenotype and the lack of testosterone leads to the default expression of the female phenotype. In fact, neurons that underlie some aspects of male sexual behavior form in both sexes and then die off in females as a result of testosterone deprivation (McCarthy and Konkle, 2005). It is in the intrauterine period that gender identity (feeling as belonging to the male or female sex) and various aspects of sexually differentiated forms of cognitive behavior arise.

Whereas sexual differentiation of the gonads occurs in the first two months post conception, sexual differentiation of the brain doesn't occur until the last half of pregnancy. This can lead to a potential mismatch of phenotypic sex and gender identity (Swaab and Garcia-Falgueras, 2009). But the story does not end here. For an adult male to engage in male typical behavior, other "activational" processes that target different neural circuitry must occur: masculinization of areas that activate male sexual behavior (e.g. the preoptic area of the hypothalamus) and defeminization of brain areas (e.g. the ventral medial nucleus of the hypothalamus) to suppress female sexual behavior (McCarthy et al., 2009).

A number of findings regarding sexual differentiation of the brain have come to light since the initial organizational/activational model was proposed. Some of the findings support the initial theory, some extend the theory to include other morphogenic factors, and yet others shed a new light on our understanding of how sexual differentiation of the brain occurs. These findings will be reviewed in the following sections.

FACTORS THAT INFLUENCE HORMONALLY-INDUCED SEXUAL DIFFERENTIATION

Building upon the foundation of the organizational/activational theory, the emphasis on gonadal hormones as the single causative factor for the origin of sex differences has significantly shifted in the last two decades. A number of factors have been identified that modulate hormone-induced sexual differentiation, including the sex chromosomes themselves, and indirect effects on hormonal expression via exposure to drugs, environmental contaminants and elements in the diet. Therefore, in contrast to the original postulates of the organizational/activational theory, it appears that various additional factors may be implicated in this dichotomy, thus expanding the original theoretical framework.

GENETIC FACTORS

Sexually dimorphic gene expression

While gonadal hormones play a significant role in sexual differentiation of the brain – as suggested by the organization/activational theory – they are unlikely to account for all sexually differentiated traits. Rather, it is believed that both genetic and hormonal mechanisms are likely involved in an interconnected, dependent manner. Gonadal hormones, therefore, function as secondary proximate factors

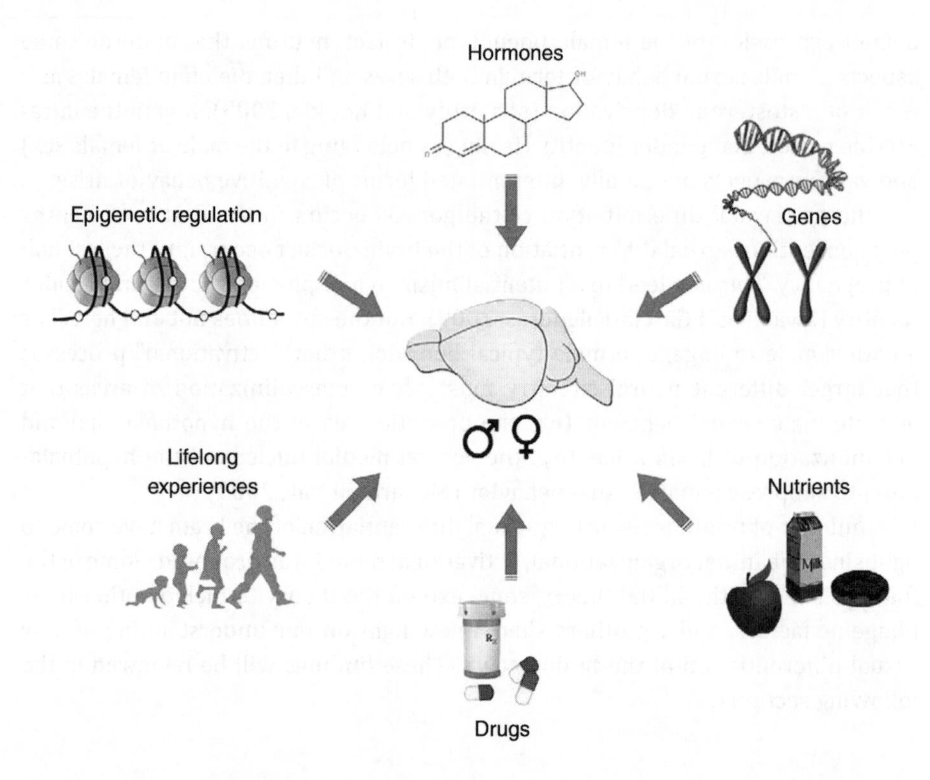

Figure 13.1 Direct and indirect factors influencing sexual differentiation of brain and behavior

that are downstream of the primary sex-specific chromosomes. Taken together, the sex chromosomes cause sex differences in the expression of gonadal hormones, which in turn produce widespread sexually differentiated effects on many tissues (Arnold, 2009). More specifically, the expression of genes residing on the sex chromosomes (X and Y) are thought to influence the sexually differentiated brain-behavior phenotype prior to the onset of gonadal hormone secretion, as well as thereafter, suggesting a complex gene–hormone interaction (Davies and Wilkinson, 2006). For instance, the *Ar* gene responsible for encoding the androgen receptor is found on the X chromosome, resulting in differential expression in male and female animals (McAbee and DonCarlos, 1999). As sex-linked genes may constitute an additional factor influencing the sexual differentiation of brain and behavior, three possible mechanisms by which this may occur include X-linked gene dosage effects, X-linked imprinting, and Y-specific gene expression.

Before examining these genetic mechanisms, however, it is important to revisit the mammalian XX/XY sex chromosome expression patterns. Sex is determined by the presence or absence of the Y chromosome. In brief, males inherit a single Y chromosome from their father and an X chromosome from their mother (XY), whereby the X chromosome is active. Females, on the other hand, inherit two X chromosomes, one from each parent (XX). One X chromosome is significantly (but not completely) silenced via X-inactivation in each cell – a process preventing a potentially toxic double dosage of X chromosome gene products in females (Heard, 2004).

As previously mentioned, sexual differentiation in neural circuitry and behavior may arise via a process called X-linked gene dosage effects, which involves regulatory mechanisms implicating X chromosome expression. Given that females inherit two X chromosomes, as opposed to males, X-linked genes that escape X-inactivation will be more highly expressed in the female brain. That is, the expression of escape genes on the X chromosome is more likely to be exhibited in females. This suggests that X-linked gene dosage effects may potentially account for some neurobiological sexually dimorphic differences (Davies and Wilkinson, 2006; Abramowitz et al., 2014).

In addition to X-linked gene dosage as a contributing factor, X-linked genomic imprinting may also produce changes, resulting in differences between males and females. Genomic imprinting is a phenomenon involving a small subset of genes that are "imprinted" or marked indicating parental origin. That is, some imprinted genes are expressed from the mother whereas others are from the father (Reik and Walter, 2001; Abramowitz et al., 2014). Given the manner by which sex chromosomes are inherited and the imbalance of X chromosomes between the sexes – namely, a maternal and paternal X chromosome inherited by females (XX) and a maternal X chromosome inherited by males (XY) – differential gene expression as a consequence of X-linked imprinting is likely to emerge. The expression of paternal X-linked imprinted genes, therefore, would be solely exhibited in the female brain, whereas the expression of maternal X-linked imprinted genes would be exhibited in both sexes, but relatively higher in males (Davis et al., 2006; Davis and Wilkinson, 2006). As such, the functional, or in some cases even dysfunctional, expression of imprinted X-linked genes may account for fundamental sex differences in brain and behavior.

Finally, Y-specific gene expression may be an additional mechanism by which sexual differentiation may occur. The Y chromosome possesses a region called the non-recombining region (NRY), where the Y chromosome cannot recombine with the X chromosome during cell division. Genes located in the NRY region are solely expressed in the male brain (Skaletsky et al., 2003), implying a male-specific predisposition. All in all, while it is evident that sex hormones may influence sexual disparity in brain development and behavior, they are unlikely to be the sole causative factor. It appears to be the case that genes are, surprisingly, influencing sexual differentiation more than originally thought.

CLINICAL POPULATIONS

In addition to early animal studies of gonadal hormone manipulations, the study of endocrine conditions in humans had been useful in assessing the contribution of gonadal hormones and sex-linked genes to sexually differentiated brain development. Here we focus on three conditions – congenital adrenal hyperplasia, Turner syndrome, and androgen insensitivity – illustrating the influence of the dynamic gene–hormone interaction on sexual differentiation.

Congenital adrenal hyperplasia (CAH)

Marked by a mismatch between gonadal hormones and sexual differentiation, congenital adrenal hyperplasia (CAH) stems from an enzymatic defect, 21-hydroxylase deficiency. This results from expression of a single autosomal recessive gene. With a prevalence rate of 1 in 15,000 (van der Kamp and Wit, 2004), CAH illustrates one of the most common conditions associated with atypical sexual differentiation. In early fetal life, high levels of adrenal androgens (testosterone and dihydrotestosterone) are produced, but low levels of cortisol. While both sexes are affected by CAH, studies in non-human species have confirmed 'male-typical' behavior in females with CAH, as a consequence of the masculinizing and defeminizing effects of excess androgens. Moreover, CAH females possess masculinized external genitalia, whereas CAH males retain male external genitalia (Ehrhardt and Meyer-Bahlburg, 1979; Cohen-Bendahan et al., 2005).

Turner Syndrome (TS)

Characterized by dysgenesis of the gonads, Turner Syndrome (TS) is a genetic developmental disorder with a population prevalence of 1 in 2000 females. Abnormal development of the gonads is the product of a chromosomal abnormality – 45, XO – where a reduction in X-linked gene dosage typically results in the complete loss of an entire X chromosome (Uematsu et al., 2002). While TS individuals appear to be female at birth, gonadal hormones appear to be lacking both pre- and postnatally. Compared to typically developing females, TS subjects are susceptible to a range of neuropsychological deficits in memory and attention, social interaction, and visuospatial skills (Nijhuis-van der Sanden et al., 2003). Moreover, alterations in structure and function have been identified in several brain regions (Davis and Wilkinson, 2006).

Androgen insensitivity (AI)

Although individuals with androgen insensitivity (AI) are genetic males (i.e. possess the male XY chromosomes) and exhibit the typical cascade of male sexual differentiation, they lack functioning androgen receptors. That is, the tissues of the body are insensitive to androgen (i.e. testosterone) and, consequently, the external genitalia differentiate in a female-typical direction (testicular feminization; Ehrhardt and Meyer-Bahlburg, 1979). This is thought to be, in part, due to gene expression on the Y chromosome. While AI individuals are reared as females and exhibit female-typical behaviors, they are typically not diagnosed until puberty, when they fail to menstruate. AI is rare, with a population prevalence of 1 to 5/100,000 (Cohen-Bendahan et al., 2005).

Environmental factors

Gonadal hormones are well positioned as important fetal secondary factors that may interact with environmental risk factors and/or experiences to asymmetrically affect the sexes. Several lines of evidence suggest that sexual differentiation is vulnerable to the effects of a variety of pre- and perinatal environmental factors, many of which possess estrogenic-like or anti-androgenic properties. Deviations from the normal pattern of androgen release by the gonads can greatly interfere with sexually differentiated brain development and, subsequently, masculinization and/or feminization of the behavioral phenotype. Maternal exposure to drugs, nutrients, and environmental contaminants are being increasingly recognized for their impact on aspects of offspring sexual differentiation, resulting in altered brain–genital development.

Here we will consider emerging evidence supporting the influence of drugs, maternal diet, and environmental contaminants on offspring sexual differentiation, via alterations in sex hormone levels during critical periods of fetal development.

DRUGS

Numerous experiments in non-human species have identified demasculinization of sexually related behaviors in male rats, as well as disrupted uptake of gonadal hormones in the brain, following prenatal cocaine exposure (Raum et al., 1990). Moreover, prenatal exposure to barbiturates, nicotine, THC, morphine, and alcohol during critical developmental periods produces a similar feminization of offspring behavior, attenuating gonadal hormone activity and thereby altering the degree of sexual differentiation (Kakihana et al., 1980; Reinisch and Sanders, 1982; Lichtensteiger and Schlumpf, 1985; Segarra and Strand, 1989).

DIET AND NUTRITION

With respect to nutrition, the effects of maternal diet and infant feeding choice influence hormone levels in the mother and offspring with potential long-term health implications (Sharpe, 2013). For example, although the mechanism remains to be determined, maternal obesity has been thought to be related to reduced fetal testosterone or androgen production, given the twofold increase in the risk of undescended testis and penile abnormalities in male offspring at birth (Akre et al., 2008; Sharpe, 2013). Reduced sperm counts have been exhibited in adult males of obese mothers, suggesting possible alterations in androgen activity in early development (Ramlau-Hansen et al., 2007).

From a postnatal perspective, the consumption of soy formula milk, as opposed to standard milk, by infant marmosets has been shown to reduce testosterone levels during the early neonatal period. This is thought to be related to the high plant estrogen content in soy milk, suggesting that the effects of infant feeding choice may also play a crucial role in influencing the production and/or modulation of sex hormones, and thereby affecting the extent to which the developing brain is masculinized or feminized (Sharpe et al., 2002).

ENVIRONMENTAL CONTAMINANTS

Recently, perinatal exposure to environmental agents and toxins has been shown to modify sex hormone production, altering the degree of masculinization and/or feminization of the developing brain. Polychlorinated biphenyls (PCBs) – used in transformers and electrical equipment – have been linked to a range of reproductive and physiological impairments (Sonnenschein and Soto, 1998). Prenatally, PCBs act on estrogen hormone receptors and interfere with hormone-dependent developmental processes, including sexual brain differentiation, reproductive maturation, and neuroendocrine gene and protein expression (Lackmann, 2002; Dickerson and Gore, 2007). A reduced gonadal hormone in newborns, and altered cognitive and behavioral function, have been correlated with prenatal PCB exposure in both humans and rats, emphasizing the possible role of environmental contaminants in sexual differentiation (Cao et al., 2008).

In addition to PCBs, humans are routinely exposed to bisphenol A (BPA), an estrogenic agonist that binds to nuclear estrogen receptors and interferes with normal estrogenic signaling. Present in a variety of beverage and food products, dental composites and manufactured goods, BPA is present at detectable levels in the human population (Calafat et al., 2005) and, more importantly, poses a significant risk for the developing human fetus (Rubin et al., 2006). Prenatal exposure to BPA induces alterations in the prostate, and the epididymis in developing male offspring, and alters vaginal opening time in females (vom Saal et al., 1998).

At a cellular level, prenatal exposure to BPA induces rapid, membrane-initiated estrogenic effects, altering the typical estrogenic signaling and, consequently, brain patterns of defeminization and/or masculinization (Nagel et al., 1999). The effects of BPA emphasizes the importance of how exposure to estrogenic environmental chemicals during gestation can alter sex hormone production and developing external genitalia, thereby altering fetal sexual differentiation. Overall, it appears that environmental risk factors and experiences are capable of altering important hormone-dependent events during critical periods of brain development, inducing altered sexually dimorphic differences in brain differentiation and function.

CHALLENGES

In the more than 50 years that have passed since the organizational/activational theory of sexual differentiation was published by Phoenix, Goy, Gerall and Young (1959), there has been an explosion of research aimed at understanding the nature of hormonal influences on sex differences. The original paper offered a new and easily understood framework in which to view the action of sex hormones. Phoenix and colleagues (1959) suggested that the link established between sex hormones and their "organizing action" extended our understanding of the role of gonadal hormones in the regulation of sexual behavior in three important ways by demonstrating:

1. The action of gonadal hormones in the prenatal period.
2. The link between gonadal hormones and neural tissue changes (similar to the link between gonadal hormones and genital tissue changes).

3. That prenatal exposure to gonadal hormones establishes "behavioral differences between the sexes."

The three points stressed by Phoenix and colleagues have not gone without challenge and controversy. Although point two has been largely supported by subsequent experimental evidence (and will therefore not be further discussed), points one and three have aroused significant debate. The current understanding of the role of gonadal hormones in the regulation of sexual behavior with respect to the two controversial points is discussed below.

THE IMPORTANCE OF GONADAL HORMONES IN THE PRENATAL PERIOD

Although a clear link has been established for the organizational role of sex hormones in the perinatal period and the activational actions of hormones in the pubertal and adult period of life, this dichotomous role of sex hormones is not as clear as was once believed. Arnold and Breedlove (1985) offer evidence from a number of experiments that challenge this view and cite the seasonal changes in the brains of songbirds as a prime example. Elaborate abundant evidence has accumulated that suggests the pubertal period of life is a time in which hormones are organizational rather than activational (Schultz et al., 2009). In addition, the finding that the production of steroid hormones (steroidogenesis) can occur in local areas of the brain has cast some doubt as to exactly when and how gonadal hormones have their major influence (Zwain and Yen, 1999). Konkle and McCarthy (2011) propose that early gonadal steroidogenesis is later replaced by local steroidogenesis, and both are important to complete the organizational process of sexual differentiation.

THE IMPORTANCE OF PRENATAL EXPOSURE TO GONADAL HORMONES IN ESTABLISHING SEX DIFFERENCES

The second controversy that arises from the organizational/activational theory is the perception that heterosexual behavior results from "normal" hormonal exposure during development resulting in male and female typical brains and bodies. This implies that homosexual behavior and transgender identity occur when the process goes awry and thus is "abnormal."

Charles Phoenix (2009) commented that the initial work published in 1959 might have been responsible for the increased interest in understanding human sexual orientation and the research that followed. MR imaging of sexually dimorphic areas in homosexual (gay and lesbian) and transgender individuals has demonstrated that sexual identity and behavior occur more on a continuum of behavior and less as a sexual dichotomy (Savic and Lindstrom, 2008; Hahn et al., 2014). Transgender issues continue to gain considerable public interest, for example with the revelation that former Olympian gold medalist, Bruce Jenner, is transgender and has embraced the female identity of Caitlyn. Although current society is more accepting of trans-gendered individuals than in the past, there is much to be done to raise public awareness of the origin of this form of gender identity.

As espoused by the Phoenix group in the late 1950s, prenatal exposure to hormones affects both gonadal and neural tissues to organize them in a particular sexually differentiated form. It is now known that the timing of differentiation of gonads is considerably different from that for differentiation of the brain, and there is no guarantee that identical hormonal conditions will prevail at both time points (Swaab and Garcia-Falgueras, 2009). The evidence that sexual behavior such as homosexuality and transgender identity is at least in part programmed prenatally by hormonal exposures (socialization and cultural influences most likely play a role as well; for a review see Hines, 2011) may be an important advance towards achieving more widespread acceptance of these and all other gender types.

CONCLUSION

More than 50 years ago, Phoenix and colleagues made a controversial claim: the brain was central to long-term, hormonally-induced changes in behavior. This work initiated a paradigm shift that gave rise to the organizational/activational theory of sexual differentiation. The influence of this work has crossed boundaries from sexual behavior to stress, cognitive function, and even mental states such as anxiety and depression (Blaustein and McCarthy, 2009). Although the conceptual framework has repeatedly been tested in the intervening years, its basic tenets have been largely supported (Arnold, 2009). "Fifty years hold no special place in the life of an idea, yet an idea that remains relevant for a half a century reflects an endurance that few ideas achieve" wrote Kim Wallen (2009) in a review of the organizational/activational theory. A notion that was viewed with considerable skepticism and debate amongst the scientific community in the middle of the twentieth century has transformed the thinking of how masculine and feminine behavior arise, and is now even conveyed as dogma in high school biology textbooks.

REFERENCES

Abramowitz, L.K., Olivier-Van Stichelen, S. and Hanover, J.A. (2014) Chromosome imbalance as a driver of sex disparity in disease. *Journal of Genomics,* 2: 77–88.

Akre,O., Boyd, H.A., Ahlgren, M. et al. (2008) Maternal and gestational risk factors for hypospadias. *Environmental Health Perspectives,* 116: 1071–6.

Arnold, A.P. (2009) The organizational-activational hypothesis as the foundation for a unified theory of sexual differentiation of all mammalian tissues. *Hormones and Behavior,* 55: 570–8.

Arnold, A.P. and Breedlove, S.M. (1985) Organizational and activational effects of sex steroids on brain and behavior: a reanalysis. *Hormones and Behavior,* 19: 469–8.

Blaustein, J.D. and McCarthy, M.M. (2009) Phoenix, Goy, Gerall, and Young, Endocrinology, 1959: 50 years young and going strong. *Endocrinology,* 150: 2501.

Calafat, A.M., Kuklenyik, Z., Reidy, J.A. et al. (2005) Urinary concentrations of bisphenol A and 4-nonylphenol in a human reference population. *Environmental Health Perspectives*, 113: 391–5.

Cao, Y., Winneke, G., Wilhelm, M. et al. (2008) Environmental exposure to dioxins and polychlorinated biphenyls reduce levels of gonadal hormones in newborns: results from the Duisburg cohort study. *International Journal of Hygiene and Environmental Health,* 211: 30–39 .

Cohen-Bendahan, C.C., van de Beek, C. and Berenbaum, S.A. (2005) Prenatal sex hormone effects on child and adult sex-typed behavior: methods and findings. *Neuroscience and Biobehavioral Reviews,* 29: 353–84.

Davies, W., Isles, A.R., Burgoyne, P.S. and Wilkinson, L.S. (2006) X-linked imprinting: effects on brain and behaviour. *BioEssays,* 28: 35–44.

Davies, W. and Wilkinson, L.S. (2006) It is not all hormones: alternative explanations for sexual differentiation of the brain. *Brain Research,* 1126: 36–45.

Dickerson, S.M. and Gore, A.C. (2007) Estrogenic environmental endocrine-disrupting chemical effects on reproductive neuroendocrine function and dysfunction across the life cycle. *Reviews in Endocrine and Metabolic Disorders,* 8: 143–59.

Ehrhardt, A.A. and Meyer-Bahlburg, H.F.L. (1979) Prenatal sex hormones and the developing brain: effects on psychosexual differentiation and cognitive function. *Annual Review of Medicine,* 30: 417–30.

Gross, R.E. and Meeker, I.A. Jr. (1955) Abnormalities of sexual development: observations from 75 cases. *Pediatrics,* 16: 303–24.

Hahn, A., Kranz, G.S., Kublbock, M. et al. (2014) Structural connectivity networks of transgender people. *Cerebral Cortex,* doi: 10.1093/ercor/bhu194

Heard, E. (2004) Recent advances in X-chromosome inactivation. *Current Opinions in Cell Biology,* 16: 247–55.

Hines, M. (2011) Prenatal endocrine influences on sexual orientation and on sexually differentiated childhood behavior. *Frontiers in Neuroendocrinology,* 32: 170–82.

Kakihana, R., Butte, J.C. and Moore, J.A. (1980) Endocrine effects of maternal alcoholization: plasma and brain testosterone, dihydrotestosterone, estradiol, and corticosterone. *Alcoholism: Clinical and Experimental Research,* 4: 57–61.

Konkle, A.T.M. and McCarthy, M.M. (2011) Developmental time course of estradiol, testosterone, and dihydrotestosterone levels in discrete regions of male and female rat brain. *Neuroendocrinology*, 152: 223–35.

Lackmann, G.M. (2002) Polychlorinated biphenyls and hexachlorobenzene in full-term neonates. *Biology of the Neonate,* 81: 82–5.

Lichtensteiger, W. and Schlumpf, M. (1985) Modification of early neuroendocrine development by drugs and endogenous peptides. In H. Parvez (ed.), *Progress in Neuroendocrinology*. Paris: VNU Science Press. pp. 153–66.

McAbee, M.D. and DonCarlos, L.L. (1999) Regulation of androgen receptor messenger ribonucleic acid expression in the developing rat forebrain. *Endocrinology,* 140: 1807–14.

McCarthy, M.M., de Vries, G.J. and Forger, N.G. (2009) Sexual differentiation of the brain: mode, mechanisms, and meaning. In A.E. Etgen and D. W. Pfaff (eds), *Molecular Mechanisms of Hormone Actions on Behavior*. New York: Academic. pp. 505–42.

McCarthy, M.M. and Konkle, A.T.M. (2005) When is a sex difference not a sex difference? *Frontiers in Neuroendocrinology,* 26: 85–102.

Nagel, S.C., vom Saal, F.S. and Welshons, W.V. (1999) Developmental effects of estrogenic chemicals are predicted by an in vitro assay incorporating modification of cell uptake by serum. *Journal of Steroid Biochemistry and Molecular Biology,* 69: 343–57.

Nijhuis-van der Sanden, M.W., Eling, P.A. and Otten, B.J. (2003) A review of neuropsychological and motor studies in Turner Syndrome. *Neuroscience and Biobehavioral Reviews,* 27: 329–338.

Phoenix, C.H. (2009) Organizing action of prenatally administered testosterone propionate on the tissues mediating mating behavior in the female guinea pig. *Hormones and Behavior,* 55: 566.

Phoenix, C.H., Goy, R.W., Gerall, A.A. and Young, W.C. (1959) Organizing action of prenatally administered testosterone propionate on the tissues mediating mating behavior in the female guinea pig. *Endocrinology,* 65: 369–82.

Ramlau-Hansen, C.H., Nohr, E.A., Thulstrup, A.M. et al. (2007) Is maternal obesity related to semen quality in the male offspring? A pilot study. *Human Reproduction*, 22: 2758–62.

Raum, W.J., McGivern, R.E., Peterson, M.A. et al. (1990) Prenatal inhibition of hypothalamic sex steroid uptake by cocaine: effects on neurobehavioral sexual differentiation in male rats. *Developmental Brain Research,* 53: 230–36.

Reik, W. and Walter, J. (2001) Genomic imprinting: parental influence on the genome. *Nature Reviews Genetics,* 2: 21–32.

Reinisch, J.M. and Sanders, S.A. (1982) Early barbiturate exposure: the brain, sexually dimorphic behavior and learning. *Neuroscience and Biobehavioral Reviews,* 6: 311–19.

Rubin, B.S., Lenkowski, J.R., Schaeberle, C.M. et al. (2006) Evidence of altered brain sexual differentiation in mice exposed perinatally to low, environmentally relevant levels of bisphenol. *Endocrinology,* 147: 3681–91.

Savic, I. and Lindstrom, P. (2008) PET and MRI show differences in cerebral asymmetry and functional connectivity between homo- and heterosexual subjects. *Proceedings of the National Academy of Sciences of the United States of America,* 105: 9403–8.

Schultz, K.M., Molenda-Figueira, H.A. and Sisk, C.A. (2009) Back to the future: the organizational-activational hypothesis adapted to puberty and adolescence. *Hormones and Behavior,* 55: 597–604.

Segarra, A.C. and Strand, F.L. (1989) Prenatal administration of nicotine alters subsequent sexual behavior and testosterone levels in male rats. *Brain Research,* 480: 151–9.

Sharpe, R.M. (2013) Perinatal effects of sex hormones in programming of susceptibility to disease. In *Nutrition and Development: Short and Long Term Consequences for Health,* edited by the British Nutrition Foundation. Abingdon: Wiley-Blackwell. pp. 86–96.

Sharpe, R.M., Martin, B., Morris, K. et al. (2002) Infant feeding with soy formula milk: effects on the testis and on blood testosterone levels in marmoset monkeys during the period of neonatal testicular activity. *Human Reproduction,* 17: 1692–1703.

Skaletsky, H., Kuroda-Kawaguchi, T., Minx, P.J. et al. (2003) The male-specific region of the human Y chromosome is a mosaic of discrete sequence classes. *Nature,* 423: 825–37.

Sonnenschein, C. and Soto, A.M. (1998) An updated review of environmental estrogen and androgen mimics and antagonists. *Journal of Steroid Biochemistry and Molecular Biology,* 65: 143–50.

Swaab, D.F. and Garcia-Falgueras, A. (2009) Sexual differentiation of the human brain in relation to gender identity and sexual orientation. *Functional Neurology,* 24: 17–28.

Uematsu, A., Yorifuji, T., Muroi, J. et al. (2002) Parental origin of normal X chromosomes in Turner syndrome patients with various karyotypes: implications for the mechanism leading to generation of a 45,X karyotype. *American Journal of Medical Genetics,* 111: 134–39.

van der Kamp, H.J. and Wit, J.M. (2004) Neonatal screening for congenital adrenal hyperplasia. *European Journal of Endocrinology,* 151: U71–U75.

vom Saal, F.S., Cooke, P.S., Buchanan, D.L. et al. (1998) A physiologically based approach to the study of bisphenol A and other estrogenic chemicals on the size of reproductive organs, daily sperm production, and behavior. *Toxicology and Industrial Health,* 14: 239–60.

Wallen, K. (2009) The organizational hypothesis: reflections on the 50th anniversary of the publication of Phoenix, Goy, Gerall, and Young (1959). *Hormones and Behavior,* 55: 561–65.

Zwain, I.H. and Yen, S.S.C. (1999) Neurosteroidogenesis in astrocytes, oligodendrocytes and neurons of cerebral cortex of rat brain. *Endocrinology*, 140 (8): 1343–52.

14 Beyond Wise et al.: Neuroleptic-induced "anhedonia" in rats: Pimozide blocks reward quality of food

Terry E. Robinson and Kent C. Berridge

INTRODUCTION TO DOPAMINE (DA)

Of all the neurotransmitters in the brain the one that is best known to the lay public, by far, is dopamine (DA). To the public, and much of the media, DA is the brain's "pleasure transmitter." By this popular view, it is a burst of DA that produces the pleasure associated with the receipt or anticipation of natural rewards – a chocolate cake, sex, even social rewards – as well as the euphoria produced by addictive drugs, such as amphetamine, cocaine or heroin. In the popular media it is often suggested that it is pursuit of this DA "rush" that leads to a variety of impulse control disorders, such as over-eating, gambling and addiction. Of all the ideas, in all of neuroscience, this is one of the most widely known, and its origin can be traced to the 1978 paper by Roy Wise, Joan Spindler, Harriet deWit and Gary Gerber, that is the topic of this chapter. This paper is an excellent example of how the interpretation of a seemingly simple experimental finding can profoundly influence the thinking and trajectory of an entire field, and the public imagination, for decades – even if it is wrong.

But before getting to that conclusion, we should step back and consider the time leading up to the publication of this extremely influential paper. It was only a little over a decade earlier that DA was first recognized as a neurotransmitter in its own right (rather than just being a precursor of norepinephrine). This came about from a series of important studies by researchers in Sweden who developed a histofluorescence method to identify and map monoamine neurotransmitters in the brain (including DA). They thus effectively created the field of chemical neuroanatomy, as well as vividly mapping the location of dopamine in the brain. They also produced useful methods to selectively lesion monoamine-containing cells with neurotoxins, such as 6-hydroxydopamine (6-OHDA), which would help reveal dopamine's psychological functions via behavioral consequences. As a result of these studies it became clear that DA-producing neurons located in the midbrain (the substantia nigra and ventral tegmental area) send dense projections to the dorsal and ventral striatum (the caudate-putamen and nucleus accumbens) in the forebrain, where DA is released. It also became clear that the loss of this dopamine projection, especially in the striatum, produces severe motor impairments similar to Parkinson's disease in humans. Indeed, early studies on the role of DA in extra-pyramidal movement disorders led to the Nobel Prize much later (in 2000) for one of these Swedish researchers, Arvid Carlsson.

Much of the early work on the functions of DA focused on its role in sensorimotor control related to Parkinson's symptoms. However, an additional role for DA in motivation and reward was revealed in 1971 when Urban Ungerstedt (one of the group of influential Swedes mentioned above) published a paper reporting that 6-OHDA-induced degeneration of nigrostriatal DA neurons produced severe aphagia and adipsia; that is, the rats no longer ate or drank. This behavioral starvation syndrome was very similar to the aphagia/adipsia described following lateral hypothalamic lesions years before by Anand and Brobeck (1951), and studied in detail by Phillip Teitelbaum and his colleagues, as well as others. It turned out that these lateral hypothalamic lesions cut the nigrostriatal DA projection, depleting striatal dopamine (in addition to destroying neurons within lateral hypothalamus), leading Ungerstedt (1971) to suggest that perhaps the lateral hypothalamic syndrome was due to loss of striatal DA. To study that, John Marshall, Steven Richardson and Phillip Teitelbaum (1972) directly compared the behavioral deficits produced by lateral hypothalamic lesions and by selective 6-OHDA lesions of the nigrostriatal DA pathway that spared hypothalamic neurons. They found that these two procedures produced very similar effects on feeding behavior, similar sensorimotor disturbances, and similar patterns of recovery of function. Importantly, they also concluded that "Like lateral hypothalamic rats, these [DA depleted] animals have persistent motivational and regulatory deficits in feeding and drinking which cannot easily be attributed to their sensorimotor disturbances." In other words, the role of dopamine in motivation for food and drink was different from its role in movement or sensorimotor function.

DOPAMINE AS THE 'PLEASURE TRANSMITTER'

The evidence above was consistent with a role of DA in the motivation to eat, and possibly even in food reward. However, the nature of the "motivational deficits" produced by DA depletion remained unclear. Just what was the role of DA in motivation and reward? This brings us to Wise et al.'s (1978) paper, as it was the first to provide a clear and concise answer to this question. Their key experiment in this paper was devilishly simple. First, hungry rats were trained to press a lever to get a food pellet, which they avidly did. After all the rats had acquired lever-pressing for food, the experimental manipulations took place over four subsequent days. In one group food was simply not provided when the lever was pressed. In these animals responding was maintained on the first day of testing, but not surprisingly, over the next few days responding decreased to very low levels (i.e. the rats underwent extinction in the absence of the reward). Rats in the other key group continued to receive food when they lever pressed; however, before the test session they were treated with a drug, pimozide, which blocked DA receptors. The behavior of these animals was identical to the group that underwent extinction – responding decreased over the four days of testing, as if the food was no longer rewarding. A number of control groups were included to eliminate the possibility that the decrease in responding under pimozide was simply due to sensorimotor dysfunction – (it was not) – as the animals treated with pimozide were clearly capable of responding, they just didn't.

A couple of years before this Wise and his colleagues had also reported that blockade of DA receptors (but not norepinephrine receptors) reduced the rewarding effects of the psychostimulant drug amphetamine, as well as that produced by rewarding electrical brain stimulation. However, it was in this 1978 paper that a reason for this was first proposed in a single powerful sentence. Tying these studies together, Wise et al. suggested that "neuroleptics [DA blockers] appear to take the pleasure out of normally rewarding brain stimulation, take the euphoria out of normally rewarding amphetamine, and take the 'goodness' out of normally rewarding food." In this one sentence, near the end of the paper, the idea was born that DA is a neurotransmitter in the brain that mediates pleasure. Wise elaborated on this idea several later papers. For example, in a 1980 *Trends in Neuroscience* review he concluded, ". . . it may yet be of heuristic value to think of pleasure centers in the brain and to think of these centers located . . . at the synapses of one or another dopamine terminal field." "The dopamine synapse might be the place where the cold information regarding the physical dimensions of a stimulus is translated into the warm experience of pleasure."

This notion – that DA is the brain's "pleasure transmitter" – immediately had an enormous impact on the field and quickly became the dominant theoretical framework guiding research on dopamine and reward, an influence still evident today, nearly 40 years later. The influence is evident in many papers published decades later (for reviews see Berridge and Robinson, 1998; Berridge, 2007). For example, as recently as 2012 it was stated, "Currently throughout the neuroscience literature dopamine is considered both a 'pleasure molecule' and an 'antistress molecule'" (Blum et al., 2012). The influence of this notion is even more evident today in the popular media. A Google search for dopamine will reveal many examples of DA presented as a "pleasure molecule." It is quite remarkable that an idea launched by one sentence in a short paper (and supported by dozens of similar experiments by Wise and others over the next decade) could have such a strong and enduring influence, and reach a level of public awareness that arguably surpasses any other conceptual advance in neuroscience.

A CHALLENGE TO THE NOTION OF DA AS A "PLEASURE TRANSMITTER"

The pervasive influence of the DA "hedonia hypothesis" is also a testament as to how entrenched a simple and seductive idea can become, even after most people working in the field no longer accept it, including the person who first proposed it. The first real challenge to the idea came from a study by us in 1989, based on what was essentially our failed attempt to find additional support for anhedonia after dopamine loss. At the time, Kent Berridge and his colleagues had published a series of papers using a "taste reactivity" procedure to assess the pleasurable impact of tastes, from sweet to bitter. It turns out that many species (from rats to monkeys to human babies) make very characteristic and quantifiable oral-facial and other movements in response to different tastes squirted directly into the mouth, which are essentially hedonic facial expressions. Many studies established that these expressions reflect affective judgments as to the "goodness" or "badness" of tastes, such as when hunger makes sweetness taste more pleasant than when full, or when a learned taste aversion after becoming nauseous makes an originally pleasant taste become disgusting. So, in the late 1980s Berridge thought this would be a good way to add one more bit of evidence to the large amount already available in support of Wise's notion that DA mediates the pleasurable effects of foods, but using a more direct measure of affective reactions. He teamed up with Terry Robinson who was studying dopamine in the brain, and who had techniques available to measure brain dopamine and to produce dopamine-depleting lesions (along with an honors undergraduate student, Isabel Venier). Our working hypothesis going into this study was that a loss of DA would render good tastes less pleasurable and this would be reflected by a decrease in positive affective reactions to the taste, and perhaps an increase in negative reactions – thus confirming Wise's idea. But that is not what we found. In this study Berridge et al. (1989) used the neurotoxin 6-OHDA to destroy DA neurons and severely deplete DA in the striatum. The depletion was so great that many rats became profoundly aphagic and adipsic and had to be kept alive by tube feeding. Nevertheless, when sweet or bitter tastes were squirted into their mouth they reacted exactly the same as control animals. The loss of DA seemed to have no influence on their experience of the tastes as good or bad, despite the fact they showed no motivation to eat. This contradicted the previous conclusion of Wise and colleagues that rewards became worthless without dopamine. However, as a key to explanation, in all previous studies by Wise and others the tests for reward involved instrumental responding, i.e. the willingness of animals to work for a reward, which of course is not a direct measure of hedonic responses.

Two years later Berridge and Valenstein (1991) used a manipulation to change DA in the opposite direction, i.e. they *increased* feeding behavior by use of rewarding electrical brain stimulation in the lateral hypothalamus that can stimulate dopamine release. Contrary to predictions of hedonic hypotheses at the time, they showed the increased motivation to eat was not because the food tasted better, again using the taste reactivity procedure. In addition, one limitation of the 1989 study was that DA levels were measured in the entire striatum, and it was possible that DA in the ventral striatum (nucleus accumbens) was relatively spared and this was why hedonic reactions were also spared – at the time it was becoming increasingly accepted that it was DA in the ventral as opposed to dorsal striatum that was especially important in reward. Therefore, in a later paper Berridge and Robinson (1998) conducted a similar study that included rats that had essentially complete (+99%) DA depletions in both the ventral and dorsal striatum. Although profoundly aphagic and adipsic, these rats also showed normal taste reactivity. In this latter study it was also shown that DA depleted rats could learn new hedonic values of a taste when a previously good taste was rendered bad by pairing it with an aversive state (taste aversion learning), which indicated that the facial reactions reliably reflected the hedonic value perceived by the entire brain.

A MOTIVATIONAL INTERPRETATION

In writing the original 1989 paper (Berridge et al., 1989) we were faced with a dilemma: DA depletion greatly decreased food reward in the sense of the motivation to feed, consistent with many studies showing that a decrease in DA neurotransmission decreases the willingness of animals to *work* for rewards (including the Wise et al., 1978 study), but this was not accompanied by any apparent change in the ability of DA depleted rats to generate normal hedonic reactions to the pleasure of sweetness, as assessed with taste reactivity. How to explain this discrepancy between the earlier anhedonia evidence and our observation of normal hedonic impact? It was done by proposing the notion of *incentive salience* attribution, as an admittedly post hoc hypothesis to make sense of these and the earlier data (guided also by other evidence and theory about how incentive motivation works). Simply put, the hypothesis we proposed was that "dopamine neurons belong to a system that assigns salience or motivational significance to the perception of intrinsically neutral events." This notion was further elaborated by Berridge and Valenstein (1991), and then more extensively in a larger review paper by Berridge and Robinson (1998), in which it was proposed that neural systems that mediate "wanting" rewards are dissociable from those that mediate "liking" rewards, and that DA is important for "wanting" rewards, the extent to which animals are motivated to obtain them, but not for "liking" them.

The idea that "wanting" is not the same as "liking" has since been investigated extensively and has become reasonably well accepted in the literature. Furthermore, additional evidence accumulated in 1990s suggesting the role of DA in reward was not to mediate pleasure. For example, studies by Wolfram Shultz established that once presentation of a reward was predicted by a cue preceding it, DA neurons discharged upon presentation of the cue, not during consumption of the food reward, when presumably they would experience the pleasure of its taste. This was consistent with many studies using microdialysis or in vivo voltammetry showing that DA is released in *anticipation* of rewards, be they food, sex or drug rewards (e.g. Phillips et al., 2008). Other studies established that DA is released and some DA neurons discharge during aversive/stressful events (like footshock) and cues that predict them, which would presumably not be experienced as pleasurable. Importantly, studies in humans emerged addressing this question (e.g. Leyton, 2010). In one example, Brauer and deWit (1997) reported that DA receptor blockade with pimozide in humans given amphetamine reduced ratings of how much they wanted more, but had no effect on how much they liked it. Indeed, 16 years after his seminal 1978 paper Wise (1994) himself wrote, ". . . my assumption was that subjective pleasure usually accompanied reward and would be blunted by treatments that blunt reward; I no longer make this assumption . . .". Nevertheless, although no longer held by most researchers in the field, the notion of DA as the brain's "pleasure transmitter" still holds sway in the public's imagination.

In the scientific literature the debate as to the role of DA in reward has not gone away, but it has shifted to consider other aspects of reward. It has become increasingly recognized that "reward" is a multifaceted concept consisting of a number of psychologically and neurobiologically dissociable processes. Berridge and Robinson (2003) argued that reward has three major components, each of which can be further subdivided: (1) *learning* about rewards and the cues that predict their availability; (2) *motivation* for these and cues associated with them (wanting and "wanting"); (3) *affective* (hedonic) responses to the actual pleasure of rewards (liking and "liking"). Thus, as it became accepted that DA is not involved in mediating the affective (pleasurable) component of reward after all, the debate has shifted as to whether DA mediates either learning about rewards or a motivational component such as incentive salience, discussed above. This debate is quite complicated and can only be very briefly discussed here, but it has been the topic of a number of comprehensive reviews (Schultz et al., 1997; Berridge, 2007, 2012).

LEARNING INTERPRETATIONS

There are a number of different ways DA has been proposed to contribute to learning. For example, it has been suggested that it may reinforce or 'stamp-in' learned associations, such as those between two stimuli (S-S associations) or between stimuli and responses (S-R associations). A more sophisticated way is the well-known idea called the "prediction error learning hypothesis" derived from elegant electrophysiological studies in monkeys during the 1990s by Wolfram Schultz. In a series of studies Schultz and his colleagues recorded from DA neurons and found that, prior to learning, those neurons discharged upon receipt of the reward. However, as the monkeys learned that a given signal predicted the reward, DA neurons came to fire upon presentation of the cue and stopped firing upon receipt of the reward. If the cue was presented but the reward omitted there was a brief dip in DA activity. Similar results have now been found using other techniques, where DA release is measured directly in DA terminal fields. This pattern of activity is consistent with what is called a "prediction error signal," which is central to most computational models of learning, and therefore it has been hypothesized that phasic DA activity comprises a "teaching signal" necessary for learning (e.g. Schultz et al., 1997). This hypothesis has been widely adopted in the computational modeling literature.

LEARNING VERSUS MOTIVATION

However, others have challenged this learning interpretation. Most notably, Berridge (2007) has argued that "dopamine is not needed for new learning, and not sufficient to directly mediate learning by causing teaching or prediction signals." For example, a number of studies suggest DA does not appear to be necessary for much new reward learning. In the paper mentioned above showing that rats with complete striatal DA depletions showed normal hedonic reactions to tastes, Berridge and Robinson (1998) also reported these rats could learn normally about changes in the value of tastes. Normal learning has also been described in mutant mice that cannot synthesize DA (e.g. Cannon and Palmiter, 2003; Robinson et al., 2005) and following DA receptor blockade (e.g. Calaminus and Hauber, 2007). In addition, DA elevation does not appear to be sufficient to enhance new learning either. Mutant mice in which DA signaling is elevated do not appear to learn new Pavlovian or instrumental associations better than controls (e.g. Cagniard et al., 2006). Similar conclusions have been reached in studies with humans, i.e. Parkinson's patients on or off their dopaminergic medication. These latter researchers concluded that "dopaminergic drug state (ON or OFF) did not impact learning", but "... the critical factor was drug state during the performance phase" (Shiner et al., 2012). Although the DA manipulations discussed above did not seem to influence learning, they did influence behavior (performance) in ways consistent with a motivational interpretation. Thus, Robinson et al. (2005) concluded, "... dopamine is not necessary for mice to like or learn about rewards but it is necessary for mice to seek (want) rewards during goal-directed behavior." Similarly, Cagniard et al. (2006) concluded, "DA directly scales behavioral performance in the absence of new learning." In their study using DA antagonists, Calaminus and Hauber (2007) concluded that DA receptor signals in the core of the accumbens "seem to be unnecessary in updating the reward-predictive significance of cues, rather, they serve to activate instrumental behavior."

Instead, dopamine elevation does more reliably scale with motivation for rewards, including even addictive or compulsive levels of motivation. Sensitized DA release in Parkinson's patients with dopamine dysregulation syndrome, leading to pathological drug use, is correlated with "compulsive drug 'wanting' but not 'liking'", and in healthy people DA release correlates well with ratings of drug wanting but not with changes in affective state (see Leyton, 2010, for a review). Finally, individuals vary in the extent to which they attribute incentive salience to reward predictive cues. DA activity is increased in response to reward cues in rats that attribute incentive salience to such cues, but not in those for whom the cue has predictive but not incentive value (Flagel et al., 2010).

All of these studies are consistent with the incentive salience or motivational interpretation of dopamine's primary role in reward described earlier. That is, DA is critical in mediating the extent to which rewards are "wanted," but not the extent to which they are "liked" or the ability to learn about them. As put by Berridge and Robinson (2003), "Incentive salience is a motivational, rather than an affective, component of reward. Its attribution transforms mere sensory information about rewards and their cues (sights, sounds and smells) into attractive, desired, riveting incentives":

> The sight of food, drugs or other incentives is merely a sensory configuration of shape and color that is not intrinsically motivating. Attribution of incentive salience to a percept or other representation is what is suggested to make it a "wanted" target of motivation. Incentive salience or "wanting", unlike "liking", is particularly influenced by dopamine neurotransmission.

As with most debates in science, consensus is never achieved quickly but only over time, with the gradual accumulation of evidence for and against various positions. And this is the case concerning the role of dopamine in reward. We can probably safely conclude that the "hedonia hypothesis" of DA function first raised by Wise et al. (1978) that is the target of this commentary is no longer held by most researchers in the field – although it lives on in the popular media. As to learning versus motivational interpretations, these are still the subject of debate. It is obvious from the above that we think the weight of the evidence favors a motivational interpretation, but it would be presumptuous in 2016 to conclude this is an overwhelmingly consensus opinion, and so this debate will likely continue until more evidence accumulates.

NEURAL BASIS OF PLEASURE

Going back to the question as to what neural systems mediate pleasurable experiences – if not DA systems, then what? This is a difficult question to address in non-human animals, but it has been a focus of investigation of one of us (KB) over the last few decades, and considerable progress has been made, especially concerning systems that mediate the sensory pleasures associated with tastes (Berridge and Kringelbach, 2013). These results suggest that hedonic "liking" is generated by a more restricted brain substrate than DA, via a set of small "hedonic hotspots" nestled within mesocorticolimbic structures. Each hotspot uses opioid (natural heroin-type neurotransmitters), endocannabinoid (natural marijuana-type neurotransmitters) and related neurochemicals to amplify "liking" reactions, but never DA. Hedonic hotspots are relatively tiny – each is only a cubic millimeter or so in rats and probably about a cubic centimeter or so in humans. For example, one hotspot constitutes about just 10% of the entire nucleus accumbens, and there are other hotspots in the ventral pallidum, in the limbic regions of the prefrontal cortex, and in the brainstem. Hedonic hotspots appear to act together as a unified system to enhance a sensory pleasure, recruiting each other into activation, and requiring unanimous activation of several simultaneously to create an intense pleasure. So, there are "pleasure centers" in the brain – they just do not use dopamine to function.

REFERENCES

Anand, B.K. and Brobeck, J.R. (1951) Hypothalamic control of food intake in rats and cats. *Yale Journal of Biology and Medicine*, 24 (2): 123–40.

Berridge, K.C. (2007) The debate over dopamine's role in reward: the case for incentive salience. *Psychopharmacology (Berl)*, 191 (3): 391–431.

Berridge, K.C. (2012) From prediction error to incentive salience: mesolimbic computation of reward motivation. *European Journal of Neuroscience*, 35 (7): 1124–43.

Berridge, K.C. and Kringelbach, M.L. (2013) Neuroscience of affect: brain mechanisms of pleasure and displeasure. *Current Opinion in Neurobiology* 23 (3): 294–303.

Berridge, K.C. and Robinson, T.E. (1998) What is the role of dopamine in reward? Hedonic impact, reward learning or incentive salience. *Brain Research Reviews*, 28: 308–67.

Berridge, K.C. and Robinson, T.E. (2003) Parsing reward. *Trends in Neurosciences*, 26: 507–13.

Berridge, K.C. and Valenstein E.S. (1991) What psychological process mediates feeding evoked by electrical stimulation of the lateral hypothalamus? *Behavioral Neuroscience*, 105 (1): 3–14.

Berridge, K.C., Venier, I.L. and Robinson, T.E. (1989) Taste reactivity analysis of 6-hydroxydopamine-induced aphagia: implications for arousal and anhedonia hypotheses of dopamine function. *Behavioral Neuroscience*, 103: 36–45.

Blum, K., Chen, A.L., Giordano, J. et al. (2012) The addictive brain: all roads lead to dopamine. *Journal of Psychoactive Drugs*, 44 (2): 134–43.

Brauer, L.H. and DeWit, H. (1997) High dose pimozide does not block amphetamine-induced euphoria in normal volunteers. *Pharmacology, Biochemistry and Behavior*, 56 (2): 265–72.

Cagniard, B., Beeler, J.A., Britt J.P. et al. (2006) Dopamine scales performance in the absence of new learning. *Neuron*, 51 (5): 541–7.

Calaminus C. and Hauber W. (2007) Intact discrimination reversal learning but slowed responding to reward-predictive cues after dopamine D1 and D2 receptor blockade in the nucleus accumbens of rats. *Psychopharmacology (Berl)*, 191 (3): 551–66.

Cannon, C.M. and Palmiter, R.D. (2003) Reward without dopamine. *Journal of Neuroscience*, 23 (34): 10827–31.

Flagel, S.B., Clark, J.J., Robinson, T.E. et al. (2011) A selective role for dopamine in stimulus-reward learning. *Nature*, 469 (7328): 53–7.

Leyton, M. (2010) The neurobiology of desire: dopamine and the regulation of mood and motivational states in humans. In M.L. Kringelbach and K.C. Berridge (eds), *Pleasures of the Brain*. Oxford: Oxford University Press. pp. 222–243.

Marshall, J.F., Richardson, J.S. and Teitelbaum, P. (1972) Nigrostriatal bundle damage and the lateral hypothalamic syndrome. *Journal of Comparative and Physiological Psychology*, 87: 808–30.

Phillips, A.G., Vacca, G. and Ahn, S. (2008) A top-down perspective on dopamine, motivation and memory. *Pharmacology, Biochemistry and Behavior*, 90 (2): 236–49.

Robinson, S., Sandstrom, S.M., Denenberg, V.H. and Palmiter, R.D. (2005) Distinguishing whether dopamine regulates liking, wanting, and/or learning about rewards. *Behavioral Neuroscience*, 119 (1): 5–15.

Schultz, W., Dayan, P. and Montague, P.R. (1997) A neural substrate of prediction and reward. *Science,* 275: 1593–99.

Shiner, T., Seymour, B., Wunderlich, K. et al. (2012) Dopamine and performance in a reinforcement learning task: evidence from Parkinson's disease. *Brain,* 135 (Pt 6): 1871–83.

Ungerstedt, U. (1971) Adipsia and aphagia after 6-hydroxydopamine induced degeneration of the nigro-striatal dopamine system. *Acta Physiologica Scandinavica,* Suppl., 367: 95–122.

Wise, R.A. (1994) A brief history of the anhedonia hypothesis. In C.R. Legg and D. Booth (eds), *Appetite: Neural and Behavioral Bases*. New York: Oxford University Press. pp. 243–63.

Wise, R.A., Spindler, J., DeWit, H. and Gerber, G.T. (1978) Neuroleptic-induced "anhedonia" in rats: pimozide blocks reward quality of food. *Science*, 201 (4352): 262–4.

FURTHER READING

Ikemoto, S. and Panksepp, J. (1999) The role of nucleus accumbens dopamine in motivated behavior: a unifying interpretation with special reference to reward-seeking. *Brain Research Reviews*, 31 (1): 6–41.

Salamone, J.D. and Correa, M. (2012) The mysterious motivational functions of mesolimbic dopamine. *Neuron*, 76 (3): 470–85.

PART 4

Brain Plasticity

15 Revisiting Krech, Rosenzweig and Bennett: Effects of environmental complexity and training on brain chemistry

Bryan Kolb

A fundamental idea of behavioral neuroscience is that experience modifies brain structure and function. The idea that experience might change the brain is not new. Darwin had noticed that the brains of domestic rabbits are smaller than those of wild rabbits and suggested that this resulted from reduced use of their "intellect, instincts, and senses." Twenty years later Ramon y Cajal (see 1928) proposed that cerebral exercise led to the expansion of "new and more extended intercortical connections." This idea was also the origin of the theory that experience-dependent changes in the brain form the basis of learning. For at least 30 years up until the mid-1950s Karl Lashley was certain that changes in the brain formed the basis of memory and called the changes "memory traces" or "engrams."

The first direct experiment to investigate the effect of experience on the brain was done by Hebb (1947). He brought laboratory rats into his home where they were treated more like family pets than research subjects. When the rats were returned to the laboratory for testing on various maze problems, they outperformed their lab-reared littermates. Hebb's conclusion was that a more stimulating environment enhanced brain function and this enhancement was the basis of improved performance on cognitive tasks.

Although Hebb (1949) suggested that the experiences of his rats must have changed circuits in the brain, it was not until a study by Krech, Rosenzweig and Bennett (1960) that direct evidence was found for the theory. They found structural changes were associated with biochemical changes in the brains of rats reared in a complex housing environment. This finding, although greeted with some skepticism at the time, proved seminal and led to thousands of papers on the topic in the following 55 years. We begin with a description of the work begun by Krech and colleagues before seeing where this leads us.

THE BERKELEY STUDIES

In their original paper in 1960, Krech et al. placed young animals in complex environments for 80 days following weaning. When they subsequently examined the rats' brains they found that the cerebral cortex was heavier and its acetylcholinesterase (AChE) activity was increased. Thus experience changed the structure of the brain and its chemistry. In a parallel study they also showed that the higher AChE levels predicted problem-solving ability. As their experiments continued, they developed an experimental protocol to try to tease out exactly what the key experiences might be (see Bennett et al., 1964). "Control" animals were housed in standard cages in groups of three and exposed to ongoing activity in the colony room but received no special treatment, a condition they labelled a "social condition." The animals receiving "enriched experience" were housed in groups of 10–12 in large cages provided with "toys" (see Figure 15.1). In addition, each day the rats were placed in an open field (90 cm x 90 cm) where there were many barriers for the animals to climb and explore. These barriers were changed daily in order to maintain the novelty of the environment. A third group, the "isolated condition," had animals placed alone in cages in which they could not see or touch other animals, although they could smell and hear them.

The increase in brain weight in the enriched animals was not due to an overall increase in body weight – the enriched animals actually weighed less than the others at the end of the experiment. Furthermore, the increase in brain weight was only seen in the cortex and there was a small, but significant, drop in the weight of subcortical regions. The increase in cortical weight was not general over the cortex. Although they did not measure all cortical regions, the weight gain was highest in the visual region (6.2%) and lowest in the somatosensory region (2.7%). For a later study the authors recruited a neuroanatomist, Marion Diamond, and she showed that the increase in cortical weight was associated with increases in cortical thickness (Diamond et al., 1964). The social control animals did not show these changes and they failed to differ from isolates on all measures.

Because the Berkeley findings were so novel the authors spent several years demonstrating the robustness of their results. They examined different strains of rats, the effects of selective breeding, and the effects of blindness and light deprivation. They also looked for other anatomical and biochemical changes. Various control experiments demonstrated that the observed changes in the enriched animals could not be attributed to differential handling, locomotor activity, stress, or rate of development. It was the novelty of the situation that was important. The commitment of the authors to their work led to the following challenge: "Because we believe that our findings demonstrate the feasibility of research on the effects of experience on the brain, and because we believe that such research offers many challenges and a wide field for investigation, we hope to see it taken up in other laboratories" (Bennett et al., 1964: 619). It was!

As important and seminal as the Berkeley experiments were, one weakness was that they did not demonstrate changes in brain organization so much as a change in brain size. It was not until the early 1970s that investigators, including the Berkeley group, began to look at neuronal organization. One method was to use the Golgi-type technique to examine the structure of individual neurons (see Figure 15.2). Golgi techniques allow visualization not only of the dendritic fields of

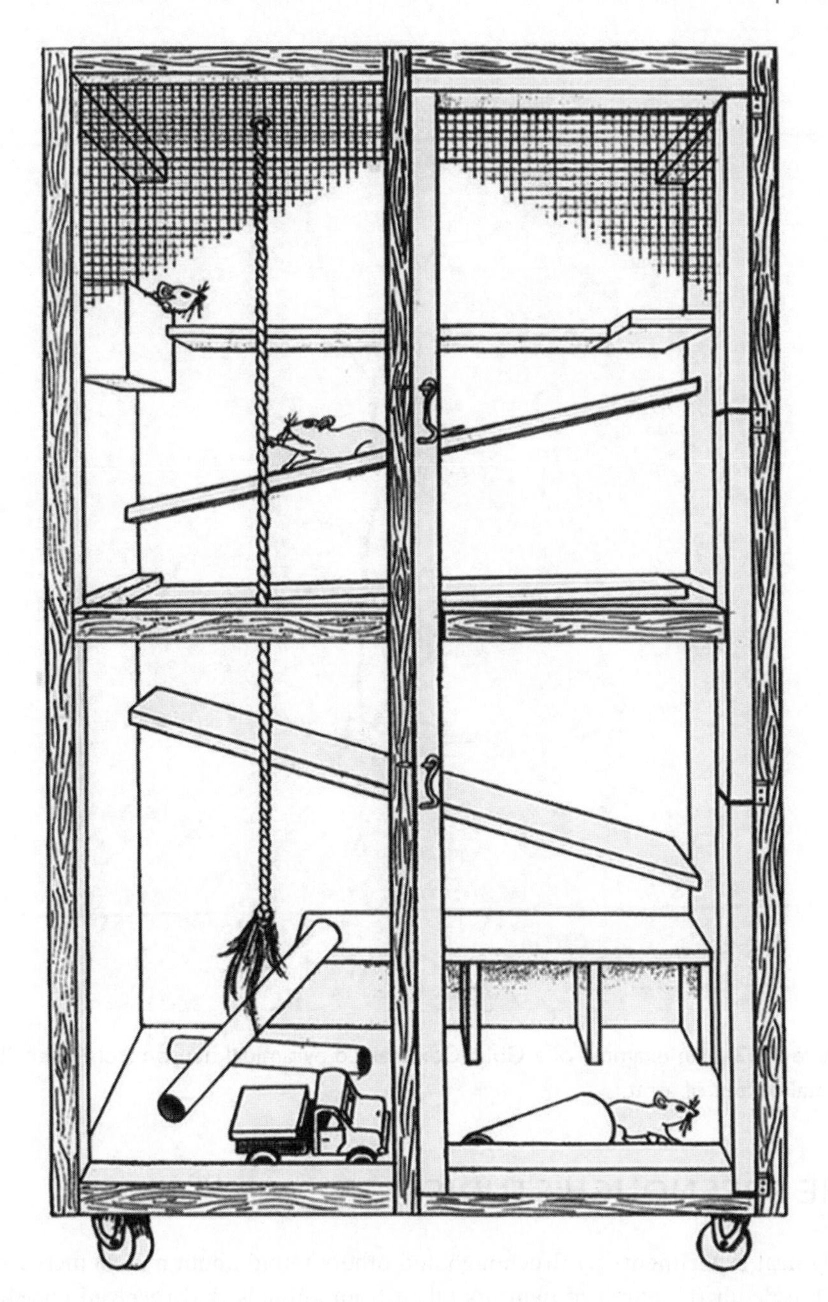

Figure 15.1 An example of an "enriched environment" used in the author's lab.

the neurons, but also of their spine density. Spine density increases meant there were more synaptic connections, and an increase in synapses in turn meant greater interneuronal communication. The most thorough morphological studies were begun in the early 1970s by Bill Greenough and his colleagues, who we turn to next.

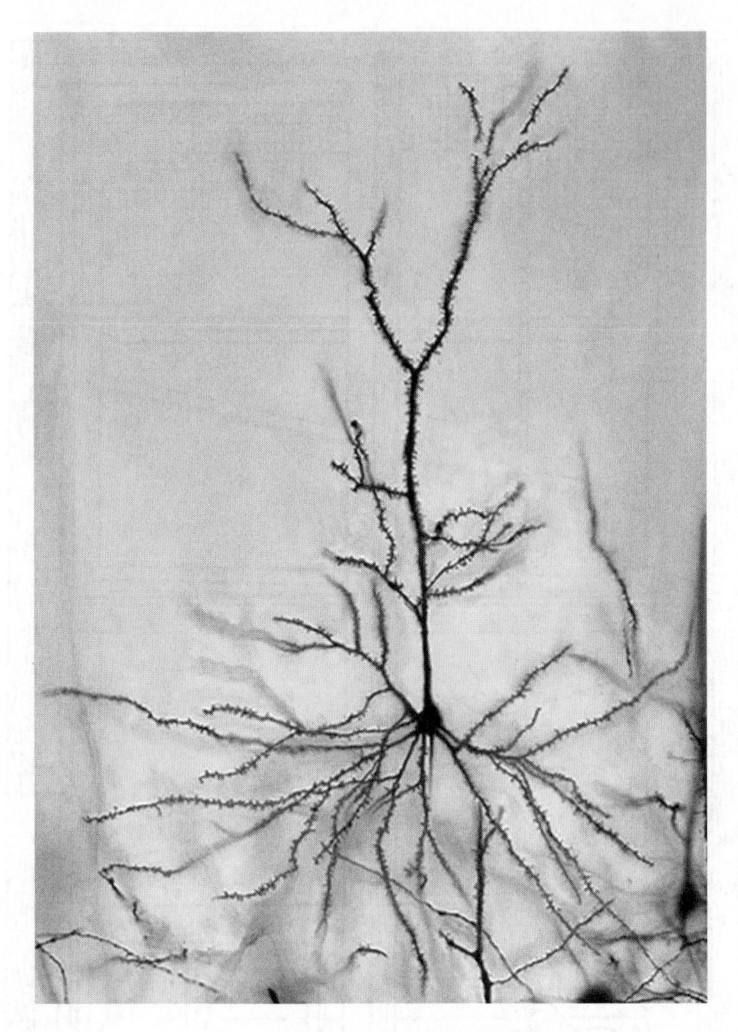

Figure 15.2 An example of a Golgi-Cox stained pyramidal neuron from layer III of parietal cortex of a rat.

THE GREENOUGH STUDIES

Typical experiments by Greenough and others found about a 20% increase in the dendritic length of neurons taken from animals that received enriching experiences (e.g. Volkmar and Greenough, 1972). Like the effects on cortical weight in the Berkeley experiments, the effects were largest in the visual cortex. Greenough's group also looked beyond the cerebral hemispheres and examined neurons in the cerebellum of animals trained on motor tasks. They found large increases in dendritic space in the Purkinje cells, the major output cells of the cerebellum, but no changes in cerebellar granule cells. This suggested that neuronal changes are not an inevitable consequence of experience but were features of only a subset of brain neurons (e.g. Floeter and Greenough, 1979).

If there is an increase in size of dendritic fields of neurons with more synapses per neuron, that would likely require more support from glial cells and blood vessels. In a heroic series of studies Sirevaag and Greenough (1987, 1988) used light and electron microscopic techniques to analyze 36 different aspects of cortical morphology in rats raised in complex or control caged-housing environments. The overall result was that there was a coordinated change in neuronal morphology as well as glial, vascular, and metabolic processes in response to different experiences. This meant that the increase in synapses per neuron was accompanied by more astrocyte material, more blood capillaries, and a higher mitochondria volume (a measure of metabolic activity) in the brains of animals with enriched experience. These many changes implied that when the brain changes in response to experience, not only are there neuronal changes but there are also metabolic adjustments to meet the requirements of increased neuronal communication.

Greenough's group also asked an important question: Is there something special about enriched housing or is it simply an example of the effects of learning? The general finding of these studies is that it is the learning that is important. When rats learn specific tasks, the learning stimulates area-dependent growth related to the specific training. Thus, for example, rats learning visual tasks show increased dendritic length in the visual cortex, whereas rats learning motor tasks show increased length in the motor cortex (see the review by Greenough and Chang, 1988). Later studies by others have shown that the learning-related changes can also be found in the prefrontal cortex when rats are trained in working memory tasks (e.g. Comeau et al., 2010).

Greenough's group addressed the issue of the relationship between novelty-induced neuronal enhancement and learning in another way. They asked whether a physiological model of learning, namely long-term potentiation (LTP), might produce changes similar to those following enriched housing or task learning. Initial studies by others suggested that there were dendritic changes associated with LTP, but Chang and Greenough (1984) performed a more thorough experimental examination. They used different types of brain stimulation and made both Golgi and EM analysis of synapse morphology. The key finding was that it was not just the activity of the neural system that drove the synaptic changes, rather that there was an optimal level and patterning of synaptic activation that was required to produce the physiological, anatomical, and biochemical features observed.

Although most of the early studies on the effects of enrichment were carried out using young, often weanling, rats, Greenough's studies also showed that young, adult, and even aged rats showed the enrichment effect, although some of the details varied by age (see below).

One important notable aspect of Greenough's studies was that he showed that changes seen when animals learn specific tasks are focal but strikingly similar to what are seen with enrichment. This finding was important because it showed that enrichment effects were not merely a response to a more normal environment versus the normally impoverished laboratory environment. The take-home point was that the experience afforded by enrichment was similar to learning experiences in profoundly changing the brain.

Chemical changes

The original Krech et al. studies showed that AChE, the enzyme that breaks down ACh, were increased by enrichment, implying an increase in acetylcholine. Over the past decades a lot of additional molecular/chemical changes have been reported, including an increased gene expression in the neocortex, increased levels of Brain Derived Neurotrophic Factor (BDNF), Fibroblast Growth Factor-2, Insulin-like Growth Factor-1, Nerve Growth Factor, NMDA receptors, signal transduction molecules, GAD 65, CREB, and many others. AChE was clearly just the very tip of the molecular iceberg associated with an enriched brain.

Beyond rodents

The original Krech et al. and Greenough studies all used rats. But how well do the rat studies generalize to other animals? Several studies have looked at enriched housing in cats and monkeys, finding similar results (e.g. Beaulieu and Colonnier, 1987; Stell and Riesen, 1987). One curious difference between monkeys and rats is that the effects of enriched housing on the visual cortex are neglible in monkeys. A possible reason is that much of monkey visual exploration does not require movement. Monkeys in relatively impoverished housing can still visually explore their environment. This visual stimulation may be sufficient to allow the development of cortical synapses. The visual system of rats has poorer acuity and is used more for guidance of movement in space and visual spatial knowledge likely requiring movement in space. Another explanation for the rat/monkey difference is that visual regions of monkeys have expanded exponentially and the "higher" visual areas may show greater experience-dependent changes. Thus, visual experiences that emphasize object exploration and object recognition may induce synaptic change in the temporal cortex, which appears to be the case (Tanaka, 1993).

The Beaulieu and Colonnier (1988) studies on enriched housing in cats not only showed similar results to those in the rat but, in addition, they found that experience increased the number of excitatory synapses per neuron and decreased the number of inhibitory ones in the visual cortex. Thus, visual experience appears to change the excitatory-inhibitory (E-I) equilibrium of the visual cortex. Indeed, more recent extensive enrichment experiences in rodents have shown similar effects on the E/I balance at critical periods during visual system development (see the review by Takesian and Hensch, 2013).

Perhaps the most remarkable studies of enriched housing of animals other than rats came from studies of fruit flies. In a study by Technau (1984) the complexity of neurons in *Drosophila melanogaster* varied with the flies' living conditions. Flies housed for three weeks either singly in small plastic vials or in groups of 200 in the "enriched" conditions of large enclosures with colored visual patterns on the walls showed about 15% more fibers in the mushroom bodies (cells in the "brain" of the fly) of the enriched flies. Clearly, the Krech et al. findings generalize well beyond rats! Indeed there are now studies across a wide range of animal taxa including insects, fish, birds, and mammals (including humans) showing the power of experience to change the brain.

NEUROGENESIS AND ENRICHMENT

Two regions of the rodent brain, the dentate gyrus of the hippocampus and subventricular zone (SVZ) of the cerebral hemispheres, continue to give birth to small numbers of new neurons in adulthood. The dentate gyrus neurogenesis has been studied in a long series of experiments by van Pragg, Kemperman, Gage and their colleagues. Beginning in the late 1990s these showed that giving animals the opportunity to run in running wheels or placing animals in enriched environments enhances the production of neurons in the hippocampus and enhances performance on learning tasks known to be dependent on the integrity of the hippocampus (e.g. Garth et al., 2016). The effects of the two experiences are somewhat different, and additive, showing that it is not simply activity that is driving the effects but also enrichment. Furthermore, Kemperman's group has shown that the effect of enriched housing requires the hippocampal neurogenesis, i.e. blocking the neurogenesis chemically prevents the enrichment benefits on behavior.

What is it about the presence of new neurons in the hippocampus that leads to the enrichment effect? The answer seems to be that the newborn dentate gyrus neurons also make new connections with other regions of the hippocampus, and presumably with neocortical regions (e.g. Bergami, 2015). Indeed, a study by Sutherland et al. (2010) showed that if rats are given unilateral hippocampal lesions, there is no enrichment effect in the ipsilateral neocortex even though the enrichment effect is normal in the contralateral (undamaged) hemisphere. It thus appears that the effect of enrichment results in connecting new hippocampal neurons to the rest of the brain and this is the same mechanism that underlies new learning.

FACTORS INFLUENCING THE EFFECTS OF EXPERIENCE-DEPENDENT PLASTICITY

Although the original Krech et al. and later Greenough studies appeared to show a fairly general effect of enriched housing on cortical structure, over the past 20 years it has become clear that things are more complicated. Consider the following examples related not just to enrichment but to other experiences as well.

THE EFFECTS OF AGE ON THE ENRICHMENT EFFECT

The original studies used young animals largely because it was expected that young brains would change more extensively than the brains of older animals. This appears true, but there are other differences. For example, Kolb et al. (2003a) placed male rats in enriched environments for three months beginning at weaning, young adulthood (four months), or in senescence (eighteen months). Although complex housing increased brain weight (~6%) and parietal cortical dendritic length and branching (~25%) at all ages, there was a surprising difference in parietal cortical spine density: spine density was ~20% *less* in neurons from enriched young animals but ~20% *greater* in adult and senescent animals (see Figure 15.3). Thus, the spines were further apart in the young animals than in the older animals. The authors speculated that this would make the younger brains more plastic to other experiences because new synapses could be added without growing additional dendritic arbor. Inspection of Figure 15.3 also shows that spine density was lower in the aged rats than in the younger groups, suggesting an effect of aging.

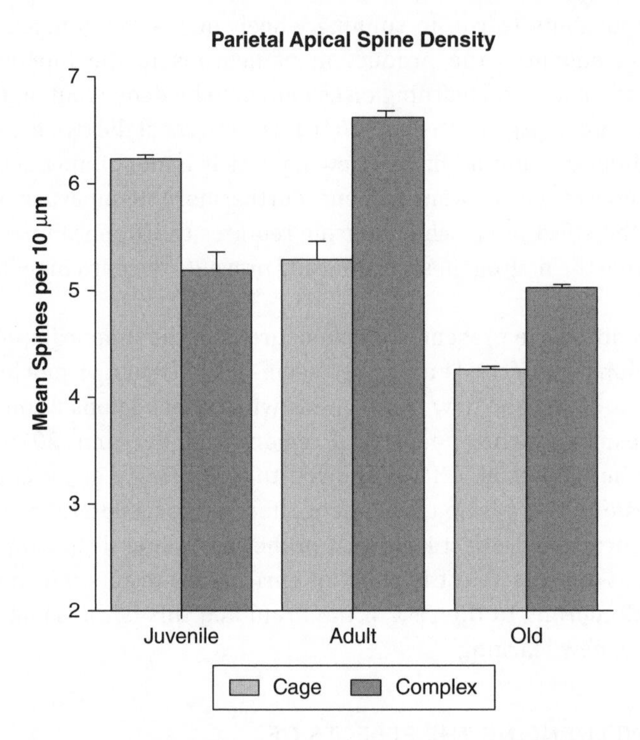

Figure 15.3 Summary of the effect of complex- versus lab-housing on the spine density in parietal cortex of male rats. (Data from Kolb et al., 2003a.)

The finding that synapse numbers could drop as well as increase in response to enriched housing was unexpected, but there are similar effects in response to learning mazes, stress, and drugs (e.g. Comeau et al., 2010). Thus, the effect of experience is not just to add synapses but to reorganize synaptic circuits.

THE EFFECTS OF SEX ON THE ENRICHMENT EFFECT

There is accumulating evidence that the male and female brains differ in their structure and respond differently to experience. Janice Juraska and her colleagues (e.g. Juraska, 1990) reported that whereas the visual cortex of young male rats shows vigorous changes to enrichment, the visual cortex of female rats does not. This is not evidence of a greater sensitivity in males, however, because her group also showed that the hippocampus shows greater enrichment-related changes in females than in males.

It is possible that age interacts with sex effects of enrichment. Kolb et al. (2003a) repeated Juraska's experiments with young and older rats and found that as she had reported, there were no changes in the visual cortex of young female rats, but there were significant changes in older female rats. This could possibly reflect differential effects of ovarian hormones at different ages. In fact, Willing and Juraska (2015) reported that female rats, but not male rats, lost significant numbers of neurons in the medial prefrontal cortex through adolescence. It is reasonable to wonder how this might affect prefrontal plasticity in the two sexes.

THE EFFECTS OF THE CORTICAL REGION ON THE ENRICHMENT EFFECT

Both the Krech and Greenough studies showed that the enrichment effects were about two times larger in the posterior cortical regions compared to the anterior cortex, but it appeared that those effects would be seen across the neocortex. Because the prefrontal cortex is often described as the brain's executive, one would predict large enrichment effects there. However the results are surprising. There are large changes in dendritic length and branching in the parietal and occipital cortex but not in the medial prefrontal or orbital prefrontal cortex in the same brains (e.g. Kolb et al., 2003c). The absence of enrichment effects is not because prefrontal neurons are not plastic. Indeed, they show robust dendritic changes in response to stress, psychoactive drugs, gonadal hormones, and task learning. One possibility is that there are transient effects of enrichment that disappear over time in the enriched environments, but these have yet to be studied in detail.

THE EFFECTS OF METAPLASTIC INTERACTIONS ON THE ENRICHMENT EFFECT

Although most experiments looking at enrichment effects treat the enrichment as a singular experience, life is not about just one experience but rather there is a "one after another" phenomenon that is important. This is referred to as metaplasticity. Consider the following example. Both environmental enrichment and experiences with psychomotor stimulant drugs modify dendrites. But if animals live in enriched environments for extended periods of time (more than three months) there is a dramatic attenuation of the effect of later exposure to psychoactive drugs such as amphetamine or nicotine (e.g. Hamilton and Kolb, 2005). The effect works the other way too. Prior exposure to psychomotor stimulants, which change dendritic length and branching in prefrontal cortex and the striatum, block enrichment effects in parietal or occipital cortex and in the striatum (e.g. Kolb et al., 2003b). It is as if there is a brain enrichment potential that can be expended but once.

CROSS-GENERATIONAL INFLUENCES OF THE ENRICHMENT EFFECT

The effects of enriched housing appear to cross generations. For example, when male rats are housed in complex environments for six weeks prior to mating with females living in standard lab caging, the brains of the adult offspring appear similar to those of animals directly housed in complex environments. Similarly, prenatal enrichment stimulates recovery from later perinatal brain injury in the offspring (e.g. Gibb et al., 2014). The likely mechanism of these effects is an epigenetic change resulting from the enriched housing.

APPLICATIONS OF THE ENRICHMENT EFFECT

In view of the powerful effects of enriched housing on so many measures of brain organization and function it is not surprising that dozens of laboratories have used enrichment as a treatment for a wide variety of therapeutic purposes. Consider several examples.

RECOVERY FROM CEREBRAL INJURY

Rats given lesions of the prefrontal cortex, motor cortex, anterior thalamus, striatum, hippocampus, entorhinal cortex or subiculum all show enhanced recovery following enriched housing. The beneficial effects occur with various types of injuries including stroke, perinatal asphyxia, surgical excisions, and neurotoxins. In addition, the positive effects are seen at all ages of brain injury ranging from neonatal to senescence, and as noted above, even gestational enrichment has beneficial effects on recovery from later perinatal brain injury. Not only are there effects of the enriched housing on behavior, but many studies have shown increased compensatory dendritic changes as well.

RECOVERY FROM THE EFFECTS OF STRESS

There is a large literature showing that prenatal exposure to stress leads to disturbances in juvenile/adolescent play behavior, increased anxiety, changes in the HPA-axis, and cognitive dysfunction in adulthood. Environmental enrichment during adolescence reverses some of these effects (Morley-Fletcher et al., 2003), but the benefits of enrichment are also seen in models of stress in adult rats. In one study, rats given a mild stressor of saline injection had a reduced hormone response if they were living in enriched housing. The authors also found a reduced effect of nicotine withdrawal in enriched rats, the effects being larger in females (Skwara et al., 2012).

REDUCED SYMPTOMS OF MODELS OF NEUROLOGICAL DISORDERS

Enriched housing reduces the motor deficits in a rat model of Parkinson Disease. Enrichment also rescues protein deficits in the striatum and hippocampus of transgenic mice expressing a human huntington transgene (a model of Huntington

Disease), leading to a delay in the onset of the disease. These studies suggest that there may be a way to modulate the pathology of at least some degenerative diseases by experiences, including enriched housing.

ENHANCING BRAIN DEVELOPMENT

Although most studies manipulating sensory system development have altered or reduced sensory experience, a study by Maffei's group in Italy (e.g. Cancedda et al., 2004) investigated visual system development in mice reared in an enriched environment. They found that raising mice this way caused earlier eye opening, and about a six-day acceleration of visual acuity measured both behaviorally and electrophysiologically. These effects were correlated with increases in BDNF and GAD 65, as well as enhanced gene expression in the visual cortex. A later study by the same group showed that enriched housing in adult rats reactivated plasticity in ocular dominance columns, which is normally only seen during the early critical period for visual system development. Hensch and Bilimoria (2012) have noted that many other experiences, ranging from genetic, biochemical, and surgical manipulations to electrical stimulation or environmental changes, also can trigger plasticity in the adult visual cortex (see Figure 15.4), and speculate that all of these manipulations trigger plasticity by a common circuit mechanism, adjusting the E/I balance.

Not only does enriched housing enhance the otherwise normal brain developmental plasticity, but Steve Dunnett's group in the UK (e.g. Dunnett et al., 1986) has shown that enriched housing facilitates the growth of neural grafts in a damaged adult striatum and this is correlated with enhanced behavioral recovery. They also showed that enrichment facilitates AChE outgrowth in acetylcholine-rich grafts and LTP in striatal grafts.

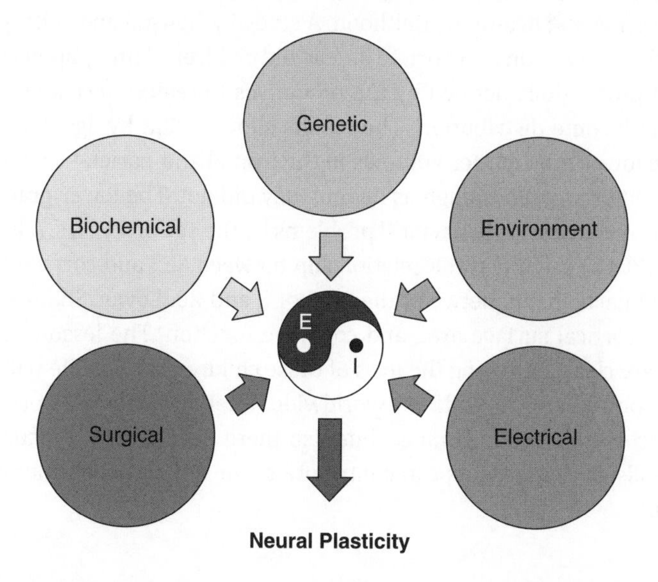

Figure 15.4 Factors that affect critical period timing. Many experimental manipulations, including enriched housing, induce juvenile-like plasticity in the visual cortex. (After Hensch & Bilimoria, 2012).

CONCLUSION

The Krech et al. study has had a profound impact on brain and behavior research over the past 50 years and more. The Berkeley group published about 30 papers (largely Bennett and Rosenzweig) in the 15 years following their first paper in 1960, fielding close to 6000 citations (or about 200 per paper). It is difficult to estimate how many papers were published by others, but a PubMed search (December 2015) using "environmental enrichment brain" yielded over 1200 papers – yet curiously the original Krech et al. paper did not show up! It is so central that it need not be cited.

The seminal observation that experience changed brain and behavior provided a compelling case for the idea that enrichment is a form of learning and that other forms of learning are based upon parallel changes in neural networks. Like the ideas of Donald Hebb (see Chapter 7) on how experience changes the brain, the work of the Berkeley group and later by Greenough's group will have a long-lasting influence on our understanding of brain plasticity and behavior.

The enrichment effect also can be put to practical purposes. It can serve as a treatment for neurological conditions, an antidote to addiction, a directive to future generations, and a curriculum for development. Early childhood education experiences such as Head Start programs are an important human application of the enrichment effect.

Finally, childhood poverty, which can be seen as the opposite of enrichment, is a health problem worldwide. The Organization for Economic Co-operation and Development (OECD) estimates that about 50 million children live below the national poverty line in industrial countries, putting them at high risk for many health conditions later in life. Thus, low socioeconomic status (SES) in childhood correlates with with poor cognitive development, language, memory, socioemotional processing, and ultimately income and health in adulthood. A study by Hanson and colleagues (2013) did repeated MRI scans on newborn to 3-year-old children, demographically balanced to represent proportions defined by the US Census Bureau in terms of gender, race, ethnicity, and income distribution. The results showed that by age 4 the lower-SES children had lower gray matter volumes in the frontal and parietal cortex than more advantaged children, even though as infants they did not. The lower gray matter volumes were correlated with behavioral problems by the age of 4 years. A later study by Noble et al. (2015) examined the relationship between SES and cortical surface area in over 1,000 participants between the ages of 3 and 20. Lower SES was associated with smaller cortical surface area and cognitive function. The lessons of the Krech et al. legacy are clear: enriching the lives of those children early in life will have a profound effect on millions of children worldwide. Interventions will not only include increased incomes for poor families, but also there needs to be a focus on helping parenting skills so that parents can contribute to the enrichment benefits (Kolb and Gibb, 2016).

REFERENCES

Beaulieu, C. and Colonnier, M. (1987) Effect of the richness of the environment on the cat visual cortex. *Journal of Comparative Neurology*, 266: 478–94.

Bennett, E.L., Kiamond, M.C., Krech, D. and Rosenzweig, M.R. (1964) Chemical and anatomical plasticity of the brain. *Science*, 146: 610–19.

Bergami, M. (2015) Experience-dependent plasticity of adult-born neuron connectivity. *Communicative and Integrative Biology*, 8: 3. e10384344.

Cancedda, L., Putignano, E., Sale, A. et al. (2004) Acceleration of visual system development by environmental enrichment. *Journal of Neuroscience*, 24: 4840–8.

Chang, F.-L. and Greenough, W.T. (1984) Transient and enduring morphological correlates of synaptic activity and efficacy change in the rat hippocampal slice. *Brain Research*, 309: 35–46.

Comeau, W., McDonald, R. and Kolb, B. (2010) Learning-induced structural changes in the prefrontal cortex. *Behavioural Brain Research*, 214: 91–101.

Diamond, M.C., Krech, D. and Rosenzweig, M.R. (1964) The effects of an enriched environment on the histology of the rat cerebral cortex. *Journal of Comparative Neurology*, 123: 111–20.

Dunnett, S.B., Whishaw, I.Q., Bunch, S.T. and Fine, A. (1986) Acetylcholine-rich neuronal grafts in the forebrain of rats: effects of environmental enrichment, neonatal noradrenaline depletion, host transplantation site and regional source of embryonic donor cells on graft size and acetylcholinesterase-positive fibre outgrowth. *Brain Research*, 378: 357–73.

Floeter, M.K. and Greenough, W.T. (1979) Cerebellar plasticity: modification of Purkinje cell structure by differential rearing in monkeys. *Science*, 206: 227–9.

Garth, A., Roeder, I. and Kempermann, G. (2016) Mice in an enriched environment learn more flexibly because of adult hippocampal neurogenesis. *Hippocampus*, 26: 261–71.

Gibb, R., Gonzalez, C. and Kolb, B. (2014) Prenatal enrichment and recovery from perinatal cortical damage: effects of maternal complex housing. *Frontiers in Behavioral Neuroscience*, 8: 223. soi 10.3389.

Greenough, W.T. and Chang, F.F. (1988) Plasticity of synapse structure and pattern in the cerebral cortex. In A. Peters and E.G. Jones (eds), *Cerebral Cortex*, Vol. 7. New York: Plenum. pp. 391–440.

Hamilton, D. and Kolb, B. (2005) Nicotine, experience, and brain plasticity. *Behavioral Neuroscience*, 119: 355–65.

Hanson, J.L., Hair, N., Shen, D.G. et al. (2013) Family poverty affects the rate of human infant brain growth. *PLoS One*, 8: e80954. doi:10.1371/journal.pone.0080954.

Hebb, D.O. (1947) The effects of early experience on problem solving at maturity. *American Pyschologist*, 2: 737–45.

Hebb, D.O. (1949) *The Organization of Behavior*. New York: Wiley.

Hensch, T.K. and Bilimoria, P. (2012) Re-opening windows: manipulating critical periods for brain development. *Cerebrum*, PMC3574806.

Juraska, J.M. (1990) The structure of the cerebral cortex: effects of gender and the environment. In B. Kolb and R. Tees (eds), *The Cerebral Cortex of the Rat*. Cambridge, MA: MIT Press. pp. 483–506.

Kolb, B. and Gibb, R. (2016) Childhood poverty and brain development. *Human Development*, DOI: 10.1159/000438766.

Kolb, B., Gibb, R. and Gorny, G. (2003a) Experience-dependent changes in dendritic arbor and spine density in neocortex vary with age and sex. *Neurobiology of Learning and Memory*, 791: 1–10.

Kolb, B., Gorny, G., Li, Y. et al. (2003b) Amphetamine or cocaine limits the ability of later experience to promote structural plasticity in the neocortex and nucleus accumbens. *Proceedings of the National Academy of Sciences, USA*, 100: 10523–8.

Kolb, B., Gorny, G., Sonderpalm, A. and Robinson, T.E. (2003c) Environmental complexity has different effects on the structure of neurons in the prefrontal cortex versus the parietal cortex or nucleus accumbens. *Synapse*, 48: 149–53.

Krech, D., Rosenzweig, M.R. and Bennett, E.L. (1960) Effects of environmental complexity and training on brain chemistry. *Journal of Comparative and Physiological Psychology*, 53: 509–19.

Morley-Fletcher, S., Rea, M., Maccari, S. and Laviola, G. (2003) Environmental enrichment during adolescence reverses the effects of prenatal stress on play behaviour and HPA axis reactivity in rats. *European Journal of Neuroscience*, 18: 3367–74.

Noble, K.G., Houston, S.M., Brito, N.H. et al. (2015) Family income, parental education and brain structure in children and adolescents. *Nature Neuroscience*, 18: 773–8.

Ramon y Cajal, S. (1928) *Degeneration and Regeneration of the Nervous System*. London: Oxford University Press.

Sirevaag, A.M. and Greenough, W.T. (1987) Differential rearing effects on rat visual cortex synapses. III. Neuronal and glial nuclei, boutons, dendrites, and capillaries. *Brain Research*, 424: 320–32.

Sirevaag, A.M. and Greenough, W.T. (1988) A multivariate statistical summary of synaptic plasticity measures in rats exposed to complex, social and individual environments. *Brain Research*, 441: 386–92.

Skwara, A.J., Karwoski, T.E., Czambel, R.K. et al. (2012) Influence of environmental enrichment on hypothalamic-pituitary-adrenal (HPA) responses to single-dose nicotine, continuous nicotine by osmotic mini-pumps, and nicotine withdrawal by mecamylamine in male and female rats. *Behavioural Brain Research*, 234: 1–10.

Stell, M. and Riesen, A. (1987) Effects of early environments on monkey cortrex neuroanatomical changes following somatomotor experience: effects on layer III pyramidal cells in monkey cortex. *Behavioral Neuroscience*, 101: 341–6.

Sutherland, R.J., Gibb, R. and Kolb, B. (2010) Unilateral hippocampal lesions block experience-dependent plasticity in the ipsilateral hemisphere. *Behavioural Brain Research*, 214: 121–4.

Takesian, A.E. and Hensch, T.K. (2013) Balancing plasticity/stability across brain development. *Progress in Brain Research*, 207: 3–34.

Tanaka, K. (1993). Neuronal mechanisms of object recognition. *Science*, 262: 685–8.

Technau, G. (1984) Fiber number in the mushroom bodies of adult Drosophila melongaster depends on age, sex and experience. *Journal of Neurogenetics*, 1: 13–26.

Volkmar, F.R. and Greenough, W.T. (1972) Rearing complexity affects branching of dendrites in visual cortrex of the rat. *Science*, 176: 1445–7.

Willing, J. and Juraska, J.M. (2015) The timing of neuronal loss across adolescence in the medial prefrontal cortex of male and female rats. *Neuroscience*, 301: 268–75.

16 Revisiting Harry Harlow: Love in infant monkeys

Stephen J. Suomi and Bryan Kolb

John B. Watson, the founder of the psychological school of Behaviorism in 1913, wanted to make psychology a purely objective part of natural science in which the goal was to predict and control behavior. For Watson the emphasis was on the observation of external behavior of people and other animals and not on inferred internal mental states. Although Behaviorism is no longer a major school of psychology, Watson had a major impact on the field and became one of the most influential psychologists of the twentieth century.

In 1928 he published a book, *Psychological Care of Infant and Child*, which became very influential, selling over 100,000 copies in the first few months (Blum, 2002). A key premise in his book was that children should be treated as small adults and that too much mothering was dangerous, i.e. too much affection undermined the development of strong character and led to the development of weak and anxious adults. Although today it is easy to wonder why this perspective became so influential, it was widely accepted and admired. Even the US government supported his ideas and warned parents of the dangers of too much rocking and playing with their children (for an interesting review of this story see Blum, 2002).

It was against the background of Watson's theory and conjecture and a lack of objective studies of the role of mothering that Harlow's studies appeared. But rather than finding that the infant–mother bond was largely a means for the infant to obtain food and drink, Harlow – as the title of his *Scientific American* review suggests – highlighted the importance of love. In addition, Harlow's studies began the scientific search for what defines love, and laid the foundation for many lines of contemporary research that search for the genetic and neural basis of love.

THE HARLOW STUDIES

The first studies on the role of mothering came from an unlikely source. In the 1940s Harry Harlow at the University of Wisconsin was studying learning in monkeys. When he began his studies there was a heavy behaviorist bias towards describing learning in laboratory rats via equations that were intended to include the variables that might influence the rate of learning and make predictions about the nature of the learning. The brain was not considered relevant in these studies, but rather the rat's brain was treated like a black box that simply made calculations based upon conditioning. Harlow believed that the various mazes of the behaviorists made the animals act like little automatons because the researchers underestimated what the rats could actually do.

For various reasons he decided to study monkeys and make the learning tasks more challenging so as to reveal their real thinking rather than automatic behaviors. However, there were no experimental self-sustaining monkey colonies in the US at that time and the importation of monkeys from Asia was fraught with difficulties. Animals often arrived in poor shape and died. Harlow decided to start a small colony at Wisconsin, and to ensure that the monkeys born in the lab were healthy they would be hand reared. There was a history of raising children in orphanages (formerly called Foundling Homes) in which the children were kept sterile and separated from others to avoid infection and disease, so Harlow believed that a similar system should work for the monkeys.

By 1956 more than 60 baby monkeys had been taken from their mothers at a few hours of age and raised individually in nurseries and fed every 2 hours. The babies grew well and were healthy. But their behavior was not right. They would clasp themselves, rock, suck their thumbs, and stare into space. When older and placed with other monkeys they were clearly not socialized and interacted poorly, if at all. Harlow and his students noticed that the babies were hugging diapers that had been placed on the cage floor to provide warmth and softness, much like Linus in the *Peanuts* cartoon strip. This was a pivotal observation leading Harlow to study the role of contact comfort in brain and behavioral development.

In the first studies baby monkeys were provided with two wire "mother" monkeys, one of which was covered with soft terrycloth but provided no food, whereas the other, plain wire, mother provided food from an attached baby bottle (see Figure 16.1). If the role of mothers was to provide food then one might expect the infants to spend more time near the food-providing mother, but instead, the baby monkeys spent significantly more time attached to the cloth mother rather than the food-providing mother. Harlow (1958) concluded "These data make it obvious that contact comfort is a variable of overwhelming importance in the development of affectional response, whereas lactation is a variable of negligible importance."

Further studies showed that contact played an important role in reducing fear and providing security. Harlow placed the baby monkeys in strange rooms in which they could explore either in the presence of the surrogate cloth mother or in her absence. The monkeys used their surrogate mother as a secure base to explore the room by leaving her briefly and then rushing back to be comforted before making longer forays around the room. In the absence of the surrogate mother the monkeys were distraught and often would crouch and clasp themselves and rock, scream and cry (Harlow, 1959).

Figure 16.1 An infant monkey with the wire and cloth surrogate models

One important question still to be answered was whether contact comfort had to come from mothers rather than other monkeys such as similar-aged peers. To answer this question, infant monkeys were housed with three peers instead of mothers for the first 6 months of life before being moved into larger social groups. The peer-reared infants established strong social bonds with each other, similar to the attachments of mother-reared infants to their mothers (see Figure 16.2).

Figure 16.2 Peer-reared monkeys

Although the peer-reared animals were much more "normal" than the surrogate-reared monkeys, they were excessively anxious and were unusually fearful in novel situations: they were reluctant to approach novel objects and tended to be shy in initial encounters with unfamiliar peers. In addition, the peer-reared animals had restricted social repertoires, likely because cage-mates had to serve both as attachment figures and playmates – a dual role that is quite abnormal (Suomi, 2002).

Longitudinal studies of peer-reared monkeys have found that not only are they more anxious, but they also exhibit more extreme adrenocortical and noradrenergic reactions to social separations than mother-reared monkeys, even after they have been living in stable social groups for extended periods. In addition, the peer-reared monkeys show lower CSF concentrations of 5-HIAA (the primary metabolite of serotonin) than mother-reared counterparts. This difference appears well before 6 months of age and persists through adolescence and into adulthood (Higley and Suomi, 1996).

The obvious question to ask about the monkeys raised without real mothers is whether the behavioral abnormalities can be reversed. To explore this idea, Harlow and Suomi (1971) placed 6-month-old isolated monkeys with 3-month-old youngsters ("therapists") for 2 hours a day. Although the isolated animals were initially fearful of the younger animals, the younger animals would cling to them, much as to their own mothers. The isolated monkeys slowly began to respond to the therapists and most were rehabilitated to a functional life.

MILD STRESS EXPERIMENTS IN RATS

As Harlow and his students were studying love in infant monkeys, Seymour Levine wondered whether contact comfort was the whole story. With Harlow's encouragement, he designed a simple study using three groups of rats. One group simply stayed with their mothers in a cage. A second group were removed from the mother for three minutes and exposed to a mild electric shock, which was thought to be a model of early trauma. A third group was removed from the mother but not shocked, a so-called "handled group." Thus, Levine could compare a sheltered infancy with a stressful one.

As the animals developed Levine conducted various behavioral tests. In one the animals were placed in a box on a grid that became the source of a mild electric shock. The animals could escape the shock simply by running away. To Levine's surprise the sheltered animals were slow to learn to escape the shock, whereas the animals in the other groups displayed less fear and quickly learned to escape the shock. In another test, much like Harlow's open field exploration test, the rats were placed alone in a large enclosure studded with toys to explore. Once again, the sheltered animals showed intense fear and showed little exploration. The handled and early stressed rats were bolder and explored more extensively.

Levine concluded that infant rats needed contact comfort from their caregiver but they needed other stimulation as well – even mild negative stimulation (e.g. Levine, 1960; Levine and Lewis, 1959). It was remarkable that a single 180-second experience in development could have such profound consequences on later adult behavior, so it is not surprising that Levine's studies were greeted with much skepticism. Over 60 years later the findings have stood the test of many replications. Handling the animals for only 180 seconds profoundly changed brain and behavioral development. (Levine later went on to study the effects of early adversity on metabolic functions in squirrel monkeys.)

A clue to how the handling could have had such profound effects came from later studies by Michael Meaney and his colleagues, who showed that the early handling altered the hypothalamic-pituitary axis (HPA), making it more efficient, allowing better recovery from stress (Meaney et al., 1985). Later studies showed that removal of the pups led to increased mothering (licking and grooming) by the dams, which influenced the development of central systems that serve to activate or inhibit the expression of behavioral and endocrine responses to stress (see the review by Caldji et al., 2000). The increased licking and grooming of the pups also altered gene expression in the infants, which in turn altered brain and behavior in adulthood (see below).

But clearly there must be a limit to the intensity of early stress that is beneficial. Indeed there is. Studies over the past 20 years have shown that prolonged gestational stress (daily stressing of mothers for 5–7 days in late gestation) also leads to anxious offspring (e.g. Vallee et al., 1997; Weinstock, 2001). In fact, the effects of stress may be indirect. For example, studies showing that stressing fathers before they mate with the mothers also leads to anxious offspring who have deficits in learning (e.g. Harker et al., 2015). Taken together, the Harlow and stressed rat studies have made it clear that early experiences have profound impact on the development of animals. For monkeys the early mother–infant bond is critical. For rats this bond is less important, but aspects of the mother–infant interaction, such as licking and grooming, play a central role in the development of brain and behavior (e.g. Caldji et al., 2000).

THE ROMANIAN ORPHANAGE STUDIES

Although the effects of being raised in orphanages have long been a topic of research interest, it was probably the example of the Romanian orphanages under the communist regime that has had the greatest impact on thinking about aversive early experiences in orphanages. In the 1960s the Romanian communist regime believed that the route to greater productivity was to have more workers, and the route to more workers was by increasing birth rate. To this end, they implemented a law that all families must have at least five children, but they did not provide sufficient resources for families to support so many offspring and instead opened orphanages where these children could be raised. However, these institutions were grim and the infants and children had a barren existence that was better than Harlow's surrogate-reared monkeys – but not much. Children were housed in groups with about a 1:25 caregiver to infant ratio, meaning that there was little one-on-one with a caregiver and little contact comfort for the infants. The result was a dramatic stunting of physical and cognitive development in these unfortunate children.

When the communist government fell, children were adopted by loving families in western countries such as the UK, US, Canada, and Australia, with the expectation that the effects of the early deprived experiences could be reversed. Unfortunately this was only true in children adopted before they were 12–18 months old. After this age the children remained severely scarred and over 25 years later they still have significant cognitive and emotional deficits, including an IQ drop of about 15 or more points, smaller brains, abnormal brain electrical activity, and a host of serious chronic cognitive and social deficits that do not appear to be easily reversed (e.g. Johnson et al., 2010; Lawler et al., 2014; Rutter, 2004).

Sadly, in hindsight, this outcome was entirely predictable from the Harlow studies. Importantly, however, the studies on surrogate and peer-reared monkeys over the past 30 years have provided important insights into the mechanisms underlying the severe effects of early adversity, as we will discuss below.

GENE–ENVIRONMENT INTERACTIONS

While we have emphasized the difference between mother-reared and peer-reared infants, there are significant individual differences in anxiety levels in both groups. Studies of normally reared monkeys, both in the wild and in captivity, have found that 15–20% of the animals are excessively fearful. These monkeys consistently respond to mildly challenging or novel situations with pronounced physiological arousal, often with prolonged activation of the HPA axis, coupled with extreme behavioral disruption (Suomi, 2002).

Taken together, the observations related to fear-provoking situations in peer-reared monkeys and the variability in normal populations provided an opportunity to examine how experience and genetics interact to shape a monkey's expression of fear and aggression. Given the effects of being peer-reared on serotonin levels noted earlier, it was reasonable to look for possible differences in the serotonin transporter gene (5-HTT). It was known in humans that there are two variants of this gene, a short allele (LS) and a long allele (LL). The LS version is associated with less efficient transcriptional efficiency of the 5-HTT, leading to the possibility that the LS allele may result in decreased serotoninergic function.

Bennett and colleagues (2002) examined the CSF 5-HIAA concentrations and found that whereas the levels did not vary with the 5-HTT type in mother-reared animals, they did in peer-reared animals. These latter animals with the LS allele had significantly lower levels of 5-HIAA than those with the LL allele. One interpretation of this finding is that the mothers are acting as a kind of "buffer" for the potentially deleterious effects of the LS gene on serotonin metabolism.

A different form of gene–environment interaction was seen in studies of alcohol consumption. Surprisingly, whereas peer-reared monkeys with the LS variant drank significantly more alcohol than those with the LL allele, the opposite was true of mother-reared monkeys. Thus, the LS allele appears to be a risk factor for the peer-reared monkeys but a protective factor for the mother-reared animals (Bennett et al., 1998).

These researchers also found a relationship between the 5-HTT gene and dominance-related assertive behavior. Across all monkeys, the peer-reared monkeys with the LL allele were the least assertive, the peer-reared LS and mother-reared LL were intermediate, and the mother-reared LS were the most assertive.

In sum, Bennett and colleagues showed a gene–environment interaction on three different measures, but the precise nature of the interaction was different on each measure. One implication of the studies is that individuals with the LS variant and poor family histories may be at risk for psychopathology, whereas the LS variant may be adaptive for those with good family histories (Suomi, 2002).

EPIGENETIC STUDIES

Epigenetics refers to changes in gene expression that are not related to changes in the DNA. The genes expressed in a cell are influenced by factors within the cell and the cell's environment. The cell's environment determines which genes are expressed and thus how the cell functions, including in the nervous system. But the cell's environment does not only include chemicals around the cell in the body, but also the environment that an organism finds itself in throughout life. Thus, our experiences can influence our genes.

Epigenetic mechanisms can influence protein production either by blocking a gene to prevent transcription or by unlocking a gene so that it can be transcribed. Chromosomes are wrapped around supporting molecules of a protein called histone, which allows many meters of a chromosome to be packaged in a small space. For any gene to be transcribed into messenger RNA, its DNA must be unspooled from the histones. Thus, one measure of changes in gene expression is histone modification (see Figure 16.3). Methyl groups (CH_3) or other molecules bind to the histones, either blocking them from opening or allowing them to open. An environmental event can either induce or remove one or more blocks, thus allowing the environment to regulate gene expression. Epigenetic mechanisms provide a mechanism whereby experience induces changes in our brain that make each of us a unique individual.

Early life experiences have been shown to alter gene expression profiles in both humans and rodents (e.g. McGowan et al., 2008; Roth et al., 2009; Weaver et al., 2003), so it was a reasonable question to ask whether there are postnatal epigenetic changes in monkeys exposed to early life adversity – and indeed there are. Provencal et al. (2012) examined whether maternal- versus peer-rearing had differential effects on DNA methylation in early adulthood in the dorsolateral prefrontal cortex or T cells. The results revealed differential DNA methylation in both the prefrontal cortex and T cells, thus supporting the hypothesis that the response to early-life adversity is system- and genome-wide and persists into adulthood. Thus, the degree of methylation at promotor sites

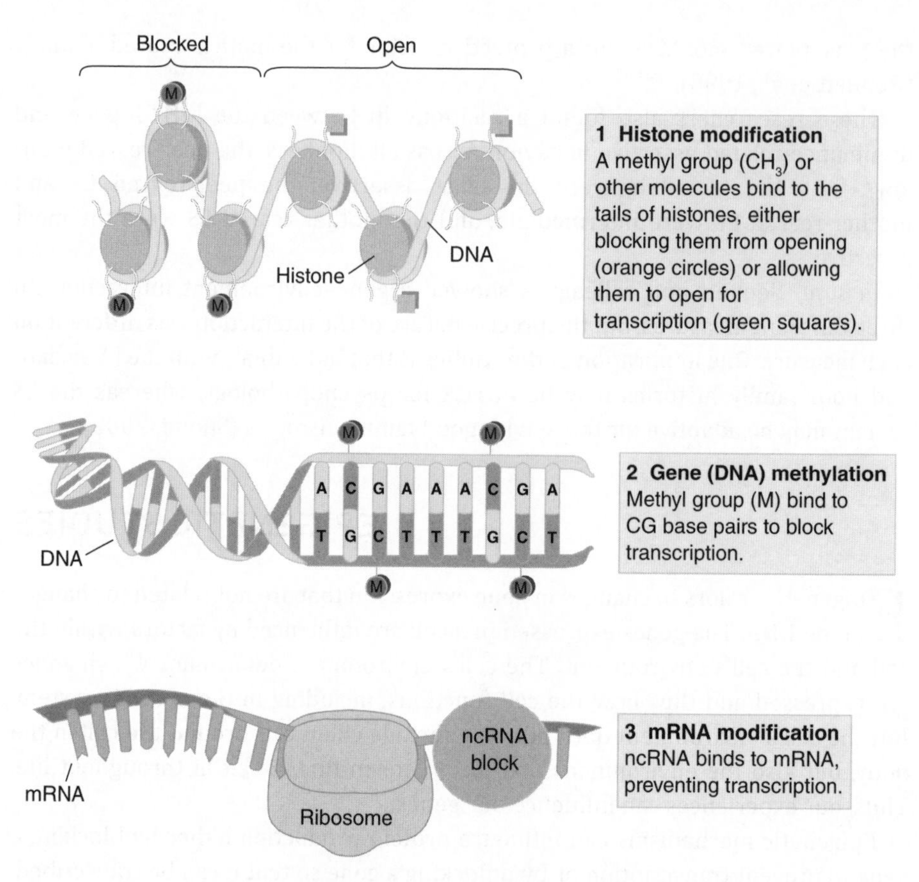

Figure 16.3 Epigenetic mechanisms (from Kolb et al., 2016: 104)

(promotors are regions of DNA that initiate the transcription of a particular gene) distinguish mother-reared from surrogate peer-reared (reared with an inanimate surrogate as well as daily socialization periods with age-mate peers) monkeys.

Parallel studies in rats have shown that not only do early aversive experiences influence methylation patterns in the medial prefrontal cortex, but so do more positive experiences such as enriched housing (see Chapter 15) (Mychasiuk et al., 2012, 2013). The complementary effects of aversive versus positive experiences lead to the question of whether housing surrogate-reared monkeys with therapists (see above) might be able to reverse some of the negative effects of early impoverished rearing.

INFLUENCE OF THE HARLOW STUDIES

Research over the past six decades since the early Harlow rearing studies has provided a remarkable demonstration that his rhesus monkey model of early childhood adversity (nursery-rearing) produces a wide array of behavioral, physiological, and neurobiological deficits that parallel those identified in human studies of early adversity (Bennett, 2010; Harlow and Harlow, 1965; Kaufman et al., 2000; Kraemer, 1992; Sackett, 1984; Suomi, 1997, 2002). But this influence goes far beyond the monkey studies.

The Harlow studies led to a broader interest in the effects of early experiences on brain and behavioral development in many species, including effects of adverse experiences (e.g. Levine, 1960; Babenko et al., 2015) as well as brain injury (e.g. Harlow et al., 1964; Kolb, 1995) and positive experiences such as enrichment (see Chapter 15) and tactile stimulation (e.g. Muhammad et al., 2011).

Finally, as noted in the introduction, the prevailing view in 1950, which stemmed from the ideas of Watson and other behaviorists, was that mother–infant interactions were largely based upon nutritional needs of the infants. The dramatic effects of the Harlow studies led directly to an alternate view of the importance of human care-givers in normal brain and behavioral development, and a revolution in both government and public views of proper child rearing. One continuing advantage of the monkey studies is that researchers can seek mechanisms that underlie the effects of early adverse experiences, with the goal of finding a way to reverse the adverse effects and give children a better life.

REFERENCES

Bennett, A.J. (2010) *Nonhuman Primate Research Contributions to Understanding Genetic and Environmental Influences on Phenotypic Outcomes Across Development*. Abingdon: Wiley.

Bennett, A.J., Lesch, K.P., Heils, A. and Linnoila, M.V. (1998) Serotonin transporter variation, CSF 5-HIAA concentrations, and alcohol-related aggression in rhesus monkeys. *American Journal of Primatology*, 45: 168–9.

Bennett, A.J., Lesch, K.P., Heils, A. et al. (2002) Early experience and serotonin transporter gene variation interact to influence primate CNS function. *Molecular Psychiatry*, 7: 118–22.

Blum, D. (2002) *Love at Goon Park*. Cambridge, MA: Perseus Books.

Caldji, C., Diorio, J. and Meaney, M.J. (2000) Variations in maternal care in infancy regulate the development of stress reactivity. *Biological Psychiatry*, 48: 1164–74.

Dettmer, A.M. and Suomi, S.J. (2014) Nonhuman primate models of neuropsychiatric disorders: influences of early rearing, genetics, and epigenetics. *ILAR Journal,* 55: 361–70.

Harker, A., Raza, S., Williamson, K. et al. (2015) Preconception paternal stress in rats alters dendritic morphology and connectivity in the brain of developing male and female offspring. *Neuroscience*, 303: 200–10.

Harlow, H.F. (1959) Love in infant monkeys. *Scientific American*, 200: 68–74.

Harlow, H.F. Akert, K. and Schiltz, K.A. (1964) The effects of bilateral prefrontal lesions on learned behavior of neonatal, infant and preadolescent monkeys. In J.M. Warren and K. Akert (eds), *The Frontal Granular Cortex and Behavior.* New York: McGraw-Hill. pp. 126–148.

Harlow, H.F. and Harlow, M.K. (1965) The effect of rearing conditions on behavior. *International Journal of Psychiatry*, 1: 43–51.

Harlow, H.F. and Suomi, S.J. (1971) Social recovery by isolation-reared monkeys. *Proceedings of the National Academy of Sciences, USA*, 59: 538–49.

Higley, J.D. and Suomi, S.J. (1996) Reactivity and social competence affect individual differences in reaction to severe stress in children: investigations using nonhuman primates. In C.R. Pfeffer (ed.), *Severe Stress and Mental Disturbance in Children.* Washington, DC: American Psychiatric Press.

Johnson, D.E., Guthrie, D., Smyke, A.T. et al. (2010) Growth and associations between auxology, caregiving environment, and cognition in socially deprived Romanian children randomized to foster vs ongoing institutionalized care. *Archives of Pediatric Adolescent Medicine*, 164: 507–16.

Kaufman, J., Plotsky, P.M., Nemeroff, C.B. and Charney, D.S. (2000) Effects of early adverse experiences on brain structure and function: clinical implications. *Biological Psychiatry*, 48: 778–90.

Kolb, B. (1995) *Brain Plasticity and Behavior*. Hillsdale, NJ: Lawrence Erlbaum.

Kolb, B., Whishaw, I.Q. and Teskey, G.C. (2016) *Introduction to Brain and Behavior*, 5th edition. New York: Worth.

Kraemer, G.W. (1992) A psychobiological theory of attachment. *Behavioral and Brain Sciences*, 15: 493–541.

Lawler, J.M., Hostinar, C.E., Mliner, S.B. and Gunnar, M.R. (2014) Disinhibited social engagement in postinstitutionalized children: differentiating normal from atypical behavior. *Developmental Psychopathology,* 26: 451–64.

Levine, S. (1960) Stimulation in infancy. *Scientific American*, 202: 81–6.

Levine, S. and Lewis, G.W. (1959) Critical period for effects of infantile experience on maturation of stress response. *Science*, 129: 42–3.

McGowan, P.O., Sasaki, A., D'Alessio, A.C. et al. (2009) Epigenetic regulation of the glucocorticoid receptor in human brain associates with childhood abuse. *Nature Neuroscience*, 12: 342–8.

Meaney, M.J., Aitken, D.H., Bodnoff, S.R. et al. (1985) The effects of postnatal handling on the development of the glucocorticoid receptor systems and stress recovery in the rat. *Progress in Neuropsychopharmacology and Biological Psychiatry*, 9: 731–4.

Muhammad, A., Hossain, S., Pellis, S.M. and Kolb, B. (2011) Tactile stimulation during development attenuates amphetamine sensitization and structurally reorganizes prefrontal cortex and striatum in a sex-dependent manner. *Behavioral Neuroscience*, 125: 161–74.

Mychasiuk, R., Zahir, S., Schmold, N. et al. (2012) Parental enrichment and offspring development: modification to brain, behavior and the epigenome. *Behavioural Brain Research*, 228: 294–8.

Mychasiuk, R., Harker, A., Ilnytskyy, S. and Gibb, R. (2013) Paternal stress prior to conception alters DNA methylation and behaviour of developing rat offspring. *Neuroscience*, 241: 100–5.

Roth, T.L., Lubin, F.D., Funk, A.J. and Sweatt, J.D. (2009) Lasting epigenetic influence of early-life adversity on the BDNF gene. *Biological Psychiatry*, 65: 760–9.

Rutter, M., O'Connor, T.G.and the English and Romanian Adoptees (ERA) Study Team (2004) Are there biological programming effects for psychological development? Findings from a study of Romanian adoptees. *Developmental Psychology*, 40: 81–94.

Sackett G.P. (1984) A nonhuman primate model of risk for deviant development. *American Journal of Mental Deficiency*, 88: 469–476.

Suomi S.J. (1997) Early determinants of behaviour: evidence from primate studies. *British Medical Bulletin*, 53: 170–84.

Suomi, S.J. (2002) How gene-environment interactions can shape the development of socioemotional regulation in rhesus monkeys. In B.S. Zuckerman, A.F. Lieberman and N.A. Fox (eds), *Emotional Regulation and Developmental Health: Infancy and Early Childhood*. New York: Johnson & Johnson Pediatric Institute. pp. 5–26.

Vallee, M., Mayo, W., Dellu, F. et al. (1997) Prenatal stress induces high anxiety and postnatal handling induces low anxiety in adult offspring: correlation with stress-induced corticosterone secretion. *Journal of Neuroscience*, 17: 2626–36.

Weaver, I.C., Cervoni, N., Champagne, F.A. et al. (2004) Epigenetic programming by maternal behavior. *Nature Neuroscience*, 7: 847–54.

Weinstock, M. (2001) Alterations induced by gestational stress in brain morphology and behaviour of the offspring. *Progress in Neurobiology*, 65: 427–51.

17 Revisiting Bliss and Lømo: Long-term potentiation and the synaptic basis of learning and memory

G. Campbell Teskey

People have long pondered the question of how we learn and remember. Neuroscientists more formally express this question something like the following: "What are the functional and structural changes that occur in the brain following an experience which store those memories and shape our future responses?" To address this important question they have relied heavily on model systems and electric stimulation. A critically important paper in the history of the neural basis of learning and memory was published in 1973 by Tim Bliss and Terje Lømo. Their landmark article entitled "Long-lasting potentiation of synaptic transmission in the dentate area of the anaesthetized rabbit following stimulation of the perforant path" in *Journal of Physiology* paved the way for an extraordinarily fruitful avenue of research. That paper was the first full-length description of long-term potentiation (LTP), broadly defined as the long-lasting enhancement of synaptic transmission following brief high-frequency synaptic stimulation. As of 2015 the article has garnered more than 3500 citations and spawned research that has not only identified critical molecular and cellular mechanisms of LTP, but has also inspired a number of groundbreaking discoveries, namely: the role of NMDA and AMPA receptors in memory and its associated plasticity; the role of intracellular messengers like calcium/calmodulin-dependent protein kinase II (CaMKII) that transduce signals from the surface receptors of cells to their nuclei and result in new RNA and protein synthesis; and the functional opposite of LTP – long-term depression (LTD). The discovery of LTP has, moreover, provided insights into the role of synaptic plasticity in learning and memory in healthy brains, and in disease states both in adults and during development.

Before we examine the influence of Bliss and Lømo's (1973) paper, it is worth exploring what relevant information was known at the time they conducted their experiments. Donald Hebb had clearly hypothesized that correlated pre- and post-synaptic neuronal activity was the basis for information storage in the central nervous system (Hebb, 1949). It was well documented that moderately lasting changes in synaptic transmission could be induced in invertebrate preparations, vertebrate neuromuscular junctions and the spinal cord, but long-lasting changes

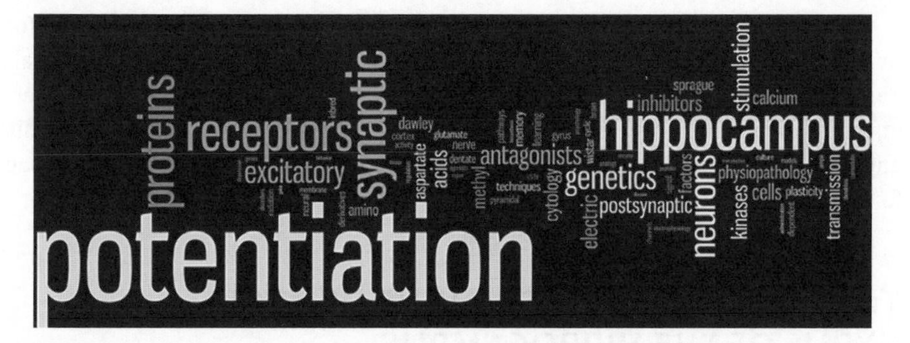

Figure 17.1 A representation of the associated words and their weighting (proportional to the size of the word) from the 5000 Medline records that have "long-term potentiation" in the subject heading

had not yet been discovered in the mammalian forebrain – yet it is the forebrain areas which are larger in mammals that were thought to underlie long-term memory. But where to look?

Per Andersen's laboratory, that housed and supported the work of Bliss and Lømo, had previously examined the electrophysiological properties of the hippocampus, a beautiful cortical structure that has a relatively simple and delineated anatomy and gives rise to large and easily recorded electrical potentials. Andersen and colleagues had determined that the recorded evoked response in the dentate gyrus that is elicited by electrically stimulating the perforant path can be understood as the combined effect of a number of single neurons behaving in close temporal synchrony (Andersen et al., 1966, 1971a, 1971b). The rising phase of the extracellular field potential corresponds with the intracellular excitatory postsynaptic potential (EPSP), whereas the extracellular population spike corresponds with the intracellularly recorded action potential. This was important because it allowed researchers to stimulate intact animals and make sense of the recorded evoked potentials.

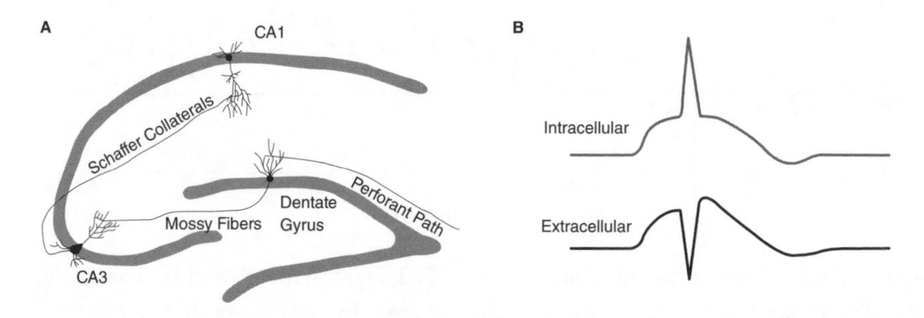

Figure 17.2 A representation of the excitatory neurons and their major synaptic connections in the rodent hippocampus (A). The intracellularly and extracellularly recorded excitatory postsynaptic potentials (EPSPs), action potentials, and population spikes in the dentate gyrus are elicited by perforant path stimulation (B).

Short-term plasticity effects, known as post-tetanic potentials, which lasted for minutes, were also known to occur (Feng, 1941; Green and Adey, 1956; Gloor et al., 1964), and Eccles in 1964 and Kandel and Spencer in 1968 discussed such after-effects as expression of synaptic plasticity. However, memories can last a long time and a clear demonstration of long-term synaptic enhancement in the mammalian forebrain had yet to be demonstrated.

THE DISCOVERY OF LTP IN THE DENTATE GYRUS OF THE HIPPOCAMPUS

Terje Lømo, who began his graduate work in Per Andersen's laboratory in 1965, examined the synaptic mechanisms and organization of the dentate area of the hippocampal formation. He published a number of articles with Andersen on the location and identification of excitatory synapses on hippocampal pyramidal cells and their mode of activation (Andersen et al., 1966; Andersen and Lømo, 1967). He also published two single-authored peer-reviewed abstracts on the potentiation of monosynaptic EPSPs (Lømo, 1966, 1968) as well as two papers on the characterization and potentiation of perforant pathway to dentate gyrus (Lømo, 1971a, b). These papers were the direct forerunners of the 1973 paper, and although the effects of the tests lasted only a few minutes, Lømo and Andersen felt they were onto something important (Lømo, 2003). Tim Bliss joined Andersen's

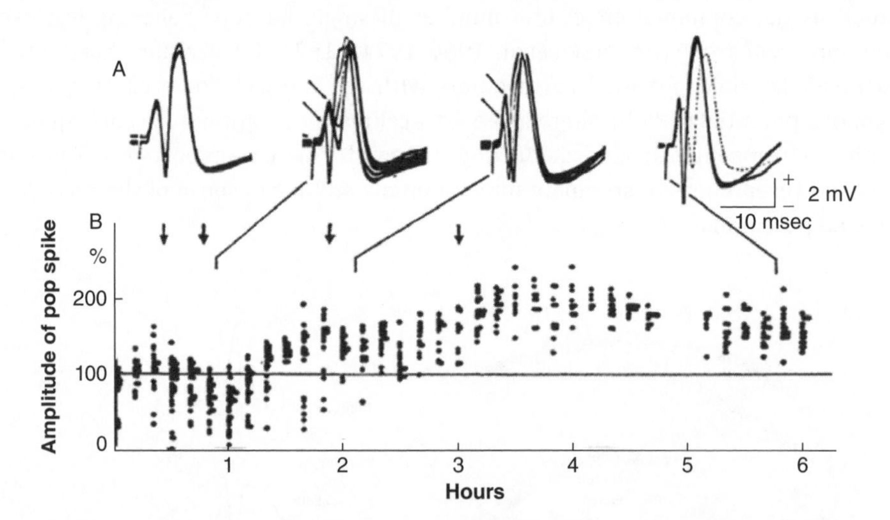

Figure 17.3 Taken from Bliss and Lømo (1973, Fig. 5. panel A and B). Extracellular recording of field potentials in the granule cell layer following high-frequency tetanization. An increase in the size of the population spike is observed that lasts for at least 6 hours – long-term potentiation (LTP). The figure illustrates the variability of the population spike with superimposed responses at the four times shown. The amplitude of the population spikes is plotted in B.

laboratory in 1968 with the intention to work on a simpler brain structure and pursue his interest in the neural basis for learning. Bliss and Lømo worked together on the perforant path to dentate system that had been fleshed out over the previous years. Lømo, in his review of the discovery of LTP, acknowledges that it was Bliss's interest in the memory mechanisms related to synaptic enhancement that really propelled the work (Lømo, 2003).

LTP: From physiology to molecular biology

Following Bliss and Lømo's 1973 publication, the first major advance was probably the demonstration that LTP could be induced in the hippocampal slice preparation (Schwartzkroin and Wester, 1975). This then allowed the powerful patch-clamping technique to interrogate the detailed physiological and molecular mechanisms operating at the single cell level in an exquisitely controlled environment. Most LTP experiments have been carried out in the circuit from CA3 via the Schaffer collateral-commissural pathway to the CA1 region where LTP is robust, but it is important to keep in mind that the "rules" learned at this synapse do not generalize to all other synaptic systems across the brain.

Ten years after the Bliss and Lømo article, Collingridge et al. (1983) determined that the NMDA receptor was required for *induction* of LTP within the hippocampus from CA3 to CA1. However, it took another 12 years before the resolution of the debate as to whether the enhanced synaptic transmission after LTP induction was due to a presynaptic increase in transmitter release or alternatively to a postsynaptic increase in AMPA receptor response. On the presynaptic side was the observation that during LTP the number of synaptic failures decreased, indicating an increase in the probability of transmitter release (Malinow and Tsien, 1990). However, opinion soon shifted to a postsynaptic mechanism when it was observed that some CA1 synapses were postsynaptically silent, but following high-frequency tetanisation that gave rise to LTP, these synapses rapidly acquired AMPA receptors and became postsynaptically active (Isaac et al., 1995). When LTP became generally understood as largely a postsynaptic phenomenon this heralded the era of studying the precise molecular mechanisms that regulate AMPA receptor trafficking to synapses.

As with most scientific fields, real progress is made when new technologies are brought to bear on old questions. The technical advance and adoption of two-photon microscopy allowed researchers to image single dendritic spines and uncage glutamate onto just those spines. This level of control coupled with electrophysiology allowed for the experiment where NMDA receptors could be activated by uncaging glutamate onto a single spine while being paired with postsynaptic depolarization. NMDA receptor activation led to a rapid increase in AMPA mediated currents, demonstrating that LTP is synapse specific as close neighboring synapses were not potentiated. Moreover the rapid increase in spine volume that persists for the duration of the experiment is now considered an important morphological component of synaptic plasticity (Matsuzaki et al., 2004).

Many prominent scientists in the LTP field then moved into the cell and asked questions like "How does the cell surface signal become internalized and result in increased AMPA receptor trafficking?" It seems that the influx of calcium through the NMDA receptor is required for LTP. Moreover, CaMKII appears to be the critical target of calcium as it is both necessary and sufficient for LTP, resulting in CaMKII being dubbed the "memory molecule." The problem is that CaMKII activation is transient, lasting about one minute (Lee et al., 2009). Thus some of the other downstream targets of CaMKII must be important to maintain LTP. The search continues.

Because the expression of LTP – at least at the CA1 synapses – is critically dependent on the highly regulated trafficking of AMPA receptors to and from synapses, the LTP field once again embraced a molecular biology approach and examined how AMPA receptor mobility was regulated. Two main models have been put forward to explain the behaviour of AMPA receptors. In one model, activity increases the number of cell surface AMPA receptors from intracellular storage pools via exocytosis. In a second model, following activity synapses capture cell surface receptors as they enter the synapse from membranes that are outside the synapse. In both models AMPA receptors become tethered to the post-synaptic density (PSD) within the synapse. There are, however, still many outstanding questions. For instance, does CaMKII change the properties of the PSD to accommodate more AMPA receptors, and what is the role of CaMKII in the maintenance of LTP?

DIVERSITY OF LTP

One of the early lessons following Bliss and Lømo's landmark paper was that there are different forms of LTP at different synapses both within and outside of the hippocampus. For instance, mossy fibre synapses – the synapses between the axons of the dentate gyrus granule cells and the proximal apical dendrites of CA3 pyramidal neurons – are mechanistically similar to forms of LTP at corticothalamic synapses and cerebellar parallel fibre synapses, and yet distinct from NMDA receptor-dependent LTP. They are distinct because triggering of mossy fibre LTP does not require activation of NMDA receptors, or for that matter ionotropic glutamate receptors. Moreover, even at one synapse – like the CA1 excitatory synapse – there are multiple forms of LTP, because the expression of enhanced synaptic strength depends on the frequency, pattern, and intensity of stimulation.

There is even LTP at inhibitory synapses manifesting as either presynaptic GABA release or the increase in postsynaptic GABA receptor number/responsiveness (Castillo et al., 2011). This is important because inhibitory synaptic plasticity, by altering the excitatory/inhibitory balance, plays an important role in the refinement of neural circuits and in many forms of experience dependent learning. The field of homeostatic plasticity showed us that there are several forms of plasticity that stabilize the properties of neural circuits. These include mechanisms that regulate neuronal excitability, stabilize total synaptic strength, and influence the rate and extent of synapse formation. Many of these mechanisms likely change with Hebbian mechanisms to allow experience to modify the properties of neuronal networks selectively (Turrigiano, 1999). Obvious counter phenomena to stimulation-induced increases in synaptic strength are both decay and use-dependent decreases in synaptic potentiation like LTD.

LTD

In the first decade of LTP research it was shown that LTP could be disrupted and reversed by certain forms of induced synaptic activity, a phenomenon termed "depotentiation" (Barrionuevo et al., 1980). Later, low-frequency stimulation was shown to induce LTD in the cerebellum (Ito, 1989) and an LTD that was NMDA receptor-dependent at CA1 synapses as well as in the neocortex. These observations are considered important because a reversal of LTP by LTD mechanisms can prevent runaway excitation that would result in unstable neuronal networks. Since the time that reliable LTD stimulation protocols were worked out, LTD mechanisms have been shown to be involved in development, learning and memory, addiction, and neurological disorders such as mental retardation and Alzheimer's disease.

In vivo

As some researchers drilled down to reveal the precise molecular mechanisms underlying NMDA-dependent LTP at CA1 synapses, other researchers preferred to pursue different but no less important questions. Studies of activity-induced facilitation of sensorimotor synapses underlying the defensive gill reflex in *Aplysia* (Bailey and Kandel, 1993) demonstrated that long-term functional and structural synaptic modifications could serve as the substrate for learning and memory at the behavioral level. In the awake behaving rodents LTP can last for several weeks (Racine et al., 1995), an observation not possible in the slice preparation, and the maintenance of LTP over these durations requires new gene transcription and protein synthesis. This is achieved via intracellular signalling pathways that link the activity that induces LTP to the nucleus. Ultimately the key transcript factor, cyclic adenosine monophosphate response element-binding protein (CREB), as well as immediate early genes, are involved. While we still know very little about the identity and operation of newly synthesized proteins that are required to maintain LTP, this also continues to be examined.

As we move up from the molecular level we see that long-term structural remodelling occurs. Morphological changes include enlargement of pre-existing spines and their associated PSDs, the splitting of single PSDs and spines into new dendritic spines, and creation of new functional synapses. There are even changes in dendritic tree structure with increases in dendritic length and number of branches following LTP, and on the functional level LTP in the motor cortex results in alterations to the topographic representations of movements (Monfils et al., 2004). Thus the relationship between LTP and structural remodelling of synaptic connections is an area of research that has gained substantial experimental attention and support.

MEMORY

LTP has a set of characteristics, such as its rapid induction and its enduring response, which makes it attractive as the synaptic basis of learning and memory. Moreover, the NMDA receptor appears to be the structural instantiation of Hebb's rule – the conjunction of pre- and post-synaptic activation – and serves as a coincidence detector, and thus plays a central role in associative learning. But is LTP really a synaptic mechanism of memory? Electrical stimulation through an electrode is uncontestedly an artificial way to increase synaptic strength, and several issues and concerns about the connection between LTP and memory have been raised (Eichenbaum, 1996).

Studying the relationship between LTP and memory can be divided into three general approaches: demonstrations of changes in synaptic physiology as a consequence of learning, known as behavioural LTP experiments; attempts to interfere with learning by saturating hippocampal plasticity using stimulation protocols that induce LTP; and attempts to block both learning performance and LTP by pharmacological or genetic manipulations. The literature has multiple examples of spectacular successes and equally spectacular failures, and the field has progressed by becoming more sophisticated in the experimental approaches, adding appropriate control conditions as well as a better theoretical foundation with the elimination of naive experiments and their interpretations.

Studies examining synaptic plasticity and memory have been pursued in many different systems, including the lateral amygdala, cerebellar Purkinje cells, hippocampus, olfactory cortex, and sensory and motor neocortex. One behavioural LTP experiment that tested the assumption that skilled-learning-induced potentiation and artificial models of plasticity (LTP and LTD) share common mechanisms was provided by Monfils and Teskey (2004). They examined the acquisition of a skilled motor task and synaptic plasticity in the sensorimotor cortex of awake freely behaving rats. Skilled motor training was previously found to induce a functional reorganization of the forelimb area (Kleim et al., 1998) and to induce an increase in synaptic efficacy, measured in vitro, on the side contralateral to the reaching forelimb (Rioult-Pedotti et al., 1998). The skilled reaching task, however, is rather complex (Whishaw et al., 1991), and involves three different stages of acquisition and proficiency levels. Acquisition of task requirements occurs when the rat learns to reach for a pellet, but it is the second stage of increase in skill proficiency that involves learning a new motor skill. In the third stage, maintenance of skill proficiency, the rat continues to reach at an asymptotic level with the reaching behaviour and success rate remains unaltered. Because the task demands vary over the stages of skilled learning it is important to disambiguate what occurs at the synaptic level by examining neocortical evoked potentials over the three stages. We measured neocortical evoked potential recordings in awake freely behaving rats to examine whether skilled training would induce changes in polysynaptic efficacy on the side contralateral to the reaching forelimb. We found that the increase in task proficiency, but not the acquisition of task requirements or the maintenance of task proficiency, induced an increase in synaptic efficacy on the side contralateral to the reaching forelimb. We also tested the hypothesis that skilled-learning-induced potentiation shares similar mechanisms to LTP and LTD by artificially manipulating polysynaptic efficacy in skilled rats with high- and low-frequency stimulation. We observed that, compared with the ipsilateral side, less potentiation but more depression could be induced on the side contralateral to the reaching forelimb. This indicated that a transient, network-based, LTP-like mechanism operates during the learning of a skilled motor task.

LTP IN HUMANS

LTP has been demonstrated in isolated human cortical tissue obtained from patients undergoing surgery (Chen et al., 1996), with identical properties to rodent preparations. In 2005 Tyler and colleagues gave the first demonstration that the rapid repetitive presentation of a visual checkerboard (a photic 'tetanus') led to a persistent enhancement of one of the early components of the visual evoked potential in neurologically normal people. The potentiated response was limited to one component of the visual evoked response in the hemisphere contralateral to the tetanized visual hemifield. Because the potentiation was selective it rules out overall brain excitability as an explanation. While that study used stimulation with light pulses, other experiments have employed electrical stimulation of human brains to induce potentiation effects.

ELECTRICAL STIMULATION AS THERAPY

While there is a centuries-old intrigue between electrical stimulation and changes in biological function – think Frankenstein – the proposal that electrical stimulation may modulate neural circuits in a wide range of disease states and allow recovery of normal circuit functions has merit. Deep brain stimulation and transcranial magnetic stimulation (TMS) have rapidly emerged as important therapeutic tools in movement disorders as well as other neurological and psychiatric diseases, although the precise underlying physiological mechanisms need to be demonstrated. However, multiple studies have shown that both TMS and direct current stimulation can impact motor and cognitive functions in healthy subjects and patients with neurological or psychiatric disorders (Demirtas-Tatlidede, 2013). Stimulation-induced activity-dependent synaptic plasticity is likely a potential mechanism of action. A recent in vitro study found that direct current activation induced synaptic potentiation via NMDA receptor activation (Fritsch et al., 2010).

CONCLUSION

Given the dazzling array of various forms of LTP and LTD at synapses throughout the brain it is a virtual certainty that the brain expresses long-lasting, activity dependent synaptic modifications as important mechanisms by which experiences modify circuit function and that these plastic modifications underlie behavioural and cognitive experience. It is likely that the mechanisms that underlie the various forms of LTP and LTD subserve an enormous set of functions both during development and in adult learning and memory.

There is certainly still a future for studies that probe the detailed mechanisms of LTP built on the foundation of Bliss and Lømo's work.

REFERENCES

Andersen, P., Blackstad, T.W. and Lømo, T. (1966) Location and identification of excitatory synapses on hippocampal pyramidal cells. *Experimental Brain Research*, 1: 236–48.

Andersen, P., Bliss, T.V.P. and Skrede, K.K. (1971a) Unit analysis of hippocampal population spikes. *Experimental Brain Research*, 13: 208–21.

Andersen, P., Bliss, T.V.P. and Skrede, K.K. (1971b) Lamellar organization of hippocampal excitatory pathways. *Experimental Brain Research*, 13: 222–38.

Andersen, P., Holmqvist, B. and Voorhoeve, P.E. (1966) Entorhinal activation of dentate granule cells. *Acta Physiologica Scand*inavica, 66: 448–60.

Andersen, P. and Lømo, T. (1967) Control of hippocampal output by afferent volley frequency. *Progress in Brain Research*, 27: 400–12.

Bailey, C.H. and Kandel, E.R. (1993) Structural changes accompanying memory storage. *Annual Review of Physiology,* 55: 397–426.

Barrionuevo, G., Schottler, F. and Lynch, G. (1980) The effects of repetitive low frequency stimulation on control and "potentiated" synaptic responses in the hippocampus. *Life Sciences*, 27: 2385–91.

Bliss, T. and Lømo, T. (1973) Long-lasting potentiation of synaptic transmission in the dentate area of the anaesthetized rabbit following stimulation of the perforant path. *Journal of Physiology*, 232: 331–56.

Castillo, P.E., Chiu, C.Q. and Carroll, R.C. (2011) Long-term plasticity at inhibitory synapses. *Current Opinion in Neurobiology*, 21: 328–38.

Chen, W.R., Lee, S., Kato, K. et al. (1996) Long-term modifications of synaptic efficacy in the human inferior and middle temporal cortex. *Proceedings of the National Academy of Sciences*, 23: 8011–15.

Collingridge, G.L., Kehl, S.J. and McLennan, H. (1983) Excitatory amino acids in synaptic transmission in the Schaffer collateral-commissural pathway of the rat hippocampus. *Journal of Physiology,* 334: 33–46.

Demirtas-Tatlidede, A., Vahabzadeh-Hagh, A.M. and Pascual-Leone, A. (2013) Can noninvasive brain stimulation enhance cognition in neuropsychiatric disorders? *Neuropharmacology*, 64: 566–578.

Eccles, J.C. (1964) *The Physiology of Synapses*. Berlin: Springer.

Eichenbaum, H. (1996) Learning from LTP: a comment on recent attempts to identify cellular and molecular mechanisms of memory. *Learning & Memory*, 3: 61–73.

Feng, T.P. (1941) Studies on the neuromuscular junction: XXVI. The changes of the end-plate potential during and after prolonged stimulation. *Chinese Journal of Physiology*, 3: 341–72.

Fritsch, B., Reis, J., Martinowich, K. et al. (2010) Direct current stimulation promotes BDNF-dependent synaptic plasticity: potential implications for motor learning. *Neuron*, 66: 198–204.

Gloor, P., Vera, C.L. and Sperti, L. (1964) Electrophysiological studies of hippocampal neurons. III. Responses of hippocampal neurons to repetitive perforant path volleys. *Electroencephalography and Clinical Neurophysiology*, 17: 353–70.

Green, J.D. and Adey, W.R. (1956) Electrophysiological studies of hippocampal connections and excitability. *Electroencephalography and Clinical Neurophysiology*, 8: 245–62.

Hebb, D.O. (1949) *The Organization of Behavior.* New York: Wiley.

Isaac, J.T., Nicoll, R.A. and Malenka, R.C. (1995) Evidence for silent synapses: implications for the expression of LTP. *Neuron*, 15: 427–34.

Ito, M. (1989) Long term depression. *Annual Review of Neuroscience*, 12: 85–102.

Kandel, E.R. and Spencer, W.A. (1968) Cellular neurophysiological approaches in the study of learning. *Physiological Review*, 48: 65–134.

Kleim, J.A., Barbay, S. and Nudo, R.J. (1998) Functional reorganization of the rat motor cortex following motor skill learning. *Journal of Neurophysiology*, 80: 3321–5.

Lee, S.J., Escobedo-Lozoya, Y., Szatmari, E.M. and Yasuda, R. (2009) Activation of CaMKII in single dendritic spines during long-term potentiation. *Nature*, 458: 299–304.

Lømo, T. (1966) Frequency potentiation of excitatory synaptic activity in the dentate area of the hippocampal formation. *Acta Physiologica Scandinavica* 68 (Suppl. 277): 128.

Lømo, T. (1968) Potentiation of monosynaptic EPSPs in cortical cells by single and repetitive afferent volleys. *Journal of Physiology*, 194: 84.

Lømo, T. (1971a) Patterns of activation in a monosynaptic cortical pathway: the perforant path input to the dentate area of the hippocampal formation. *Experimental Brain Research*, 12: 18–45.

Lømo, T. (1971b) Potentiation of monosynaptic EPSPs in the perforant path–dentate granule cell synapse. *Experimental Brain Research*, 12: 46–63.

Lømo, T. (2003) The discovery of long-term potentiation. *Philosophical Transactions of the Royal Society of London*, 358: 617–20.

Malinow, R. and Tsien, R.W. (1990) Presynaptic enhancement shown by whole-cell recordings of long-term potentiation in hippocampal slices. *Nature*, 346: 177–80.

Matsuzaki, M., Honkura, N., Ellis-Davies, G.C. and Kasai, H. (2004) Structural basis of long-term potentiation in single dendritic spines. *Nature*, 429: 761–6.

Monfils, M.H. and Teskey, G.C. (2004) Skilled-learning-induced potentiation in rat sensorimotor cortex: a transient form of behavioral LTP. *Neuroscience*, 125: 329–36.

Monfils, M.H., VandenBerg, P.M., Kleim, J.A. and Teskey, G.C. (2004) Long-term potentiation induces expanded movement representations and dendritic hypertrophy in layer V of rat sensorimotor neocortex. *Cerebral Cortex*, 14: 586–93.

Racine, R.J., Chapman, C.A., Trepel, C. et al. (1995) Post-activation potentiation in the neocortex. IV. Multiple sessions required for induction of long-term potentiation in the chronic preparation. *Brain Research*, 702: 87–93.

Rioult-Pedotti, M.S., Friedman, D., Hess, G. and Donoghue, J.P. (1998) Strengthening of horizontal cortical connections following skill learning. *Nature Neuroscience*, 3: 230–4.

Schwartzkroin P.A. and Wester, K. (1975) Long-lasting facilitation of a synaptic potential following tetanization in the in vitro hippocampal slice. *Brain Research*, 89: 107–19.

Teyler, T.J., Hamm, J.P., Clapp, W.C. et al. (2005) Long-term potentiation of human visual evoked responses. *European Journal of Neuroscience*, 21: 2045–50.

Turrigiano, G.G. (1999) Homeostatic plasticity in neuronal networks: the more things change, the more they stay the same. *Trends in Neurosciences*, 22: 221–27.

Whishaw, I.Q., Pellis, S.M., Gorny, B.P. and Pellis, V.C. (1991) The impairments in reaching and the movements of compensation in rats with motor cortex lesions: an endpoint, videorecording, and movement notation analysis. *Behavioral Brain Research*, 42: 77–91.

18 Beyond Pons et al.: Massive cortical reorganization after sensory deafferentation in adult macaques

Theresa A. Jones

INTRODUCTION

The somatosensory cortex contains maps of skin surfaces, muscles and joints that can be detected by measuring the response of cortical neurons to sensory stimulation. For example, the region of cortex in which neurons are most responsive to touch of a particular skin surface is considered to be the somatosensory representation for that skin surface. The entire collection of representations makes a human resembling map, the homunculus (little person) in the somatosensory cortex.

In the late 1970s through to the 1980s there was major excitement surrounding findings that cortical maps of body surfaces changed in response to sensory manipulations. Old concepts about the adult brain being fixed and hard wired were still being overturned at the time. In this context, it was remarkable to realize that something as concrete as the somatosensory homunculus in the adult brain continued to change in response to new sensory experiences. This was suggested by the findings that somatosensory representations shifted their boundaries in response to manipulations such as behavioral training, peripheral nerve stimulation, nerve transections and amputations.

Merzenich and colleagues characterized how digit (finger) representations in the primary somatosensory cortex (S1) of monkeys reorganized in response to such manipulations (Jenkins et al., 1990) (see Figure 18.1A). They found that if sensory inputs from one or two digits were removed, the cortical representations of remaining digits expanded. If two digits were attached to one another, such that the monkeys used them together like a single digit, the normal sharp border between the representations of the two digits disappeared, leaving a fused cortical representation. If monkeys were trained to maintain contact on a rotating disk with the tips of one or two digits in order to receive a food reward, the representations of the trained digits enlarged.

What were the mechanisms of the map changes? There was growing evidence around this time that new neural connections could be added to the adult brain,

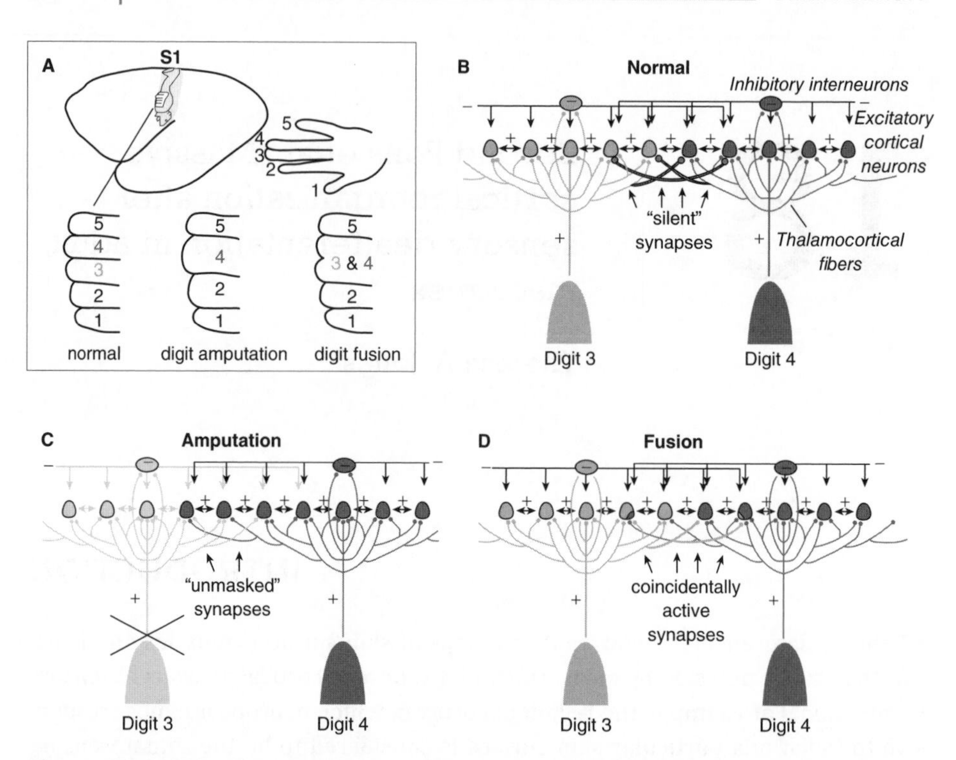

Figure 18.1 (A) Examples of how representations of the digits (fingers) in the primary somatosensory cortex of monkeys rapidly change in response to peripheral sensory manipulations (adapted from Jenkins et al., 1990). Digit 3 amputation results in a spread of surrounding digit representations into the digit 3 region. If two digits are sewn together, there is a partial merging (fusion) of their cortical representations. **(B–D)** Putative mechanisms of rapid cortical reorganization. **(B)** The terminations of axonal projections from individual thalamic neurons carrying sensory information to the cortex partially overlap in the cortex. These fibers excite both excitatory and inhibitory cortical neurons, which in turn excite and inhibit surrounding cortical neurons. Inhibition can overcome excitation near the borders of a representation, resulting in functionally silent thalamocortical synapses. **(C)** The expansion of surrounding digit representations after amputation could be explained by a release of the inhibition that normally occurs due to stimulation of the missing digit. **(D)** The fusion of digit representations could be explained by making previously silent synapses more efficacious as a result of the coincident activation of the overlapping thalamocortical projections of adjacent representations.

for example as a result of behavioral training or enriched environment housing. However, the somatosensory representations started to change very rapidly in response to sensory manipulations, which made it seem unlikely that their reorganization involved new synapse formation. In addition, the changes were always rather small. In Merzenich's studies above, the borders of representations typically shifted less than 1 mm. The maximum shift in representations was within a range corresponding to the axonal projection territory of individual thalamic neurons that carry sensory information to the cortex. These axonal projections of thalamic neurons overlap in the cortex (Figure 18.1B). The fact that plasticity of sensory maps in adults seemed to be constrained to the size corresponding to individual thalamocortical projections implied that the map changes could be due to shifts in the activity of existing excitatory and inhibitory synapses in the cortex.

The expansion of a representation could be accomplished by a release of inhibition near the outer domain of the thalamocortical projections to that representation (see Figure 18.1C). This would allow pre-existing thalamocortical synapses that were previously ineffective in activating cortical neurons to become effective. This general idea came to be referred to as the "unmasking of silent synapses" (Buonomano and Merzenich, 1998). Silent synapses could also be unmasked via Hebbian synaptic plasticity mechanisms: the coincident activation of the projections of adjacent skin surfaces could strengthen previously silent synapses, resulting in the fusion of adjacent representations (Figure 18.1D)

Rasmusson, Turnbull and Leech (1985) were the first to uncover evidence for cortical reorganization that was too large in extent to be explained by the above mechanisms. They found a raccoon in the wild that was missing a forepaw, presumably due to a prior accident. The completely healed state of the wound suggested that it probably occurred long ago. Upon mapping the forepaw region of the racoon's somatosensory cortex, they discovered many neurons that were activated by tactile stimulation of the stump of the forearm as well as by stimulation of the *hindpaw*. The differences relative to normal raccoon maps far exceeded the projection territory of individual thalamocortical neurons. In retrospect, these findings were predictive of those to come by Pons et al., as reviewed below. At the time, however, it was hard to know what to make of findings from a single wild animal with an uncertain life history, and assumptions about the limits of cortical reorganization continued.

PONS ET AL.'S FINDINGS

Pons et al.'s (1991) findings overturned the conception that cortical map plasticity was limited to very small shifts in the borders of representations. They examined a region of S1 that contained representations of the upper limb in four macaque monkeys that had, 12 years earlier, lost all sensation of their upper limb due to peripheral nerve deafferentation. Normally the upper limb region of macaques extends 10–14 mm in SI and is bordered by representations of the face and trunk. The face representations of the chin and lower jaw most directly border the upper limb area (see Figure 18.2A). In monkeys with upper limb deafferentation, Pons et al. found that every single one of the neurons in the upper limb region was responsive to tactile stimulation of the face. The new face representation consisted entirely of chin and lower jaw portions of the face. It appeared as if the directly adjacent portion of the face representation was stretched across the denervated upper limb region of cortex.

How could this have happened? Pons et al. considered that such large-scale reorganization could be accomplished by the sprouting of new thalamocortical projections, or it could be due to reorganization at the level of the thalamus or brainstem. Subsequent studies supported the latter possibility. Jones and Pons (1998) found in monkeys with the same long-term limb deafferentation that thalamic neurons that are normally responsive to upper limb stimulation were mostly responsive to stimulation of the lower face. This suggested that the thalamic neurons that normally carry upper limb information into the cortex began to instead carry face information (see Figure 18.2B). Reorganization at the level of the brainstem may contribute to the thalamic reorganization. The brainstem nuclei that receive face and upper limb sensory input (trigeminal and cuneate nuclei, respectively) are near one another. After spinal cord injury, face afferents to the trigeminal nucleus sprout new axonal projections into the cuneate nucleus (Jain et al., 2000) and it becomes responsive to face stimulation (Kambi et al., 2014).

Pons et al. pointed out that their findings raised many new questions about the mechanisms and function of the cortical reorganization. They predicted that answering these questions could lead to the ability to harness cortical plasticity mechanisms for therapeutic applications. Although the work to do so continues, the authors' prediction about potential therapeutic applications of cortical reorganization appears to be in the process of being fully realized. The remainder of this chapter focuses on some of the clinically-relevant research directions that have emerged since their landmark discovery.

PHANTOM LIMB PHENOMENA

Most people who have had a limb amputated experience phantom sensations of their missing body part. Phantom sensations are sometimes elicited by touching the amputation stump. The individual feels both the touch to the stump and the phantom sensation at the same time. Subsequent to Pons et al.'s study, Ramachandran (1993) found that phantom sensations in some patients with amputations also could be elicited by touching the face. Like touching the stump, the phantom sensations and the sensation that the face is being touched are perceived together.

Several other studies found evidence for major reorganization of somatosensory cortex after upper limb amputation in humans (e.g. Flor et al., 1995). Resembling the findings by Pons et al., face representations shift into former upper limb territory of the somatosensory cortex in humans with limb amputation. There is also a parallel reorganization of the motor cortex and other cortical areas. Together with Ramachandran et al.'s findings, this suggests that the cortical reorganization is reflected in changes in sensory perceptions, though not normal ones. The simultaneous perception of touch to the missing limb and to the face suggests that the original functions of the brain circuits responsible for upper limb perception do not simply disappear in parallel with the shifts in cortical representations.

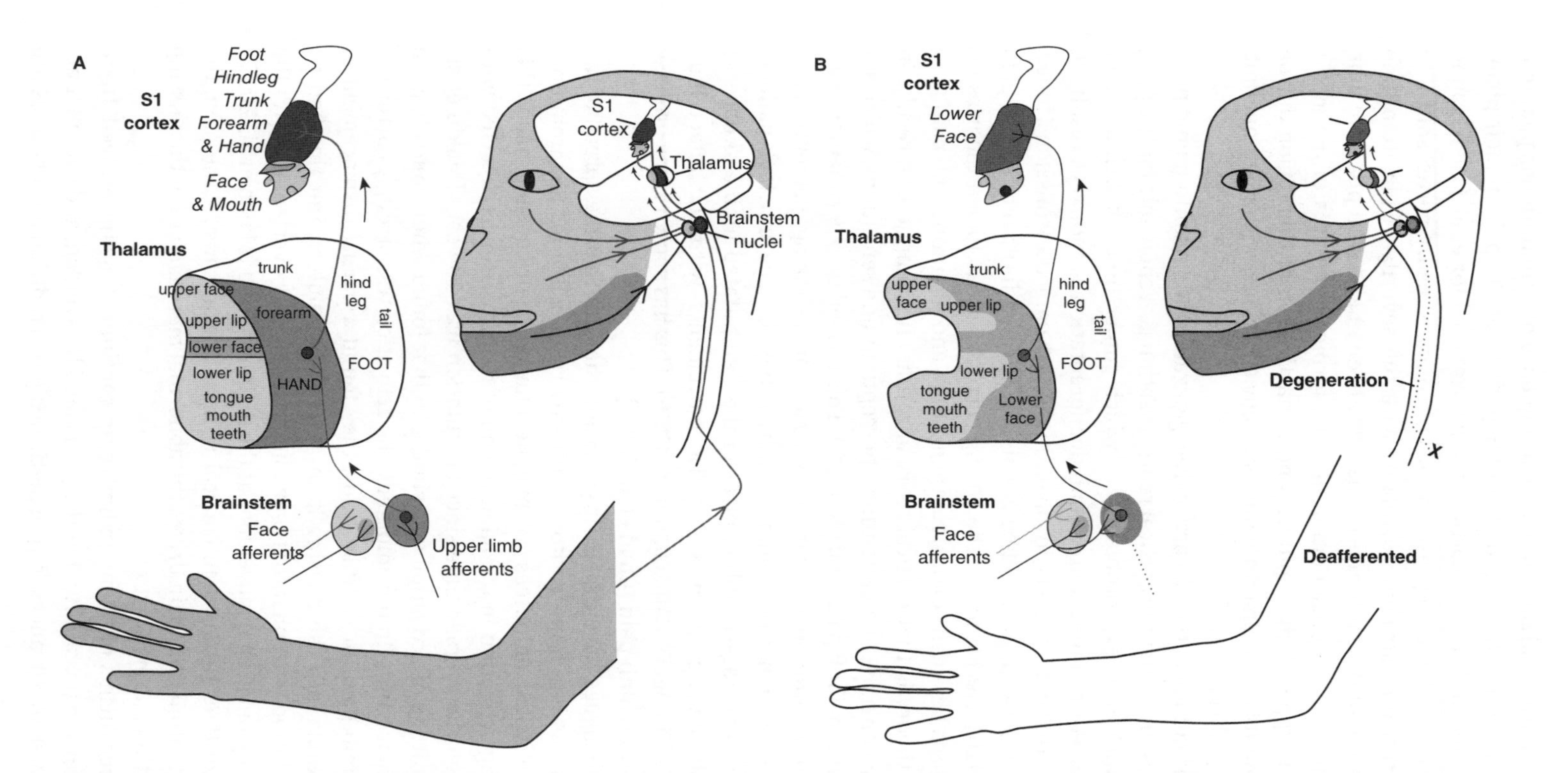

Figure 18.2 Putative mechanisms of massive cortical reorganization after sensory deafferentation of the upper limb as shown in simplified somatosensory pathways of normal monkeys **(A)** and after upper limb deafferentation **(B)**. Pons et al. found that all of the normal upper limb territory of primary somatosensory cortex (S1) was responsive to sensory stimulation of the lower face in monkeys 12 years after limb deafferentation. Other monkeys with the same manipulation were later found to have a reorganized thalamus, with neurons in the upper limb territory of the thalamus being mostly responsive to lower face stimulation. Sprouting of face afferents into the brain stem nucleus of the upper limb can contribute to thalamic and cortical reorganization. The illustration of the thalamus is adapted from Jones and Pons (1998).

Unfortunately, another all too common consequence of amputation is phantom limb pain. Many amputees develop debilitating perceptions of pain in their phantom limb. This is also linked with cortical reorganization. Flor et al. (1995) found that the degree to which face representations of somatosensory cortex shift into the upper limb territory strongly correlates with the severity of phantom limb pain. Reorganization of the motor cortex has also been linked with phantom limb pain. A spread of mouth representation into the former hand area of the motor cortex is found in upper limb amputees, and the activation of this region during imagined movements of the phantom limb strongly correlates with phantom limb pain (Lotze et al., 2001).

Given the correlations between cortical reorganization and phantom limb pain, it is tempting to assume that the cortical reorganization is responsible for the pain, but there is a chicken or egg problem (i.e., which comes first) as well as other potential culprits. For example, upper limb deafferentation in monkeys results in abnormalities in thalamic regions involved in central pain transmission (Jones, 2000). Peripheral nerve abnormalities are also implicated in phantom limb pain (Weeks et al., 2010). Nevertheless, the strong correlation has fueled theories on causal relationships between cortical reorganization and phantom limb pain.

A prominent theory is that cortical reorganization after amputation results in mismatches between movement intentions, perceptions and sensations, which are perceived as phantom limb pain (Harris, 1999). This has led to experimental treatments for phantom limb pain that are focused on relinking perceptions of movement with sensation. One example is mirror box therapy, in which patients view an image of their intact limb in place of their amputated limb and observe this virtual limb making mirror-symmetric movements with the intact limb (Ramachandran and Rogers-Ramachandran, 1996). This has shown some promise in reducing phantom limb pain (Chan et al., 2007).

Makin and colleagues have called into question the idea that the spread of the face representations into limb territory is directly responsible for phantom limb pain. When they asked amputees to imagine moving their missing hand, this resulted in strong activation in the hand region of motor cortex as detected with fMRI, even though the average time since amputation was 18 years (Makin et al., 2013). The magnitude of activation was similar to that found when controls were asked to move their hands. In the amputees, the activation was strongly correlated with phantom limb pain. In another study, they found a partial displacement of face representations into the missing limb region in amputees, but this had no significant correlation with phantom pain and it was mostly overlapping with the region activated by imagining movement of the phantom limb (Makin et al., 2015). They hypothesize that, rather than cortical reorganization causing the pain, the experience of phantom limb pain drives the abnormal *maintenance* of the missing limb representations in the cortex.

While the mechanisms of phantom limb pain continue to be unresolved, there is excitement over the possibility that the pain could be avoided, and overall functionality optimized, by advances in prosthetic designs. Traditional prostheses are moved by remaining body parts, via cables, switches or buttons. Myolectric prostheses are controlled by electromyography (EMG) sensors connected to the patient's own muscles, allowing them to be moved by movement intentions. The ability to control a prosthesis with one's own movement intentions increases the feeling of "prosthetic embodiment", the sense that the prosthesis is part of one's own body. Greater use of myolectric prostheses by patients is linked with reduced phantom limb pain as well as reduced spread of the lip representation into the arm region of the motor cortex (Lotze et al., 1999). This suggests that phantom limb pain and cortical reorganization might be countered by prosthetic embodiment.

Ongoing directions in prosthetic designs are moving ever closer to mimicking the original limb. Targeted reinnervation, the rerouting of the motor and sensory nerves from an amputated limb to other muscles and skin surfaces, provides an interface between prosthesis and the neural pathways of the original limb. Connecting biosensors to the muscles and biofeedback devices to the skin surfaces of the re-routed nerves enables prostheses that are both moved and felt by the user. These developments are very promising for promoting prosthetic embodiment. They will undoubtedly lead to new questions about how cortical organization is influenced by the replacement, rather than the loss, of a limb.

FOCAL HAND DYSTONIAS

Writer's cramp, musician's cramp and golfer's cramp ("the yips") are examples of focal hand dystonias. Focal hand dystonias can develop with highly repetitive performance of skilled tasks. They typically manifest as a subtle loss of dexterity, progressing to uncontrollable and excessive muscle contractions, abnormal postures and tremor during performance of the skilled tasks. Among professional musicians, the very practice leading to full mastery of their musical instrument can lead to focal hand dystonia, often robbing them of their ability to perform at the peak of their profession.

Pons et al.'s finding raised the possibility that other types of peripheral sensory changes, such as dystonia, could involve large-scale cortical reorganization. Byl, Merzenich and Jenkins (1996) were the first to find that focal hand dystonia involved major cortical reorganization. They trained owl monkeys on a hand grip task, giving them a banana-flavored food pellet as a reward for every perfect hand-hold. Most monkeys will work hard for banana pellets, and these did, performing 300 trials per day. Over days of training, the monkeys reached a peak of performance (>240 pellets/day!), but after more days of training, that performance began to decline. The monkeys had developed symptoms of focal hand dystonia. Upon mapping the hand area of the primary somatosensory cortex, Byl et al. found a massive disorganization of its normal topography (see Figure 18.3). Receptive fields (the area of skin where touch activates the recorded neuron) were 10–20 times larger than normal, sometimes extending across the entire surface of a digit or across multiple digits. The normally sharp segregation of representations of the ventral (palm side) and dorsal (hairy side) surfaces of the hand was also lost. A subsequent study by the same group found that, with continued practice of the task, the disorganization of the somatosensory cortex was progressive over time (Blake et al., 2002). A later study using rats to model of focal hand dystonia found that the primary motor cortex had huge increases in regions in which more than one movement could be elicited by a discrete stimulation (Coq et al., 2009). The cortical changes described in these studies are reminiscent of the finger fusion effects described above (see Figure 18.2), but they are on a much more massive scale.

Elbert et al. (1998) found that focal hand dystonia in humans is linked with altered somatosensory cortical organization. In musicians with focal hand dystonia, there was reduced distance between finger representations, suggesting a fusion or overlap of their representations. Byrnes et al. (1998) found abnormalities

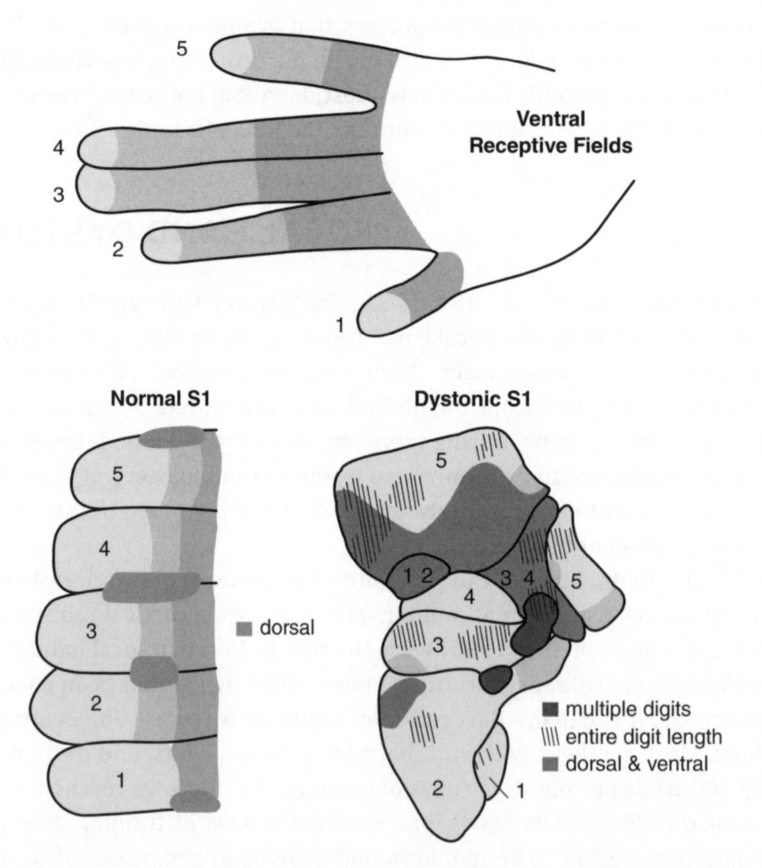

Figure 18.3 Hand dystonia is associated with derangement of somatosensory representations. The dystonic S1 is a simplified adaptation from Byl et al. (1996). They found that training on a task with highly repetitive hand movements disrupted the orderly topographic maps of the digits in S1. Regions responsive to stimulation anywhere on the digit, to either side of the digit and to multiple digits appeared.

in the primary motor cortex associated with hand dystonia. This study used trancranial magnetic stimulation (TMS) to map the motor cortex, in which pulses of TMS are used to excite different areas of the motor cortex as the motor-evoked potentials (MEP) in muscles are measured using EMG recordings. This revealed that the representations of index finger and thumb movements were displaced in participants with writer's dystonia relative to the representations of the unaffected hand in the contralateral cortex.

Findings from other TMS studies suggest that focal hand dystonia is associated with a loss of surround inhibition between the fingers. The property of surround inhibition in the motor system is detected in the tendency for the muscle activity related to a discrete movement to be coupled with inhibited muscle activity for other movements. For example, when vibration is applied to a single finger in healthy participants, this increases TMS-evoked activity in the muscles used to move that finger but decreases it in muscles that move other fingers of the same hand (Rosenkranz et al., 2005). In contrast, in musicians with focal hand dystonia, vibrating one finger increases, rather than decreases, MEP amplitudes of the muscles of other fingers. A similar effect is found when participants are asked to imagine moving one finger. In healthy controls, this increases MEP amplitudes only in those muscles involved in the imagined finger movement. In participants with writer's dystonia, this increases the MEP amplitudes of multiple fingers (Quartarone et al., 2005).

These and related findings have led to the idea that excessive repetition of synchronous movements, such as moving two fingers at the same time, can drive the cortex to merge sensory and motor representations, losing some capacity for separate sensorimotor processing of each. This impairs the ability to separately move the fingers and continued practice further disassembles cortical organization, exacerbating dystonia symptoms in a feedback loop.

It follows from the feedback loop idea that interrupting the cycle with sufficient asynchronous experience might alleviate dystonia and normalize cortical organization. Schabrun et al. (2009) tested this by providing asynchronous electrical stimulation to the muscles involved in moving the thumb and index finger for one hour in participants with writer's or musician's dystonia. Before the treatment, the motor representations of the thumb and index finger were larger and closer to one another than normal. After the treatment, the thumb and index finger representations were smaller and had moved further apart and some dystonia symptoms were also improved. Similar effects were found using a task requiring participants to pay attention to a change in vibration frequencies applied to individual fingers in random order (Rosenkranz et al., 2009).

Emerging from this line of research are new treatments for focal hand dystonia that are based on restoring segregated sensorimotor processing of the affected hand. A prominent example is sensorimotor retuning therapy, in which the healthy fingers are splinted while the affected fingers engage in intense practice of sequential (never simultaneous) movements. This therapy has been found to both improve symptoms of focal hand dystonia and normalize cortical organization (Candia et al., 2002).

STROKE

Stroke is an interruption of the blood supply to the brain resulting in tissue damage. Hemiparesis (impaired movement in one body side) is a common consequence of stroke due to the tendency for strokes to impact the vascular supply to sensory and motor systems of the brain. Upper limb impairments are particularly common after stroke. The typical response to these impairments is to compensate by relying more on the unaffected hand and arm. This results in disuse of the affected upper limb, which exacerbates its functional loss.

It so happens that the monkeys that were studied by Pons et al. received the sensory deafferentation to begin with in order to better understand this phenomenon of disuse (Taub, 1976). The monkeys lost sensation in the upper limb as a result of the peripheral deafferentation procedure, but maintained the ability to move this limb. Nevertheless, they stopped using the limb altogether. The monkeys could be trained to use the limb, and they would use it if the other limb was constrained, but on their own they would not. Taub and colleagues hypothesized that the disuse developed in response to the animals' repeated experience with the ineptitude in the limb (reviewed in Taub and Uswatte, 2006). That is, they had *learned* to not use the limb, a phenomenon referred to as "learned nonuse." However, a prolonged period of immobilization of the normal upper limb, in order to force practice in using the deafferented limb, can overcome this learned nonuse.

These observations of learned nonuse led to the development of a new treatment approach for stroke: constraint-induced movement therapy (CIMT) (Taub et al., 1993, 2014). During CIMT, over several days and for most waking hours, a large mitt is worn on the unaffected hand to render it useless. During the same time period, intense rehabilitative training of the paretic limb is provided, to return more functional use to this limb. Large clinical trials established that CIMT is effective in improving upper limb function after stroke (Wolf et al., 2006, 2008).

Cortical reorganization has been linked with the functional improvements resulting from rehabilitative training, including CIMT. An example of rehabilitative training changing the brain map was described by Nudo and colleagues (1996a, b) in squirrel monkeys that had had small strokes in the hand area of the motor cortex. In the absence of training, the remaining hand representations surrounding the stroke disappeared, replaced either by neighboring representations or by nonresponsive zones (regions in which no movements are evoked by stimulation). Training the monkeys daily to use the paretic hand to reach and grasp food pellets with their digits prevented the further loss of movement representations and promoted their expansion into neighboring representations of the primary motor cortex. Similar effects have been found in rodent stroke models (see Figure 18.4), as well as in clinical stroke populations. Liepert et al. (1998, 2000) found that when stroke patients were treated with CIMT, the hand area of motor cortex is enlarged, as detected with TMS.

Rehabilitative training is usually insufficient to restore normal function in the paretic upper limb, and the modest improvements that are achieved can require a

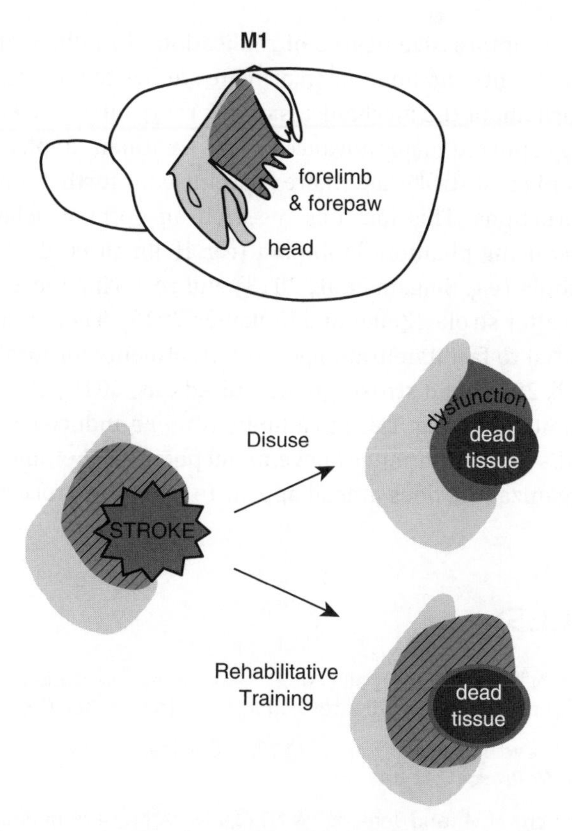

Figure 18.4 Effects of stroke in the upper limb region of primary motor cortex (M1). In the absence of interventions, sufficient stroke damage results in disuse of the affected upper limb and a loss of its remaining representations in the cortex surrounding the damage. Rehabilitative training focused on the impaired limb promotes the maintenance and reorganization of forelimb representations. The illustration is of a rodent motor cortex.

great deal of effortful training. This has led to efforts to facilitate the effects of rehabilitative training. Studies in animal stroke models have found that delivering electrical stimulation over the motor cortex during rehabilitative training results in faster improvements and greater enlargement of remaining representations of the paretic limb in the motor cortex (reviewed in Jones and Adkins, 2015). There is now a major ongoing effort to determine whether non-invasive stimulation, using TMS or transcranial direct current stimulation (TDCS), could be similarly effective in clinical stroke populations.

CONCLUSION

The years since Pons et al.'s landmark discovery have seen an explosion of research on the role of cortical reorganization in clinical health problems.

There has been an impressive degree of replication of findings on cortical reorganization from rodents and nonhuman primate studies to humans. While there is still much to learn about the mechanisms of this reorganization, Pons et al.'s findings have already spurred major advances in the treatment of phantom limb pain, focal hand dystonia and stroke, and there is promise for further advances in ongoing research directions. This includes research on optimal behavioral training strategies for reducing phantom limb pain (e.g. Hellman et al., 2015), reversing focal hand dystonia (e.g. Singam et al., 2013) and restoring more effective use of the paretic limb after stroke (Zeiler and Krakauer, 2013). There is also much interest in the potential to facilitate training-based treatments for focal hand dystonia (Kimberley et al., 2015) and stroke (Jones and Adkins, 2015) by combining these with cortical stimulation (e.g. TMS) to amplify training-induced cortical reorganization. Pons et al.'s prediction, noted above, about potential therapeutic applications of cortical reorganization, does indeed appear to be in the process of becoming a reality.

REFERENCES

Blake, D.T., Byl, N.N., Cheung, S. et al. (2002) Sensory representation abnormalities that parallel focal hand dystonia in a primate model. *Somatosens Mot Res,* 19 (4): 347–57.

Buonomano, D.V. and Merzenich, M.M. (1998) Cortical plasticity: from synapses to maps. *Annu Rev Neurosci.* 21: 149–86.

Byl, N.N., Merzenich, M.M. and Jenkins, W.M. (1996) A primate genesis model of focal dystonia and repetitive strain injury: I. Learning-induced dedifferentiation of the representation of the hand in the primary somatosensory cortex in adult monkeys. *Neurology*, 47 (2): 508–20.

Byrnes, M.L., Thickbroom, G.W., Wilson, S.A. et al. (1998) The corticomotor representation of upper limb muscles in writer's cramp and changes following botulinum toxin injection. *Brain*, 121 (Pt 5): 977–88.

Candia, V., Schafer, T., Taub, E. et al. (2002) Sensory motor retuning: a behavioral treatment for focal hand dystonia of pianists and guitarists. *Arch Phys Med Rehabil.* 83 (10): 1342–8.

Chan, B.L., Witt, R., Charrow, A.P. et al. (2007) Mirror therapy for phantom limb pain. *N Engl J Med.* 357 (21): 2206–7.

Coq, J.O., Barr, A.E., Strata, F. et al. (2009) Peripheral and central changes combine to induce motor behavioral deficits in a moderate repetition task. *Exp Neurol.* 220 (2): 234–45.

Elbert, T., Candia, V., Altenmuller, E. et al. (1998) Alteration of digital representations in somatosensory cortex in focal hand dystonia. *NeuroReport,* 9 (16): 3571–5.

Flor, H., Elbert, T., Knecht, S. et al. (1995) Phantom-limb pain as a perceptual correlate of cortical reorganization following arm amputation. *Nature,* 375 (6531): 482–4.

Harris, A.J. (1999) Cortical origin of pathological pain. *Lancet,* 354 (9188): 1464–6.

Hellman, R.B., Chang, E., Tanner, J. et al. (2015) A robot hand testbed designed for enhancing embodiment and functional neurorehabilitation of body schema in subjects with upper limb impairment or loss. *Front. Hum. Neurosci.,* 9: 26.

Jain, N., Florence, S.L., Qi, H.X. and Kaas, J.H. (2000) Growth of new brainstem connections in adult monkeys with massive sensory loss. *Proc Natl Acad Sci USA,* 97 (10): 5546–50.

Jenkins, W.M., Merzenich, M.M. and Recanzone, G. (1990) Neocortical representational dynamics in adult primates: implications for neuropsychology. *Neuropsychologia,* 28 (6): 573–84.

Jones, E.G. (2000) Cortical and subcortical contributions to activity-dependent plasticity in primate somatosensory cortex. *Annu Rev Neurosci.* 23: 1–37.

Jones, E.G. and Pons, T.P. (1998) Thalamic and brainstem contributions to large-scale plasticity of primate somatosensory cortex. *Science,* 282 (5391): 1121–5.

Jones, T.A. and Adkins, D.L. (2015) Motor system reorganization after stroke: stimulating and training towards perfection. *Physiology,* 30 (5): 358–70.

Kambi, N., Halder, P., Rajan, R. et al. (2014) Large-scale reorganization of the somatosensory cortex following spinal cord injuries is due to brainstem plasticity. *Nat Commun.* 5: 3602.

Kimberley, T.J., Schmidt, R.L., Chen, M. et al. (2015) Mixed effectiveness of rTMS and retraining in the treatment of focal hand dystonia. *Front. in Hum. Neurosci.,* 9: 385.

Liepert, J., Bauder, H., Wolfgang, H.R. et al. (2000) Treatment-induced cortical reorganization after stroke in humans. *Stroke,* 31 (6): 1210–16.

Liepert, J., Miltner, W.H., Bauder, H. et al. (1998) Motor cortex plasticity during constraint-induced movement therapy in stroke patients. *Neurosci Lett.* 250 (1): 5–8.

Lotze, M., Flor, H., Grodd, W. et al. (2001) Phantom movements and pain: an fMRI study in upper limb amputees. *Brain,* 124 (Pt 11): 2268–77.

Lotze, M., Grodd, W., Birbaumer, N. et al. (1999) Does use of a myoelectric prosthesis prevent cortical reorganization and phantom limb pain? *Nat Neurosci.* 2 (6): 501–2.

Makin, T.R., Scholz, J., Filippini, N. et al. (2013) Phantom pain is associated with preserved structure and function in the former hand area. *Nat Commun,* 4: 1570.

Makin, T.R., Scholz, J., Henderson Slater, D. et al. (2015) Reassessing cortical reorganization in the primary sensorimotor cortex following arm amputation. *Brain,* 138 (Pt 8): 2140–6.

Nudo, R.J. and Milliken, G.W. (1996a) Reorganization of movement representations in primary motor cortex following focal ischemic infarcts in adult squirrel monkeys. *J Neurophysiol.* 75 (5): 2144–9.

Nudo, R.J., Wise, B.M., SiFuentes, F. and Milliken, G.W. (1996b) Neural substrates for the effects of rehabilitative training on motor recovery after ischemic infarct. *Science,* 272 (5269): 1791–4.

Pons, T.P., Garraghty, P.E., Ommaya, A.K. et al. (1991) Massive cortical reorganization after sensory deafferentation in adult macaques. *Science,* 252 (5014): 1857–60.

Quartarone, A., Rizzo, V., Bagnato, S. et al. (2005) Homeostatic-like plasticity of the primary motor hand area is impaired in focal hand dystonia. *Brain,* 128 (Pt 8): 1943–50.

Ramachandran, V.S. (1993) Behavioral and magnetoencephalographic correlates of plasticity in the adult human brain. *Proc Natl Acad Sci USA,* 90 (22): 10413–20.

Ramachandran, V.S. and Rogers-Ramachandran, D. (1996) Synaesthesia in phantom limbs induced with mirrors. *Proc Biol Sci.* 263 (1369): 377–86.

Rasmusson, D.D., Turnbull, B.G. and Leech, C.K. (1985) Unexpected reorganization of somatosensory cortex in a raccoon with extensive forelimb loss. *Neurosci Lett.* 55 (2): 167–72.

Rosenkranz, K., Butler, K., Williamon, A. and Rothwell, J.C. (2009) Regaining motor control in musician's dystonia by restoring sensorimotor organization. *J Neurosci.* 29 (46): 14627–36.

Rosenkranz, K., Williamon, A., Butler, K. et al. (2005) Pathophysiological differences between musician's dystonia and writer's cramp. *Brain,* 128 (Pt 4): 918–31.

Schabrun, S.M., Stinear, C.M., Byblow, W.D. and Ridding, M.C. (2009) Normalizing motor cortex representations in focal hand dystonia. *Cereb Cortex,* 19 (9): 1968–77.

Singam, N.V., Dwivedi, A. and Espay, A.J. (2013) Writing orthotic device for the management of writer's cramp, *Front. Neurol.,* 4: 2.

Taub, E. (1976) Movement in nonhuman primates deprived of somatosensory feedback. *Exerc Sport Sci Rev.* 4: 335–74.

Taub, E., Miller, N.E., Novack, T.A. et al. (1993) Technique to improve chronic motor deficit after stroke. *Arch Phys Med Rehabil.* 74 (4): 347–54.

Taub, E. and Uswatte, G. (2006) Constraint-Induced Movement therapy: answers and questions after two decades of research. *NeuroRehabilitation,* 21 (2): 93–5.

Taub, E., Uswatte, G. and Mark, V.W. (2014) The functional significance of cortical reorganization and the parallel development of CI therapy. *Front Hum Neurosci.* 8: 396.

Weeks, S.R., Anderson-Barnes, V.C. and Tsao, J.W. (2010) Phantom limb pain: theories and therapies. *Neurologist,* 16 (5): 277–86.

Wolf, S.L., Winstein, C.J., Miller, J.P. et al. (2006) Effect of constraint-induced movement therapy on upper extremity function 3 to 9 months after stroke: the EXCITE randomized clinical trial. *J Am Med Assoc,* 296: 2095-104.

Wolf, S.L., Winstein, C.J., Miller, J.P. et al. (2008) Retention of upper limb function in stroke survivors who have received constraint-induced movement therapy: the EXCIOTE randomised trail. *Lancet Neurol,* 7: 33-40.

Zeiler, S.R. and Krakauer, J.W. (2013) The interaction between training and plasticity in the poststroke brain. *Curr. Opin. Neurol.,* 26 (6): 609–16.

19 Revisiting Roland: How does the human brain produce complex motor behaviours? Insights from functional neuroimaging

Jenni M. Karl and Jody C. Culham

Roland and colleagues (1980) were among the first to image the functional activity of the human brain during the performance of a motor task. They injected a radioactive tracer into the internal carotid arteries of awake participants and then used a specialized camera array placed on the head to measure the breakdown of the tracer inside the brain while participants performed different motor tasks. The technique was based on Roy and Sherrington's (1890) hypothesis that brain areas engaged in a behavioural task would metabolize glucose and oxygen more quickly and this would drive increased blood flow to these areas. Thus, breakdown of the radioactive tracer could be used to measure blood flow to an area and, by proxy, the extent of neuronal and glial activity in that area. Using this particular positron emission tomography (PET) approach, Roland and colleagues would discover that when all aspects of motor performance are considered, such as motor planning, learning and execution, almost every cortical area can contribute to motor performance.

In their seminal experiment, Roland and colleagues (1980) calculated the change in blood flow, or brain activity, as healthy participants performed three different motor tasks compared to when they rested quietly. Surprisingly, each motor task was associated with a different pattern of brain activity! When participants imagined a complex sequence of finger movements without actually executing them, activity increased only in the supplementary motor area (SMA; see Figure 19.1A); when they actually executed the finger sequence, activity increased in the SMA but also in primary motor cortex (M1; see Figure 19.1B); and when they performed a simpler motor task, repeated flexions of a single finger, activity increased in M1 but *not* SMA (Figure 19.1C). Thus, Roland and his colleagues concluded that M1 can generate simple movements with a single body part, but additional input from SMA is required to produce complex motor sequences involving multiple body parts. Specifically, SMA appears to play a specialized role in constructing movement sequences, which are then passed on to M1 during execution.

Roland immediately recognized the potential of neuroimaging approaches like PET (reviewed in Posner and Raichle, 1997; Roland, 1993) and would continue to apply these in his studies of sensorimotor control. Subsequent work would reveal increasingly impressive features of the cortical motor system and largely concentrate on two important questions: 'How is the cortical motor system organized?' and 'How do we learn motor skills?' Findings from Roland's work in the 1980s would foreshadow the outcomes of experiments conducted more than 30 years later using advanced neuroimaging techniques that would not be developed until the twenty-first century.

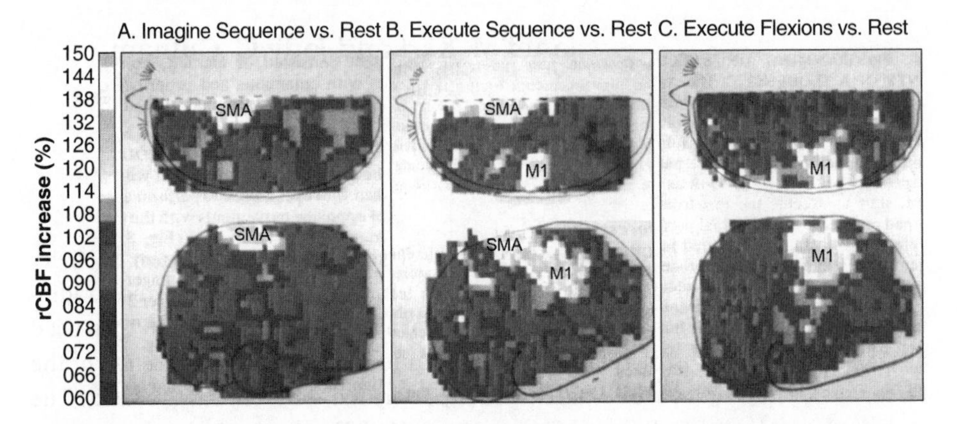

Figure 19.1 Increases in regional cerebral blood flow (rCBF) compared to rest during the perfomance of three different motor tasks. rCBF increased **(A)** in only the SMA during the imagination of a complex finger tapping sequence, **(B)** in both the SMA and M1 during the execution of a complex finger tapping sequence, **(C)** in only M1 during repeated flexions of a single digit. The top row shows a view from above and the bottom row shows a lateral view of the left hemisphere (inferior temporal cortex was not imaged). Adapted from Roland et al., 1980.

WHAT HAS PET REVEALED ABOUT HOW THE CORTICAL MOTOR SYSTEM IS ORGANIZED?

Roland and others used PET to identify a large network of cortical areas involved in producing more naturalistic behaviours like reaching or pointing to a point in space (Kawashima et al., 1994, 1995), manipulating objects with the fingers and hand (Roland et al., 1989; Seitz et al., 1991), and grasping objects with tools such as tongs (Inoue et al., 2001). They found that traditional motor areas, like M1 and the premotor cortex, were active during all three tasks while other areas were only active during certain tasks. For example, the primary and secondary somatosensory cortices (S1 and S2) were much more active during grasping, whereas the ipsilateral intraparietal sulcus (IPS) appeared to be uniquely active during tool use. Thus, they concluded that some cortical areas contributed to a range of different movements, whereas a smaller subset of areas specialized in generating specific movements with a particular body part such as the arm, hand or a tool.

An important aspect of naturalistic behaviours is the ability to coordinate the actions of multiple body parts. Consider the outfielder attempting to catch a fly ball. To succeed she must coordinate the movements of her eyes, arms, hands and legs, so as to position herself just beneath the falling ball. To explore how the cortical motor system might coordinate the actions of multiple body parts Roland and his students asked participants to perform isolated wrist movements, isolated ankle movements, or simultaneous movements of the wrist and the ankle at the same time while undergoing PET (Ehrsson et al., 2000). Surprisingly, neural activity in the premotor and parietal cortices was similar regardless of whether movements were made with only the ankle, only the wrist, or both at the same time (see Figure 19.2A). These findings suggested that certain cortical areas may encode a particular action, irrespective of the body part used to perform that action. This overlap in neural control could facilitate the coordination of movements across multiple bodily effectors.

Another important aspect of naturalistic behaviours is motor resonance – the ability to use different body parts to accomplish the same goal. Again, consider the baseball player who wishes to steal a base. To do this, the runner must run and reach out and touch the next base before the baseman can tag him with the baseball. Importantly, the runner can reach out and touch the base with either the arm or the leg. Grafton and colleagues (1992) used PET to explore how the cortical motor system might produce the same action with a number of different body parts. They asked PET participants to track a visually moving object using the index finger, tongue, toes or eyes and found that a common network of SMA, premotor and parietal areas was activated regardless of which body part was doing the tracking. Only M1 was differentially activated depending on body part: toe tracking activated superior M1 whereas finger tracking activated inferior M1. These findings were among the first to suggest that overlapping body part representations in the premotor and parietal cortices may allow us to either coordinate the movements of two or more body parts at the same time or execute a specific action with more than one bodily effector.

Timing is another important aspect of naturalistic behaviours. Consider the importance of timing for our baseball players. The outfielder must time her movements so as to position herself beneath the falling ball at exactly the right moment, while the base runner must decide when to run and when to reach for the base so as to avoid a tag by the baseman. To examine how the cortical motor system times complex movements, Larsson et al. (1996) asked participants to tap their fingers along with either an external beat or a silent internally-generated beat while undergoing PET. Most motor areas were similarly active during both tasks; however, the SMA was more active when tapping along to the internally-triggered beat and the premotor cortex was more active for the externally-triggered beat. These results suggested that most movements (even simple ones) activated the majority of cortical motor areas, but the SMA may play a special role in producing internally-triggered movements, whereas the premotor cortex may play a larger role in generating movements in response to external sensory stimuli.

In sum, early work by Roland and others suggested that individual cortical areas contribute to a variety of different motor behaviours. Specifically, the SMA contributes to the production of complex motor sequences regardless of which body parts actually perform the sequence. It also plays a special role in generating voluntary self-initiated movements. In contrast, premotor and parietal areas appear to be especially important for processing external visual, auditory and somatosensory stimuli that can guide movements made by a range of different body parts. Finally, M1 is consistently involved in the actual execution of movements and appears to be at least partially organized in a topographic manner – according to body parts. That individual cortical areas can contribute to a variety of motor behaviours provides at least a partial explanation for why overlapping 'body part' representations can be found throughout the cortical motor system.

WHAT HAS PET REVEALED ABOUT HOW WE LEARN MOTOR SKILLS?

Another line of research in Roland's lab would focus on how the cortical motor system learns different motor skills. Motor learning is known to consist of at least three stages. Initially, motor improvements are large and rapid; then improvements become smaller, slower and more incremental; finally the new skill is mastered, but ongoing practice may be required to maintain optimal performance. Early work from Roland's lab showed that different brain areas are involved in the early versus late stages of motor learning.

Seitz and Roland (1992) used PET to examine the early versus late stages of learning a complex finger tapping sequence. They found that motor and premotor areas were consistently active throughout all stages of learning, but activity in parietal and motor speech areas decreased with learning (see Figure 19.3A), while subcortical activity increased. They proposed that in the early learning stages participants may rely more heavily on somatosensory feedback from the parietal cortex and internal counting from the motor speech areas, but in later stages the motor sequence has been internalized and is mediated largely by subcortical activity.

Similar but varied results were subsequently reported by a number of different research groups. Jenkins et al. (1994) reported increased activity in the prefrontal, premotor, parietal and cerebellar areas when participants performed an unfamiliar finger tapping sequence as compared to increased activity in the SMA when they performed a well-learned sequence. Penhune and Doyon (2002) found greater activity in the basal ganglia and frontal lobe on the fifth day of training on a motor sequence task versus greater activity in M1, premotor and parietal cortices at recall four weeks later. Finally, Grafton et al. (1994) found that when participants rapidly acquired a new visuomotor skill early activity in the occipitotemporal cortex (visual processing areas) was gradually replaced by increased activity in the prefrontal, premotor and cingulate motor areas, possibly reflecting a rapid shift from a visually guided motor strategy to an internally generated motor representation.

In sum, PET studies from a number of different labs suggested that different brain areas are involved in the early versus late stages of learning a new motor skill. This shift in neural control may reflect an early dependence on online sensory and cognitive guidance of individual movements to the subsequent internalization and automatization of a specific motor sequence after it has been mastered. Nevertheless, the exact brain regions and their specific contributions at different stages of learning remained far from fully resolved.

TRADITIONAL fMRI

While extremely useful, PET continues to face some limitations for studying brain activation. The advent of functional magnetic resonance imaging (fMRI) in the early 1990s (Kwong, 1992; Ogawa et al., 1992) provided a non-invasive alternative that did not require injecting participants with a radioactive substance. In fMRI, the radioactive tracer is replaced with an *in vivo* contrast agent – the participant's own blood. Specifically, fMRI measures the blood oxygenation level dependent or BOLD signal. Neural activity leads to local increases in blood flow that flushes away deoxygenated hemoglobin, leading to an increase in the BOLD signal (typically on the order of 0.1–4%). Advantages of fMRI include its better spatial resolution (mm not cm), temporal resolution (sec not min), and sensitivity. fMRI also allows experimenters to examine activation patterns in individual participants and across a larger number of conditions. This produces more precise control conditions that can better isolate cognitive functions. Nonetheless, like PET, traditional fMRI methods aim to answer *where* in the brain certain functions are located by subtracting the BOLD signal during a control condition from the BOLD signal during an experimental condition.

WHAT HAS TRADITIONAL fMRI REVEALED ABOUT HOW THE CORTICAL MOTOR SYSTEM IS ORGANIZED?

Traditional fMRI studies initially focused on examining brain activation patterns while participants performed a variety of different eye, arm and hand movements. These studies seemed to suggest that the cortical motor system might be organized into distinct circuits – or representations – each dedicated to the control of a specific body part (for reviews see Culham and Valyear, 2006; Culham et al., 2006). For example, when participants made quick eye movements or continuously tracked a moving object with their eyes, the frontal eye fields, supplementary eye fields and the lateral intraparietal sulcus became active, implicating their involvement in a dedicated 'eye circuit'. Similar activation of the SMA, dorsal premotor cortex and superior parietal cortex occurred during real, planned, imagined and observed reaching and pointing movements, and seemed to be indicative of a dedicated 'arm circuit'. Finally, activity in the ventral premotor cortex and anterior IPS during real, imagined, planned and observed grasping and manipulatory actions was interpreted as a dedicated 'hand circuit'.

While there is a lot of support for this classic view, more recent fMRI studies have identified both distinct and partially overlapping body part representations. Cavina-Pratesi et al. (2010) found distinct representations for reaching with the arm and grasping with the hand throughout most of the parietofrontal cortex, but the two representations did partially overlap in the dorsal premotor cortex and SMA. Partially overlapping reach and grasp representations were also found in the parietal cortex in a later study by Konen et al. (2013). Other studies (Heed et al., 2011; Jastorff et al., 2010) found that overlapping hand and foot representations in the premotor and parietal cortex diverge in M1/S1 so that the hand is represented lateral to the foot (see Figure 19.2B). Finally, a number of studies (Astafiev et al., 2003; Beurze et al., 2009; Connolly et al., 2007; Levy et al., 2007) have suggested that the parietofrontal cortex is organized according to a reversible gradient of overlapping body part representations, with occipital and posterior parietal areas preferentially encoding eye movements, anterior parietal areas

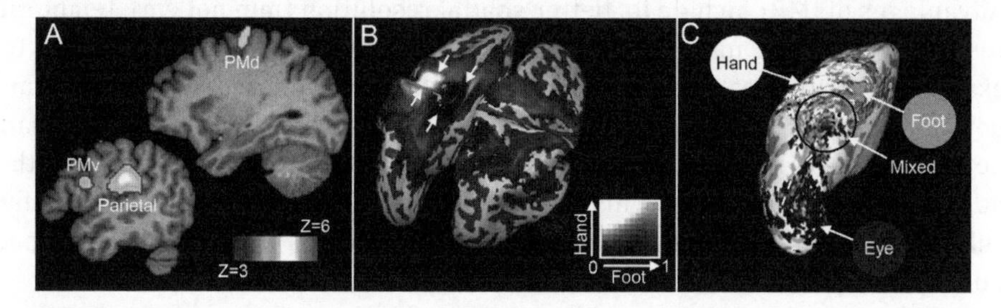

Figure 19.2 Overlapping body part representations. **(A)** PET activation in PMd, PMv and parietal cortex was similar regardless of whether the participants moved the ankle, the wrist, or both at the same time (rendered on parasagittal slices through the left hemisphere); adapted from Ehrrson et al., 2000. **(B)** Traditional fMRI analyses reveal overlapping hand and foot representations (medium gray) throughout most of the parietal and premotor cortices; however, distinct representations of the hand (white) and foot (black) are found in the primary sensorimotor cortex (rendered on partially inflated surfaces of the two hemispheres shown from a right posterior view); adapted from Heed et al., 2011. **(C)** Advanced fMRI analyses reveal primarily effector-specific (Hand-white, Foot-light gray, Eye-black) representations in primary sensory and motor cortices and a mixture of effector-specific (Hand-white, Foot-light gray, Eye-black) and action-specific (dark gray) representations in the parietal cortex (Mixed), (rendered on a partially inflated surface of the left hemisphere); adapted from Leoné et al., 2014.

encoding both eye and arm movements, M1 and SMA preferentially encoding arm movements, and then premotor areas again encoding both arm and eye movements.

Thus, while earlier fMRI studies tended to suggest that the cortical motor system might be organized into a set of segregated circuits, each dedicated to generating movements with a specific body part, later fMRI studies suggested that premotor and parietal areas, which contain overlapping body part representations, encode specific actions that can be made with more than one body part. In this way premotor and parietal areas may preferentially encode the particular action being performed whereas primary sensorimotor areas may preferentially encode the specific body part used to execute it.

WHAT HAS TRADITIONAL fMRI REVEALED ABOUT HOW WE LEARN MOTOR SKILLS?

As with PET, traditional fMRI studies found that repeated practice of a motor skill seems to alter neural activity in a manner that reflects functional reorganization to support the learned skill. Again, the specific pattern of neural activation was found to depend on the stage of motor learning (see Floyer-Lea and Matthews, 2005; Patel et al., 2013; Ungerleider et al., 2002; Yang et al., 2014).

A common theme to emerge from fMRI studies of motor learning was that as participants practised and became more skilled at a specific motor task, activity in primary sensorimotor areas would increase while activity in secondary motor areas decreased. As an example, Petersen et al. (1998) trained people to use a single finger to trace their way through a maze while undergoing fMRI. They found that premotor and parietal areas were more active during initial unskilled performance of the task whereas SMA was more active during later skilled performance. Similar findings have been reported from a variety of labs such that activity in primary sensorimotor areas increases when participants perform a well-learned task (see Figure 19.3B; Karni et al., 1996; Hlustík et al., 2004; Shmuelof et al., 2014), while activity in secondary sensorimotor areas decreases (Steele and Penhune, 2010). Interestingly, the nature of the neural activity change may depend on the type of task being learned. For example, practice on both skilled and gross motor tasks can lead to increased activity within the M1 finger representation; however, the representations of individual fingers overlap more extensively after training on the skilled task compared to the gross motor task (Hlustík et al., 2004). Thus, despite similar levels of activation for gross and skilled motor learning, the nature of the activation may be more complex when learning a skilled motor task.

In sum, work from traditional fMRI studies suggested that many cortical areas might contribute to the initial formation of a new motor skill when sensory, cognitive, and attentional demands are highest. Subsequent decreases in secondary motor activity suggested that as the skill becomes automatized these areas no longer contribute to skill performance. At the same time, increased activity in primary sensorimotor areas could reflect the emergence of a highly efficient neural representation, which can execute the skill without inputs from higher level cortical areas. Finally, the nature of the neural representation in M1 could be related to the complexity of the motor task, with highly skilled motor tasks producing more complex patterns of activity.

ADVANCED fMRI

In traditional fMRI analyses each voxel is treated as an independent piece of data and the goal is to determine whether that voxel responds more in some experimental conditions compared to others. Two limitations are: 1) the data from a large number of voxels are typically averaged together in order to determine whether the activity in one brain region is different from another; and 2) focusing on activation magnitude can obscure differences in the way information is encoded by separate neural populations (or voxels) within a single brain region. For example, if a brain region is equally active in two different conditions, e.g. upward versus downward reaches, the region is deemed to be insensitive to reach direction even though the upward and downward reaches may be encoded by different but comparably sized and spatially overlapping populations of neurons within the same region. Increasingly advanced fMRI techniques have allowed neuroimagers to begin to move beyond these limitations to explore how individual brain areas might encode more than one type of information.

One technique that has been particularly useful is Multivoxel (or Multivariate) Pattern Analysis (MVPA) in which the activity of multiple voxels is not averaged together, but rather the *pattern* of activity among them is examined to determine whether different activity patterns coincide with different experimental conditions, even if the average activity of those voxels is equivalent across different experimental conditions (Haxby et al., 2001; Tong and Pratte, 2012). Initially, MVPA was mainly used to make binary comparisons, i.e. to determine whether the M1 activity patterns for upward and downward reaches were more different than would be expected by chance. However, Kriegeskorte and colleagues (2008) showed that it was possible to examine MVPA activity patterns across scores of conditions and then arrange those conditions on a spectrum according to the extent to which their corresponding activity patterns were similar or dissimilar, now known as representational similarity analysis (RSA).

WHAT HAS ADVANCED fMRI REVEALED ABOUT HOW THE CORTICAL MOTOR SYSTEM IS ORGANIZED?

While traditional fMRI studies discovered overlapping body part representations in the parietal and premotor cortex, more advanced fMRI studies, particularly using MVPA, have been able to distinguish, within these areas of ambiguous activation, sub-areas that use a *common* neural activity pattern to encode a particular action, regardless of which body part executes it (action-specific encoding), versus sub-areas that use *unique* neural activity patterns to encode the same action executed by different body parts (effector-specific encoding).

As one example, Leoné et al. (2014) found that parietal areas previously thought to encode only a specific body part also contained weaker representations for alternate body parts. Thus, these areas actually represent a specific action, such as grasping, but within this 'grasping' representation, the primary effector (the hand) is represented most strongly while potential alternate effectors (the mouth and foot) are represented more weakly (see Figure 19.2C).

A similar approach has been used to further characterize how the cortical motor system enables tool use. In cortical areas where traditional fMRI studies could not discriminate between hand and tool representations, MVPA was able to identify areas that were effector-specific (i.e. hands and tools are represented by different neural activity patterns) and cortical areas that are action-specific (i.e. hands and tools are represented by a common neural activity pattern). While anterior parietal areas represent hands and tools with different patterns of neural activity, and are thus effector-specific, posterior parietal and premotor areas produce the same pattern of neural activity regardless of whether the action is carried out with the hand or the tool, and are thus action-specific (Gallivan et al., 2013).

Traditional fMRI studies also found that a number of cortical areas were similarly active regardless of whether people observed, imagined or executed a particular action. Filimon et al. (2014) used MVPA to show that, within these areas of ambiguous activation, distinct neural activity patterns emerge for action observation, imagery and execution. Thus, the cortical motor system can distinguish between observed, imagined and executed movements, not only based on which cortical areas are activated in each condition, but also according to the specific neural activity pattern produced within cortical areas that are commonly activated across all three conditions.

In sum, these intermingling action- and effector-specific representations in parietal and premotor areas may enable these areas to evaluate the spatial demands of a task, select the specific action needed to accomplish that task, and then flexibly execute the action with the effector that is most appropriate in the given context (Cui and Andersen, 2011; Medendorp et al., 2005). For example, if a coin is located on a table in front of you, you would most likely reach for it using the arm and hand; however, if the coin rolls off and underneath the table, it may be necessary to reach for it with the leg and foot! Activation of the weaker 'leg' representation within the larger 'reaching' representation in the parietal and premotor cortices would likely subserve this action. Furthermore, as predicted by Roland, the encoding of multiple body parts within a single cortical area could also support the coordination of two different effectors in achieving a common goal. For example, overlapping arm and hand representations in SMA and the dorsal premotor cortex may allow us to coordinate the actions of these two body parts when reaching to grasp objects (Cavina-Pratesi et al., 2010).

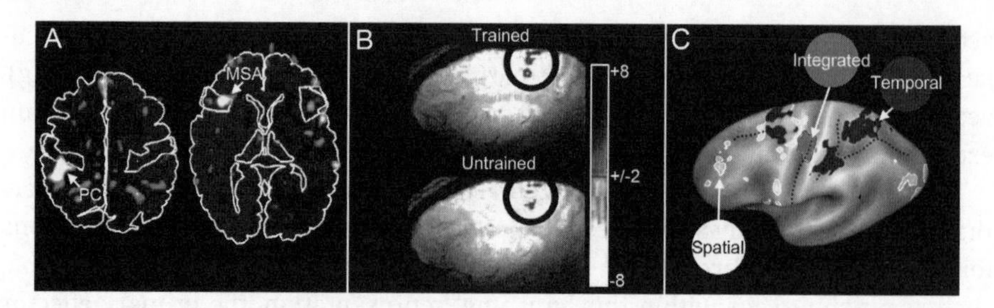

Figure 19.3 Changes in neural activation in response to motor learning. **(A)** Highlighted regions indicate decreased PET activity in motor speech areas (MSA) and the parietal cortex (PC) after participants had learned and performed a specific finger tapping sequence (rendered on horizontal slices); adapted from Seitz & Roland (1992). **(B)** Traditional fMRI analyses reveal increased activity in primary motor cortex when participants perform a well-trained finger tapping sequence (top) as compared to an untrained sequence (bottom) (rendered on parasagittal slices); adapted from Karni et al., 1995. **(C)** Advanced fMRI analyses reveal that SMA, parietal, and premotor areas encode the spatial (white) and temporal (black) features of a learned motor sequence in distinct representations, while primary sensorimotor areas integrate (gray) the spatial and temporal features of a learned motor sequence into a single, executable, representation (rendered on inflated surfaces of the left hemisphere); adapted from Kornysheva and Diedrichsen, 2014.

WHAT HAS ADVANCED fMRI REVEALED ABOUT HOW WE LEARN MOTOR SKILLS?

Advanced fMRI techniques have also made leaps and bounds in understanding how the cortical motor system supports the learning of motor skills. One particularly curious finding from traditional fMRI studies was that motor learning is often associated with both increases and decreases in neural activity. Wiestler and Diedrichsen (2013) hypothesized that the learning of a new motor skill might be mediated by the refinement of a skill-specific neuronal circuit, and the emergence of a highly distinct neural activity pattern, despite more global increases or decreases in averaged cortical activity.

To test their idea they trained participants on five different finger tapping sequences and then used MVPA to see whether they could identify unique neural activity patterns for each sequence and whether the activity patterns for trained sequences were more distinguishable (i.e. more refined) than for untrained sequences. In general, finger tapping produced activation in primary, premotor and parietal motor areas, but activity in secondary motor areas such as the dorsal premotor and parietal cortex was reduced when performing trained versus untrained sequences. Nevertheless, sequence-specific activity patterns could be identified for both trained and untrained sequences throughout most cortical motor areas. The patterns for trained sequences, however, were easier to distinguish from one another (i.e. more refined) compared to untrained sequences, especially in the SMA and pre-SMA, suggesting that these regions might encode longer highly specific sequences, whereas other motor areas that discriminate both trained and untrained sequences, but with less accuracy, may encode shorter motor elements that are used in more than one sequence.

Highly skilled motor behaviours, such as playing a musical instrument, dancing a tango or maneuvering a race car, involve activating a precise set of muscles, in a particular order, with specific timing. Consider learning to play a new piece of music on the piano. Not only must you learn which fingers to press and in which order (the spatial structure of the sequence), you must also master the timing and rhythm of the finger presses (the temporal structure of the sequence). A major question in motor learning is: when you learn a motor skill are the spatial and temporal structures of the sequence encoded in a single representation or are they encoded in separate representations? If it is the first, then you would have to learn and refine two separate neural representations in order to produce the same finger sequence with two different rhythms. If it is the second, then you could simply recombine separate previously learned spatial and temporal structures and this would allow you to play a particular musical piece at different speeds, tempos or rhythms with relative ease.

In an ingenious study, Kornysheva and Diedrichsen (2014) trained participants to produce nine different finger sequences which actually consisted of three different temporal structures uniquely paired with three different spatial structures. They then used MVPA to determine which cortical areas contained nine unique neural activity patterns – one for each finger sequence. The only cortical area to do so was M1, which suggests that M1 does bind the temporal and spatial structures of each motor sequence into a single, unique, representation. Next, they used MVPA to see if there were cortical areas that encoded the temporal and spatial structures of the sequences independently, the idea being that cortical areas that encode only spatial features would contain only three neural activity patterns – one for each spatial structure with no regard for temporal structure, and vice versa for areas that encode only temporal features. Amazingly, it worked! SMA, the dorsal premotor and parietal cortex all contained neural activation patterns specific only to spatial structure, while SMA, the dorsal premotor, ventral premotor, anterior and posterior cingulate, and extrastriate visual areas contained neural activity patterns specific only to temporal structure. Thus, there were even some cortical areas, SMA and the dorsal premotor cortex, that encoded both the temporal and spatial features of the sequence, but in separate representations (see Figure 19.3C).

Together, these findings have important implications for understanding how the cortical motor system learns motor skills. In the traditional view, decreases in overall activity in secondary motor areas might lead to the conclusion that highly trained motor skills are largely encoded in primary motor areas with relatively little input from secondary motor areas. In contrast, the combined results from Diedrichsen's lab suggest that multiple mechanisms support motor learning. First, almost all cortical motor areas contribute to skill learning through the refinement of specialized neural representations for each new skill. Second, this refinement is accompanied by increases in neural efficiency, especially in secondary motor areas, and decreased average activity in these areas. Third, the decomposition of skilled movements into their spatial and temporal structures in parietal and premotor areas likely increases the efficiency and flexibility of the motor system, because these representations can then be flexibly combined to produce a complex motor action appropriate for any given situation.

CONCLUSION

In sum, functional neuroimaging has revealed that huge swathes of cortex are involved in processing motor information, even areas not previously thought to participate in motor control including those in the ventral stream (Gallivan and Culham, 2015). Although initially confusing and elusive, the cortical motor system does appear to be organized in a manner that allows us to efficiently coordinate multiple body parts in order to produce naturalistic movements and learn different motor skills. This organization seems to strike a balance between premotor and parietal areas that preferentially encode the goal of an action, and traditional sensorimotor areas that preferentially encode the body part used to execute the action.

In learning a new motor skill, repeated practice seems to lead to the refinement of skill-specific neural representations throughout the brain. Importantly, the spatial and temporal features of the skill are encoded in separate representations throughout supplementary, parietal and premotor areas, whereas they are integrated into a single, executable, representation in M1. The presence of multiple copies of these representations suggests that each cortical area may modify or utilize the representation in a different way. Copies in the parietal cortex may allow us to modify the motor skill based on incoming visual, auditory and somatosensory information, copies in the premotor cortex may allow us to modify the motor skill based on higher-level cognitive inputs, and copies in SMA may allow us to voluntarily execute the skill in the absence of other external cues. Thus, once a motor skill is learned, the flexible recombination of these structural and temporal motor representations likely allows us to adapt this single behaviour to almost any imaginable situation.

Finally, while neuroimaging techniques have improved significantly since Roland's first PET experiments in the 1980s there are still a number of hurdles to overcome. First, any functional neuroimaging technique is correlational. Thus, although we are now able to identify motor-related activity in most brain areas, we cannot be certain of the particular role each area plays in actually producing that behaviour. In the future, the combination of functional neuroimaging with temporary perturbation techniques, such as transcranial magnetic stimulation (TMS), should help to clarify some of the confusion. Second, perhaps the biggest drawback for functional neuroimaging is that it is highly susceptible to motion within the scanner. Even the slightest movements can corrupt the data and render these unusable. This greatly limits the types of actions and experimental populations that can be studied using fMRI. Thus, there is a strong need to develop novel methods for dealing with motion artifacts – especially when trying to study motor behaviour in patient populations, developing infants and non-human species. Given the dramatic increase in our knowledge about sensorimotor control in the 35 years since Roland's seminal publication, we can only imagine the possibilities to come.

REFERENCES

Astafiev, S.V., Shulman, G.L., Stanley, C.M. et al. (2003) Functional organization of human intraparietal and frontal cortex for attending, looking, and pointing. *Journal of Neuroscience,* 23 (11): 4689–99.

Beurze, S.M., de Lange, F.P., Toni, I. and Medendorp, W.P. (2009) Spatial and effector processing in the human parietofrontal network for reaches and saccades. *Journal of Neurophysiology,* 101: 3053–62.

Cavina-Pratesi, C., Monaco, S., Fattori, P. et al. (2010) Functional magnetic resonance imaging reveals the neural substrates of arm transport and grip formation in reach-to-grasp actions in humans. *Journal of Neuroscience,* 30 (31): 10306–23.

Connolly, J.D., Goodale, M.A., Cant, J.S. and Munoz, D.P. (2007) Effector-specific fields for motor preparation in the human frontal cortex. *NeuroImage,* 34: 1209–19.

Cui, H. and Andersen, R.A. (2011) Different representations of potential and selected motor plans by distinct parietal areas. *Journal of Neuroscience,* 31 (49): 18130–6.

Culham, J.C., Cavina-Pratesi, C. and Singhal, A. (2006) The role of parietal cortex in visuomotor control: What have we learned from neuroimaging? *Neuropsychologia,* 44 (13): 2668–84.

Culham, J.C. and Valyear, K.F. (2006) Human parietal cortex in action. *Current Opinions in Neurobiology,* 16 (2): 205–12.

Ehrsson, H.H., Naito, E., Geyer, S. et al. (2000) Simultaneous movements of upper and lower limbs are coordinated by motor representations that are shared by both limbs: a PET study. *European Journal of Neuroscience,* 12: 3385–98.

Filimon, F., Rieth, C.A., Sereno, M.I. and Cottrell, G.W. (2014) Observe, executed, and imagined action representations can be decoded from ventral and dorsal areas. *Cerebral Cortex,* Epub ahead of print, doi: 10.1093/bhu110.

Floyer-Lea, A. and Matthews, P.M. (2005) Distinguishable brain activation networks for short- and long-term motor skill learning. *Journal of Neurophysiology,* 94: 512–18.

Gallivan, J.P. and Culham, J.C. (2015) Neural coding within human brain areas involved in actions. *Current Opinion in Neurobiology,* 33: 141–9.

Gallivan, J.P., McLean, D.A., Valyear, K.F. and Culham, J.C. (2013) Decoding the neural mechanisms of human tool use. *Elife,* 2: e00425. Doi: 10.7554/eLife.00425.

Grafton, S.T., Mazziotta, J.C., Woods, R.P. and Phelps, M.E. (1992) Human functional anatomy of visually guided finger movements. *Brain,* 115: 565–87.

Grafton, S.T., Woods, R.P. and Tyszka, M. (1994) Functional imaging of procedural motor learning: relating cerebral blood flow with individual subject performance. *Human Brain Mapping,* 1 (3): 221–34.

Haxby, J.V., Gobbini, M.I., Furey, M.L. et al. (2001) Distributed and overlapping representations of faces and objects in ventral temporal cortex. *Science,* 293: 2425–30.

Heed, T., Beurze, S.M., Toni, I. et al. (2011) Functional rather than effector-specific organization of human posterior parietal cortex. *Journal of Neuroscience,* 31 (8): 3066–76.

Hlustík, P., Solodkin, A., Noll, D.C. and Small, S.L. (2004) Cortical plasticity during three-week motor skill learning. *Journal of Clinical Neurophysiology,* 21 (3): 180–91.

Inoue, K., Kawashima, R., Sugiura, M. et al. (2001) Activation in the ipsilateral posterior parietal cortex during tool use: a PET study. *NeuroImage,* 14: 1469–75.

Jastorff, J., Begliomini, C., Fabbri-Destro, M. et al. (2010) Coding observed motor acts: different organizational principles in the parietal and premotor cortex of humans. *Journal of Neurophysiology,* 104: 128–40.

Jenkins, I.H., Brooks, D.J., Nixon, P.D. et al. (1994) Motor sequence learning: a study with positron emission tomography. *Journal of Neuroscience,* 14 (6): 3775–90.

Karni, A., Meyer, G., Jezzard, P. et al. (1996) Functional MRI evidence for adult motor cortex plasticity during motor skill learning. *Nature,* 377 (14): 155–8.

Kawashima, R., Roland, P.E. and O'Sullivan, B.T. (1994) Fields in human motor areas involved in preparation for reaching, actual reaching, and visuomotor learning: a positron emission tomography study. *Journal of Neuroscience,* 14 (6): 3462–74.

Kawashima, R., Roland, P.E. and O'Sullivan, B.T. (1995) Functional anatomy of reaching and visuomotor learning: a positron emission tomography study. *Cerebral Cortex,* 2: 111–22.

Konen, C.S., Mruczek, R.E.B., Montoya, J.S. and Kastner, S. (2013) Functional organization of human posterior parietal cortex: grasping- and reaching-related activations relative to topographically organized cortex. *Journal of Neurophysiology,* 109: 2897–908.

Kornysheva, K. and Diedrichsen, J. (2014) Human premotor areas parse sequences into their spatial and temporal features. *eLife,* 3: e03043. Doi: 10.7554/eLife.03043.

Kriegeskorte, N., Mur, M. and Bandettini, P. (2008) Representational similarity analysis – connecting the branches of systems neuroscience. *Frontiers in Systems Neuroscience, 2*: 4. Doi:10.3389/neuro.06.004.2008.

Kwong, K.K., Belliveau, J.W., Chesler, D.A. et al. (1992) Dynamic magnetic resonance imaging of human brain activity during primary sensory stimulation. *Proceedings of the National Academy of Sciences, USA,* 89: 5675–9.

Larsson, J., Gulyás, B. and Roland, P.E. (1996) Cortical representation of self-paced finger movement. *NeuroReport,* 7 (2): 463–8.

Leoné, F.T.M., Heed, T., Toni, I. and Medendorph, W.P. (2014) Understanding effector selectivity in human posterior parietal cortex by combining information patterns and activation measures. *Journal of Neuroscience,* 34 (21): 7102–12.

Levy, I., Schluppeck, D., Heeger, D.J. and Glimcher, P.W. (2007) Specificity of human cortical areas for reaches and saccades. *Journal of Neuroscience,* 27 (17): 4687–96.

Medendorp, W.P., Herbert, C.G., Crawford, D. and Vilis, T. (2004) Integration of target and effector information in human posterior parietal cortex for the planning of action. *Journal of Neurophysiology,* 93: 954–62.

Ogawa, S., Tank, D.W., Menon, R. et al. (1992) Intrinsic signal changes accompanying sensory stimulation: functional brain mapping with magnetic resonance imaging. *Proceedings of the National Academy of Sciences, USA,* 89: 5951–5.

Patel, R., Spreng R.N. and Turner, G.R. (2013) Functional brain changes following cognitive and motor skills training: a quantitative meta-analysis. *Neurorehabilitation and Neural Repair,* 27 (3): 187–99.

Penhune, V.B. and Doyon, J. (2002) Dynamic cortical and subcortical networks in learning and delayed recall of timed motor sequences. *Journal of Neuroscience,* 22 (4): 1397–1406.

Petersen, S.E., Van Mier, H., Fiez, J.A. and Raichle, M.E. (1998) The effects of practice on the functional anatomy of task performance. *Proceedings of the National Academy of Sciences, USA,* 95: 853–60.

Posner, M.J. and Raichle, M.E. (1997) *Images of Mind.* New York: Henry Holt and Company.

Roland, P.E. (1993) *Brain Activation.* New York: Wiley-Liss Inc.

Roland, P.E., Eriksson, L., Widén, L. and Stone-Elander, S. (1989) Changes in regional cerebral oxidative metabolism induced by tactile learning and recognition in man. *European Journal of Neuroscience,* 1 (1): 3–18.

Roland, P.E., Larsen, B., Lassen, N.A. and Skinhøj, E. (1980) Supplementary motor area and other cortical areas in organization of voluntary movements in man. *Journal of Neurophysiology*, 43 (1): 118–36.

Roy, C.S. and Sherrington, C.S. (1890) On the regulation of the blood-supply of the brain. *Journal of Physiology,* 11: 85–108.

Seitz, R.J. and Roland, P.E. (1992) Learning of sequential finger movements in man: a combined kinematic and Positron Emission Tomograph (PET) study. *European Journal of Neuroscience,* 4: 154–65.

Seitz, R.J., Roland, P.E., Bohm, C. et al. (1991) Somatosensory discrimination of shape: tactile exploration and cerebral activation. *European Journal of Neuroscience,* 3: 481–92.

Shmuelof, L., Yang, J., Caffo, B. et al. (2014) The neural correlates of learned motor acuity. *Journal of Neurophysiology,* 112 (4): 971–80.

Steele, C.J. and Penhune, V.B. (2010) Specific increases within global decreases: a functional magnetic resonance imaging investigation of five days of motor sequence learning. *Journal of Neuroscience,* 30 (24): 8332–41.

Tong, F. and Pratte, M.S. (2012) Decoding patterns of human brain activity. *Annual Review of Psychology,* 63: 483–509.

Ungerleider, L.G., Doyon, J. and Karni, A. (2002) Imaging brain plasticity during motor skill learning. *Neurobiology of Learning and Memory,* 78: 553–64.

Wiestler, T. and Diedrichsen, J. (2013) Skill learning strengthens cortical representations of motor sequences. *eLife,* 2: e00801. Doi: 10.7554/eLife.00801.

Wu, T., Kansaku, K. and Hallett, M. (2003) How self-initiated memorized movements become automatic: a functional MRI study. *Journal of Neurophysiology,* 91: 1690–8.

Yang, J. (2014) The influence of motor expertise on the brain activity of motor task performance: a meta-analysis of functional magnetic resonance imaging studies. *Cognitive, Affective, & Behavioural Neuroscience,* 15 (2): 381–94.

Index